The Tree of Knowledge

Claudio Ronchi

The Tree of Knowledge

The Bright and the Dark Sides of Science

 Springer

Claudio Ronchi
Karlsdorf-Neuthard
Germany

Original Italian Edition L'Albero della Conoscenza, Editoriale Jaca Book, Milano, 2010.
© Editoriale Jaca Book 2010. All rights reserved.
Copyright for the English translation by Springer Verlag, 2014.

ISBN 978-3-319-34652-6   ISBN 978-3-319-01484-5   (eBook)
DOI 10.1007/978-3-319-01484-5
Springer Cham Heidelberg New York Dordrecht London

Printed on acid-free paper

Springer is part of Springer Science+Business Media (www.springer.com)

To my wife Luciana

Preface to the English Edition

The story of this book is somewhat unusual. A few years ago I started working on a monograph, where I intended to collect a number of problems concerning ancient and modern physics. I had on my desk a pile of assorted notes taken during many years and during the course of many lectures and discussions, which would have provided enough material to fill a thick volume. Yet, when, after my retirement, I maintained more regular contact with my birth country, I was encouraged by old friends to first publish a short version of the book, addressed to a broader circle of Italian readers. This implied selecting suitable subjects and establishing the depth of their treatment at a level suitable for and palatable to non-specialists. After discussing with the publisher's advisors, I was asked to keep the text within what they considered "standard limits" for this specific book market. I first accepted the suggestion and did not worry about the consequences. As the work progressed, however, I began fearing not only to have almost deprived the book's skeleton of its necessary substance, but also to have weakened the force of the argumentation on delicate and debated questions. It has been indeed a rather lengthy and hesitant work that finally led to the publication by Jaca Book of the essay entitled *L'Albero della Conoscenza*. However, when, some months after publication, I re-read this slim book, I realised that it did indeed present some positive traits. Of these, the best was a necessary distance to so vast a subject, at least in terms of approach that, anyhow, even several weighty tomes would have hardly been sufficient to expose in any detail even the most pertinent matter. I thus resumed working on the English version by maintaining the same height of flight over the subjects as in the Italian edition. I only inserted some further details in the central part of the book and a new section regarding the p-adic numbers—which few people are familiar with—since they represent today a most promising instrument for invigorating the mathematics of quantum physics. The last chapters, dealing with anthropological aspects of science remained substantially unchanged.

The book, at any rate, required short but fluent expositions of sometimes complex arguments and, in this respect, I should like to gratefully acknowledge the invaluable, competent contribution of Tina Saavedra who polished my English so that this book perhaps makes for pleasant reading and at least reduces the reader's fatigue which the original text might have caused.

In the end, the *Tree of Knowledge* turned out to be much more similar to its Italian brother than to the monograph I had initially in mind. However, I myself am satisfied: since the book is essentially a *plaidoyer* for simplicity and clarity in science I find its size and contents comply with these attributes.

Neuthard, July 2013   Claudio Ronchi

Contents

Prologue

When, half a century ago I was beginning my studies, the cultural climate was characterised by an absolute confidence in the progress of science and technology. The war, having ended a few years ago, had shown that the victory of the democracies over the dictatorships had been obtained by those countries which had invested more in promoting the intellectual resources of their people and in their ability to acquire and to employ new knowledge. During the subsequent reconstruction, the conviction steadily grew in strength that the applications of binomial science-technology was changing the world for the better, as had never occurred in previous history. The political conflicts of the post-war period, even through a continuous succession of international crises, indicated a common concern to safeguard a social progress that seemed to depend more on this paired combination of science and technology than on any specific political form of government; indeed, it was believed that diffusion of knowledge would in and of itself serve as the most effective guarantee against repeating the totalitarian aberrations of the recent past. "Knowledge is good, whatever its object might be", was proclaimed as a slogan from every chair or podium, and, we must admit that, in the so-called advanced countries this value was indeed widely distributed among all population levels. Starting as a privilege of the elite, knowledge quickly became a public patrimony and everyone was free to employ it for the common good. Capitalist and socialist countries were both inclined to see in the scientific and technological progress achieved a test bench of the validity of their respective political systems. A framework of questions was, obviously, considered concerning the ethical value of knowledge, but they concerned its potentially negative applications and uses. Furthermore, it was believed that even the threat of the destructive power of the new atomic weapons, made possible by the progress of science, in the future would have rendered useless and, therefore, improbable devastating wars, like that in which the entire world had been involved. However, although the events of the following years confirmed, at least partially, this prophecy, today we must admit that it would be ingenuous to expect from the generations of the last post-war period a more solidly founded ethical behaviour than that of the antecedent generations, and that their moral stature has grown so much as to protect them from the temptation to use science for the purpose of conquest and oppression of the adversary.

It is, however, undeniable that man has at present assumed a more critical attitude with regard to the social and cultural system in which he lives. Though being aware of the constructive potential of the recent technical and scientific progress, he is far from thinking that this implies a moral and intellectual superiority with respect to his ancestors, preserving him from the errors of the past. Indeed, concerning questions regarding the role of knowledge in our society, the tenet that this is an absolute good is today frequently under discussion, if not in serious doubt. One has never spoken so much as one presently does about the ethical problems related to the admissibility of any type of scientific research, and current opinions are deeply divided in a field in which the exercise of total freedom has been always regarded as one of the more important benchmarks of our civilisation. On the other hand, in the present society, characterised by the disintegration of its traditional cultural texture and by the loss of authority of civil and religious moral forces, it is difficult to establish a clear reference for an ethical appraisal of private and collective behaviours, in particular those regarding the proper use of reason. The arguments on ethics in knowledge are founded on the one hand on a number of utilitarian criteria and, on the other, on more or less rigorous scientific reasoning. Although, from a more practical view, the contrasting theses do not consider the value of knowledge itself, but rather of its possible applications, it is a common belief of all contenders that the two aspects are *not* separable—which is plausible within the context of modern society. This renders convergence of opinions almost impossible, since the question is placed simultaneously on two different planes that intersect on the labile and inconstant line of what we call common interest.

The strained relation between ethics and knowledge had been perceived from a religious standpoint many centuries before modern science was born. In this regard, the Biblical passage that has often provoked exegetes of the Christian–Jewish religion is contained in the second chapter of the Book of Genesis where it is said that God produced *lignum vitae in medio paradisi lignumque scientiae boni et mali* and then admonished Man: *Ex omni ligno paradisi comede; de ligno autem scientiae boni et mali ne comedas; in quocumque enim die comederis ex eo, morte morieris* (Gen. 2, 7–8).[1] Why this interdiction on knowledge, which, a priori, has an ethically neutral value, in as much as it encompasses good as well as evil?

---

[1] The text of *Vulgata* contains the expression *in quocumque die … morte morieris* whose deep significance is lost in most of the current translations: *You are free to eat from any tree in the garden; but you must not eat from the tree of the knowledge of good and evil, for if you eat from it, you will certainly die.*

Eating of the fruit of this tree, that is "knowing" anytime it happens (*in quocumque die*), does it contain in itself the germ of death? The answer of the exegetes is almost unanimous: the Biblical text does not speak about abstract knowledge as we mean it today, but about operating beyond the established limits, as represented by the fenced garden of paradise. In particular, the knowledge of evil is not aprioristic, but is only revealed after having experienced it. The transgression is, therefore, in asserting one's freedom against that of the Creator. Just for this reason, *the tree of life* and *the tree of knowledge* were planted *in medio* of paradise, but the fruits of the latter excluded those of the former. Knowledge did not guarantee life and life did not necessarily imply knowledge. They were essentially the terms which asserted the finiteness of man in the face of the work of creation and at the risk of acting freely. The root of the Hebrew word (*Deh'et*) = *knowledge* is used in the same book of Genesis in order to indicate the intercourse between man and woman that brings about procreation. It is amazing that even in the Indo-European linguistic domain the root *$g'n$ is the same to indicate both *to know* and *to generate/to be born*.[2]

This nexus presupposes in ancient man the deep psychological perception of the parallel between the risk of knowledge and that of birth and of the consequent death. In the Biblical report, the price of Adam's transgression was his irreversible departure from the closed Garden of Eden, to undertake his journey in the open cosmos, where knowledge-procreation has become his reason for being; but on gaining new knowledge, the ambivalence of its fruit is always present: good or evil? A question which concerns all of us. It seems, however, to be impossible to reach a unanimous answer, since today everyone feels free to define what is good and what is evil. Yet, reason and religion agree on asserting the existence of an objective criterion of judgement, which is only a function of the proposed final goal. In other words, the progress of knowledge has only an ethical sense if put into effect in the awareness of man's destiny.

The considerations contained in this book examine the development of human knowledge and its various implications. It is not a monograph on the history of science, but rather a series of reactions on some intellectual aspects, which, in the history of mankind, have started a causal chain the inexorable effects of which, only now we are starting to perceive.

---

[2] For instance, Greek γιγνώσκειν/γίγνεσθαι and Latin *cognoscere/gignere*.

Introduction

Approximately 200,000 years ago, *homo sapiens* began a journey which led to the colonisation of our planet. His first step was to replace, wherever he settled, other similar species, descendants from *homo erectus*, who inhabited the earth for more than 2 million years. It is probable that, in the beginning, only genetic factors such as fertility, longevity and resistance to diseases will have played an important role in enabling him to overwhelm his antagonist cousins. In fact, until 40,000 years ago, his manual abilities and the techniques in his possession were not so superior as to ensure him absolute dominance in the territories where he lived. It is only at the end of the quaternary era that, under the most difficult environmental conditions, modern *homo sapiens* exhibited a behavioural jump, that opened the door to a totally new associative system, generally identified with the term "civilisation". The motor of the process of civilisation has been the development of powerful instruments of communication and memorisation: language, graffiti and, finally, writing.

It is believed that the apparition of language is contemporary with the most ancient parietal images dated to around 35,000 B.C. Towards 30,000 B.C. the first recording signs are found on bones and served a mnemotechnical function. The first testimony of written pictography appears in the Mesopotamian lowland towards 3,300 B.C. (Uruk IVb) (Fig. 1).[3] Egyptian hieroglyphic writing is dated to 3,100 B.C. Small objects used for calculation have been recovered from archaeological layers in Susa from the same period. The Phoenician alphabetical writing system appears, in a consonantal form, around 900 B.C. with the vocalic Greek alphabet beginning around 800 B.C..

The effectiveness of communication instruments lies fundamentally in their ability to exempli-/simpli-fy complex situations and behaviours. Learning based on concrete experience is, in fact, hardly economic: it presupposes occasional, often uncontrollable and mostly not immediately accessible circumstances.

---

[3] We should mention here the case of the neolithic Vinča culture which flourished in the Balkans between the 6th and the 3rd millennium B.C. Archaeological sites have revealed clay tablets with pictographs, which might indicate a proto-writing system. The matter is, however, debated. The Vinča culture declined and finally disappeared, probably under the pressure of Indo-European tribes coming from central Asia. Yet, it would be strange that such an important achievement as writing was not taken over by these conquering peoples.

Fig. 1 Sumerian tablet with archaic cuneiform characters found in Uruk (3,100 B.C.) representing the first historical evidence of numerical writing (*British Museum, London*). The Sumerian numeration system was hexadecimal, thus the number 1 was given the name, *gesch*, the same as for 60. Hence, one needed 60 different names and symbols to identify the numbers from 1 to 60. This was very impractical and, therefore, an auxiliary decimal system was introduced with 10 names for the numbers from 1 to 10, where 10 was called *u*. The next power of 60 ($60^2 =$ 3,600) was called *schar* and smaller numbers were composed with the help of decimals: for instance, 600 was called *gesch-u* (60×10) and 36,000 was *schar-u* ($3,600 \times 10$). The next power of 60, 60^3, was called *shar-gal* (=*large schar*) [1]

Moreover, the emotional implication of the observer (fear, wrath, attraction, etc.) as well as the real time in which the experience occurs is not always favourable to what we could call the analysis of the situation. Learning through direct experience is, therefore, slow and expensive; furthermore, repetition of the same experience is, in the majority of the cases, unaccompanied by any new knowledge. On the contrary, the possibility to transmit the contents of any experience using language in a favourable space and time context, renders learning suitable to comparison, deduction and abstraction.

How it could happen that language attained so quickly the necessary precision and complexity remains a fundamental enigma. Specialists in linguistics, psychologists and anthropologists have formulated various theories on the origin of language, starting from various evolutionist, structuralist and behaviourist hypotheses, but the problem is too complex to allow us to provide an exhaustive and convincing answer.[4] If it is verisimilar that language has evolved from a form of elementary communication present in other animal species, the fact that human language has developed in a few thousands of years to an extremely complex structure cannot but arouse wonder, in the same way one wonders at the speed with which a child gets hold of the complex articulations of speech before still being

---

[4] It is worth remembering that in the 1860s the French *Societé Linguistique* decided to stop publishing further works concerning the origin of language, since the argument, then very fashionable, was finally considered merely speculative and lacking scientific value. One hundred years later, towards the end of the 1960s, the study of the origin of language experienced a revival leading to a proliferation, this time unlimited, of theories that extend to various fields, from logic to neurology.

able to exert them. It was just this process of learning that induced Noam Chomsky [2, 3] to propose his "botanical model" of language acquisition, according to which learning happens on a genetic base, like the growth of a plant, where the elements of its structure correspond to the modules of a sort of universal grammar. How then language has taken control of abstract intuitions like, for instance, number, space and time remains totally obscure. We do not know what the degree of complexity of language was at the birth of civilisation, but one can believe that in a relatively short time language had acquired sufficient flexibility to describe the interactions with the surroundings with which man was in direct contact.

From this point, the transmission of information was increasingly freed from the limits of occasional observations. In other words, members of a homogeneous group could be instructed even on circumstances and behaviours in which they did not have any direct experience. The enormous advantage of this method of transmitting within a set of complex cognitive experiences is founded on two pillars. The first, an objective one, is the reliability and the consistency of the contents of the collective patrimony of knowledge. The second, subjective, is the confidence shared by all members of the clan in a system of beliefs that far exceed the possibilities of verification by the individuals. It is on these premises that science was born. It does not, therefore, astonish us that, with an increasing concern to organise and to transmit the common patrimony of knowledge, which soon assumed a character of intangibility, especially manifested in the tradition of techniques and arts around which were consolidating the first specialised professional activities (ceramics, metallurgy, hunting/war, agriculture, preparation of the aliments, medicine).

Specialisation has been, in fact, the successive step that led to an increase in the cognitive patrimony in an organic way, without creating hiatuses or imbalances in the collective life.

Progress was closely channelled and limited in the consortia of productive activities, the members of which answered directly for the quality of their performance or their products. In theory, under normal conditions, regression was improbable, since knowledge in a certain field, rigorously acquired and transmitted, represented the ever-present foundation and the stone of comparison for every innovation. In any art, innovations occurred after long intervals of time, sometimes lasting many generations. Their origin could be a more or less accidental discovery, or could be gained from contacts with other groups.

For millennia, until modern times, the evolution of knowledge has followed a path marked by a sequence of bifurcations, the development of which has produced a progressive ramification of specialisation, with a rigorous hierarchical order, involving the subordination of the more peripheral activities to the parental ones. This structure has endured until the contribution of the primary fields to the social economy compensated for the burden of all their ancillary activities. However, the phenomenon of the hypertrophy of a field at the expense and damage of others has always represented an immanent weakness of this outline of progress, which has actually manifested itself in abnormal forms of some civilisations, whose downfall was due to serious imbalances in their patrimony of knowledge.

Examination of historical civilisations clearly demonstrates that their rate of development was never constant. A rapid initial flowering is always followed by a period of stagnation and decline, in which an old civilisation, provides the base (or the humus) on which (or from which) a new civilisation was born. The reason for the decline is always due to the tireless repetition of schemes of behavioural legacies and to a lack of innovation and competition with outside forces. Birth, apogee and decline of a civilisation are generally measured in terms of political power, economic productivity and cultural patrimony. Of these realities, only the third one can be separated from its historical context and be entirely transmitted from one dying civilisation to an emerging new one. But this operation has very rarely happened without having the old poisons simultaneously inoculate the young organisms.

Knowledge allows a civilised society to prevail on its enemies, to overcome adversities, to control their surroundings and to improve the standard of living. A desire for security and well-being represents the psychological force that pushes man on the arduous path of trying and searching. Security and well-being do not have an absolute value, but they are always related to a reasonable satisfaction, situated between what is considered possible and what is believed to be ineluctable. What happens when this state is reached? When, once the difficulties have been removed and the necessities assured,[5] does nothing remain but the perspective to exist and survive? One may answer that the natural curiosity of man has always represented a stimulus to increase his knowledge. However, curiosity (from Lat. *cur* = why?) is born from restlessness, from expectations not having been met. Very rarely do we ask ourselves the cause of favourable events. There is, indeed, a ludic aspect to knowing, but, except for exceptional cases, it does not explain the immense social effort that acquiring new knowledge demands. For this reason, in a society where the patrimony of knowledge is sufficient to ensure security and well-being, a civilisation can die of tedium and disinterest. The idea that, independent of any given social context, the transmission of a large patrimony of knowledge can have only positive effects is, therefore, fundamentally dubious. Knowledge, as an instrument of human progress, can become, like the Gorgon, a goddess that transmutes anyone who gazes upon her into stone.

---

[5] Modern studies indicate that prehistoric man enjoyed more comforts and resources than was commonly believed. Also, his fight for survival was probably less arduous than in some later historical periods.

Chapter 1
Time of Growth

"Si omnia a veteribus inventa sunt,
hoc semper novum nobis erit,
usus et inventorum ab aliis
scientia ac dispositio."[1]

(L. A. Seneca, *ad Lucilium VII, II*)

1.1 Orient and Greece

Geometry and mathematics constitute a kind of knowledge, which is directly founded on these two pillars that support our "thinking thought". The former is related to the perception of the unity of space, the latter to the multiplicity of things. The principles of non-contradiction and of identity, which represent the foundations of logic, are, in some way, a lingual translation of these primary perceptions. In mathematics the basic concept of cardinal numbers, defined by the relation between different sets (equal, greater, smaller), is bound to cerebral elaborations of visual images and is in part also present in other advanced animal species.

Cardinal numbers can be directly perceived by their characteristic nature (singularity, duality, trinity, etc.) only for a few units, generally less than five. For greater numbers we are forced "to count", i.e., to execute a sequence of visualisations, in which images of singular objects or group of objects take part. The process of counting does not require a large base. It has been sometimes observed that primitive peoples, who have not succeeded in conceiving numbers greater than three, are still able to count up to relatively large figures, relying on memorisation based on pebbles, carvings or to references to parts of the body.

---

[1] *"Even if everything had been discovered by our ancestors, this will always remain new for us: the application as well as the understanding and ordering of what was invented by others"*

C. Ronchi, *The Tree of Knowledge*, DOI: 10.1007/978-3-319-01484-5_1,
© Springer International Publishing Switzerland 2014

However, a much greater power of abstraction is demanded in defining the concept of ordinal number, that is to say, a number positioned in a given sequence of numbers. This necessarily implies the possession of a perfectly structured language in which the notion of "following" can be incorporated. The same can be said for geometrical shapes, the definition of which is based on the ability to mentally translate and overlap the images of the objects perceived in space, where their order is defined by metric (extension, distance) and topological (orientation, symmetry and connection) properties.[2] It is an astonishing fact that mathematics and geometry already reached a high degree of development in the primeval period of human civilisation, advancing in complexity to that of the empirical sciences. Their history marks the path of development for all civilisations, which, in a five thousand year history, have gradually channelled this knowledge to a common stream.

In this chapter, we will briefly review the main stages of this course, taking as a point of reference our Western civilisation, but remaining aware that the river of our scientific thought has been continuously nourished by affluent currents flowing from distant sources scattered through the entire Asian continent.

1.1.1   Geometry Called Mother of Mathematics

Geometry was born from the elaboration of problems concerning astronomy (calendar) and land-surveying (extension and borders of land property) because geometry could offer methods to solve these problems in such a way that a general consensus could be reached, an agreement which was of vital importance in society.

As for astronomy, archaeological finds pertaining to all known civilisations show that, since prehistoric times, astronomical observations have always been an object of interest and particular concern. The proof of the assiduousness and the precision with which they were performed is supplied by the remains of astronomic observatories and by the orientation of important buildings and places of cult as well as by the complexity of some calendars used in antiquity.

We don't know how the first astronomical observations were translated into numerical measurements. The oldest evidence is supplied by two Assyrian tablets dated to around 650 B.C. containing the names and the co-ordinates of about sixty stars and constellations. Modern astronomers, however, have demonstrated that the original measurements, from which these data originate, were made in the region of Nineveh in the second half of the second millennium B.C.. We must, therefore, conclude that the data of these tables were copied and then handed down for centuries in the Mesopotamian region. Furthermore, one can see that ancient Greek astronomers (Eudoxus, Aratus) accepted them virtually unquestioned until the IV century B.C.:

---

[2] We must here emphasise that our perception of the world is first defined by our sight, i.e., by the detection of photons emitted by surrounding objects. The eye is itself a simple array detector which reacts to electromagnetic waves with wavelengths somewhere between approximately 390 and 750 nm, but it is the task of the brain to elaborate the perception of space and distance. This extremely complex elaboration results from a combination of images in movement. We can, therefore, understand why in our perception of reality space and time are essentially interrelated.

Fig. 1.1 Part of an Egyptian papyrus of the XVII century B.C. found two hundred years ago during non-professional excavations near Luxor. The complete papyrus is approximately 5 m long and contains the description of about 100 problems of mathematics and geometry, still not all deciphered (*Rhind Papyrus, British Museum, London*)

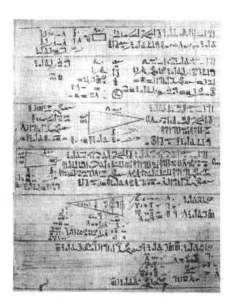

evidently, new measurements of a comparable precision had not yet been executed; otherwise they would have shown remarkable variations from the ancient ones due to the precession of the equinoxes. This indicates that these Assyro-Babylonian astronomical measurements represented the best that existed for centuries. They had certainly demanded a considerable expertise as well as permanent astronomical observatories and instruments of measurement not commonly found, meaning they could not be repeated frequently.

As for land-surveying, goniometric triangulations and measurements were vitally important in Egypt, where every year, after the Nile floods, the borders of lands had to be restored on the basis of cadastral maps. Numerous papyri exist, showing that, starting from the third millennium, Egyptian land-surveyors had already mastered methods for the correct calculation of lengths and geometric areas (Fig. 1.1).

It is likely that these notions and techniques were spread throughout all the countries of the Mediterranean Orient and in the colonies of *Magna Graecia*. How many of these notions were of local origin and how many had been imported from the East will always remain contested, but it is certain that they were expressed in scientific terms for the first time in Greek regions, terms which are still perfectly valid today.

Geometry, as the abstract science of shapes, freed from the chains of particular practical applications, bloomed in classical Greece and developed, without any break in continuity, until now. The undisputable authority, with which Greek geometers, from the VII century B.C. until the Alexandrine age, demonstrated their theorems on the sole basis of imperfect graphical representations, demonstrates, even from a modern point of view, the exactness of this science. In fact, the corpus of the principles of geometry increased incessantly over a period of more than two thousand years, surviving successive cultures and various schools, but always maintaining its own perfect coherence .

Geometric figures and the concepts of their translation and superimposition are primordially intuitive but imply a profound ability for and constant training in abstract thinking (today a critical analysis of these aspects would carry us much farther away). Still more problematic are the relations between the different elementary objects of geometry: point, line and plane, in the intuitive representation of which are placed the specific assumptions that define the property of space—for example, that two points are enough to define a straight line, that in a plane two parallel straight lines never meet, that, in a plane containing a straight line L, it exists only one perpendicular to L, and other similar propositions, the validity of which can be demonstrated only by assuming an equivalent one. It was Euclid, the greatest geometer of classical antiquity,[3] who defined the principles by which the property of geometric figures could be deduced using a rigorous method. His main work, entitled "*Elements*", represented the foundation of geometry for more than two millennia and the space defined by its principles was thought to be the sole one corresponding to real space (Figs. 1.2 and 1.3).

Euclid founded his system on five postulates that are clearly formulated in the 1st book of his *Elements*. The first two concern the property of the straight line and its extension to infinity. The third deals with the properties of the circle, the fourth those of right angles. The fifth, the most problematic one, and repeatedly reformulated in different ways, asserts the uniqueness of the parallel to a straight line L passing through a point P and lying in a plane containing L.

Fig. 1.2 Tablet of a text in cuneiform characters, from the library of King Ashurbanipal (VII century B.C.), belonging to a collection containing the name and the co-ordinates of 66 stars and constellations with their rising and sunset times. They represent a total of 190 experimental data of observation, which, when analysed using modern theory [3] reveal that the original measurements date back to the XIII century B.C. (*Tablet Mulapin, British Museum, London*)

---

[3] In spite of the importance and transmission of his work, which has come down to us almost in its entirety, we know very little about his life. Even his place and time of birth are unknown. From different quotations by later authors we infer that he flourished around 300 B.C., sometime between Aristotle and Archimedes.

All five postulates are the fruits of abstraction, the modern analysis of which leads to profound aspects. Already the uniqueness of a straight line joining two points A and B presupposes a problematic metric reference, that allows the definition of the minimal distance between A and B, but its extension to points infinitely distant exceeds our ability to represent it; its intuition presupposes a mental activity that, in some sense, *precedes* the infinite production of a line in space [4] by predicating somehow its properties. In the end, in order to understand the concept, it is necessary to describe the structure of our senses and mind and their patterns of operation.[5]

1.1.2 The Construction of "Euclidean Space"

The ancient Greeks were conscious that geometry was not only the science of abstract forms, but also the science of real space; yet they realised that the concept of space was, at the same time, a creation of geometry. However, they were never disturbed by the doubt that one could conceive various geometries. From this conviction, they attempted to produce proofs of the "truth" of the Euclidean space over a long period of time, in which ancient physicists were organising the laws of the newborn natural sciences. The effort proved, however, to be in vain. For instance, the thesis that a straight line L and a point P outside it define one plane and in this plane there is only one straight line parallel to L and passing through P, definitively failed

Fig. 1.3 Fragment from the *Elements of Geometry* of Euclid found among the papyri of Oxyrhynchus, dated to around the end of the first century A.D.. (*Museum of Archaelogy and Anthropology at the University of Pennsylvania - E2748 in the Museum catalogue*) A comparison with the Egyptian papyrus of Fig. 1.1, still reveals the unique Greek genius despite the poor quality of the drawing which is at odds with the logical rigour of the text

---

[4] Aristotle repeatedly address this problem and he solved it by accepting infinity as a *potency*, but not as an *act*. In his physical model the universe is, however, finite and, for instance, an infinite straight line cannot exist as a real entity.

[5] One should recall that Plato's first authenticated definition of a straight line is "*the line of which the middle always covers both ends*" (*Parmenides*, 137 E), whereby he appeals here to our sense, implying that the line of sight is straight. Euclid's definition is as follows: "*a straight line is that which lies evenly with the points on itself*". This can be interpreted that among the lines having the same extremities the straight line is the least.

after innumerable endeavours lasting several centuries and was finally accepted as a postulate.

The fifth Postulate *"of the parallels"* is of the greatest importance, as it represented the starting point for speculations that finally led to modern non-Euclidean geometries as well as to concomitantly new concepts of space after more than two millennia. Let us examine the postulate as expressed in Euclid's *Elements*:

– *"If a straight line falling on two straight lines makes the interior angles on the same side less than two right angles, the two straight lines, if produced indefinitely (ἄπειρον, ápeiron), meet on that side on which the angles are less than the two right angles."*

Euclid was aware that this proposition was not demonstrated. The problem was clearly stated by Proclus (412–485 A.D.), the last eminent geometer of Greek civilisation [91]:

– *"The fact that some lines exist, which approach each other indefinitely, but yet remain non-secant (asymptotes), ... although paradoxical is nevertheless true. May not then the same thing be possible in the case of (parallel) straight lines? Indeed until the statement in the (fifth) Postulate is secured by proof, the facts shown in the case of other lines may direct our imagination in the opposite way .... It will be necessary at that stage to show that its obvious character does not appear independently of proof, but is turned by proof into a matter of knowledge"*[6]

Thomas L. Heath emphasises [92, vol. I, p. 207] that we have here the source of the idea which was later further developed by Lobachevsky namely that in a plane the lines issuing from a point P can be divided, with reference to a given line L, into two classes: *secants* and *non-secants* and the lines which divide the secant from the non-secant classes can be defined as parallel to L.

In the centuries to follow many illustrious geometers and mathematicians, from Ptolomaeus to Gauss, have tried in vain to obtain the proof of the fifth postulate. In his definitive commentary to Euclid's *Elements* Heath [92] examines a dozen of these attempts, reporting the various re-formulations of this postulate, which mark the road towards modern non-Euclidean geometries. The demonstrations, obviously all faulty, are sometimes at first glance so convincing that Heath felt it necessary to underline the points where invalid arguments were used.

Since its original formulation, the fifth postulate establishes a link between the notion of parallelism and the properties of the triangle, the simplest polygon.

---

[6] It is worthwhile reporting Proclus' criticism on the fifth Postulate.
"It is impossible to assert without some limitation that two straight lines produced from angles less than two right angles do not meet. On the contrary, it is evident that some straight lines do meet, although the argument must be proven that this property belongs to all straight lines. For one might say that the lessening of the two right angles being subject to no limitation, with such and such an amount of lessening, in excess of this they meet." [91], p. 37.

Gerolamo Saccheri (1667–1733) a professor at the university of Pavia, made significant progress in these attempts. He approached the fifth Postulate by constructing a plane quadrilater $\begin{smallmatrix} C & D \\ A & B \end{smallmatrix}$ with AC and BD equal and perpendicular to the base AB [94]. The Postulate entails that the angles $A\hat{C}D$ and $C\hat{D}B$ are both right angles. But Saccheri examines the cases that these angles be, respectively, *obtuse* and *acute* and deduces that in the former case the sum of the angles α, β, γ of a triangle is greater than two right angles and in the latter case the sum is less than two right angles. Johann Heinrich Lambert (1728–1777) made further progress by showing that the area Δ of a plane triangle in the hypothesis of an *acute* angle is:

$$\Delta = k(\pi - \alpha - \beta - \gamma)$$

And in the case of an *obtuse* angle is:

$$\Delta = k(\alpha + \beta + \gamma - \pi),$$

where k is a positive constant.

Furthermore, Lambert made the important observation that Δ is the area of a spherical triangle over a sphere of radius r (with $k = r^2$) in an acute-angle case. But in the case of the obtuse angle the formula represents the area of a spherical triangle with an imaginary radius of $r = \sqrt{-1}$, a concept that will be fully developed in Lobachevsky's *Imaginary Geometry*.

The specific "subjectivity" of Euclidean geometry has deep roots. Historically, its postulates were expressed in various forms, but all indicate that some *a priori* properties are arbitrarily attributed to the concept of infinite and infinitesimal magnitudes, according to which space can respectively "extend" or "contract" indefinitely.

To have an idea of the depth of these problems we can do no better than to consider a formulation of the fifth Postulate as elaborated by the great mathematician Johann Carl Friedrich Gauss (1777–1855):

- " *If I could prove that a rectilinear triangle can be given, the extension of which is greater than any given area, I am in a position to prove perfectly the whole of geometry*" [93].

Here the crux of the matter is touched upon: actually, as reported above, the properties of the triangle, and, in particular, the value π of the sum of its angles represents the distinction between Euclidean space and curved spaces.

When, in the XIX century, geometries were developed, in which the fifth postulate was *not* assumed, a definitive distinction between the *intuitive vision* and the *mathematical description* of the geometric space was definitively sanctioned. This distinction undermined the foundations of the metaphysical principle of the perfect correspondence between Euclidean space and real space. Nevertheless, we cannot but admire the acumen of the Greek geometers who, at the same time as they were laying down the foundations of geometry, recognised in Euclid's postulates the cornerstone that established the junction of space with a description of physical phenomena.

However, the suspicion that infinity may not have predictable properties represents, from a logical point of view, an inherent thorn in the heart of geometry. In fact, throughout the development of geometry, the non-finite, the ἄπειρον, has continuously revealed elusive aspects. These concepts not only introduced serious problems when limitlessly large extensions were being considered, but also when the geometrical shapes were confronted with the infinitely small, the infinitesimal, involving a fundamental dilemma of the most elementary morphologic intuitions. Thus, for instance, the circle, unequivocally defined in a plane as the locus of the equidistant points from a fixed point, appeared, from a physical perspective, as the limit of an infinite succession of polygons with an increasing number of sides of decreasing lengths. Many geometers could not conceive how such figures could be realised in real space without passing through this infinite succession of polygons. Therefore, the suspicion that curves, like a circle, were a sort of chimerae not only rose from metric considerations, but also from their incompatibility with conceptions of matter based on atomic theories. If, on one hand, the *continuum's* properties of geometric space were, from the beginning—unconsciously or explicitly—accepted, on the other, in the various philosophical schools, they became the subject of a controversy, which was nevermore to be settled (Fig. 1.4).

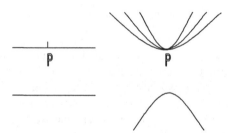

Fig. 1.4 It was only in the XIX century that Euclid's fifth postulate (of the parallels) was rejected and a new geometry was consequently developed by Nikolai Ivanovich Lobachewsky (1792–1856). The language with which Lobachewsky's and other non-Euclidean geometries were formulated is not visually intuitive, but based on abstract mathematical entities. In the sketch on the *left-hand side*, a straight line and its unique parallel passing through a point P in a given plane containing the line are designated in Euclidean space. On the *right-hand side*, Lobachewsky's curved space is sketched, where it can be seen that the straight line is transformed into a curve and an infinity of lines pass through point P that do not intersect it (i.e., by definition, parallel lines). On this subject, it must be noted that a curved space is not an absurd concept, since today it forms part of fundamental models of modern physics

1.1.3 From Forms to Numbers

Actually, Greek geometers soon understood what a portentous instrument the assumption of simple figures of continuous curvature was, like the circle. In fact, the study of the intersections of quadric and conic curves with straight lines and planes turned out to be enormously productive already in the Alexandrine age, allowing graphical solutions of quadratic and cubic equations which, in a mathematical form, would have required algebraic language and the notion of real or even complex numbers. Actually, the solutions could be obtained by overcoming or circumnavigating the conceptual difficulties and the formal limits of the mathematics of that time. The whole projective geometry, with its broad developments and applications, is based on the supposed presence of one of these curves to the operator's inner eye.[7] Without a doubt we can assert that, in classical Greece, geometry was centuries ahead of mathematics in solving complex problems. Mathematics had, in fact, a longer gestation period. During the age of the first *floruit* of geometry, mathematics was limited to practical measurements and methods of calculation. Given the mediocre precision of practicable measurements in usual applications, integer numbers were mostly used (Fig. 1.5). The concept of the fraction and of rational numbers was, however, clear from the very beginning and was applied in order to solve simple geometric problems by means of numerical equations. However, the concept of *continuum* and of real numbers always remained relegated to the frontier of the field of operation, so

 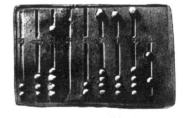

Fig. 1.5 On the *left-hand side* Representation of a Persian civil servant collecting tribute in order to finance the military expedition of King Darius against Greece. He is executing numerical calculations using a small abacus, holding it in his left hand. The number written on the table, probably the total figure collected, is expressed in "*Mirioi*" (10,000) and corresponds to 7.543.733. (*Detail of the so-called "Vaso di Dario", manufactured probably around 340–320 B.C. and found in 1851 near Canosa di Puglia, Italy, now in the National Archaeological Museum of Naples, cat. H3253*). On the *right-hand side* an abacus dating to Roman times similar to that depicted on the vase

---

[7] It is known that, in order to completely liberate theory from a visual concretisation of geometrical curves, in 1847 Karl Georg Christian von Staudt, a German geometer, published a treaty of projective geometry [2] in which he did not use any figures.

that the Greeks called the numbers that are not expressible as fractions of integer numbers, ἄλογοι (*álogoi*), that is literally translated into the modern appellation of *irrationals,* definitively establishing a verbal warning on conflicts with reason their use could possibly entail.

However, the application of rational numbers very quickly reached their limit. Pythagoras (582–507 B.C.) demonstrated that the length of the hypotenuse of a triangle rectangle cannot be generally expressed as a fraction of the lengths of the catheti.

But the reaction of Greek mathematicians in the face of this serious drawback was typically conditioned by a *forma mentis* that refused to accept any concept that was not clearly defined, and with regard to numbers, they were of the opinion that the integers were the only ones to be clearly defined. Therefore the concept of incommensurability remained bound to geometry, and never extended to mathematics.

In the many schools of classical Greece the analysis of the concept of limitless (ἄπειρον, *ápeiron*), considered as infinitesimal or infinite, converged in a synthesis advanced by Aristotle, who maintained many points common to the different doctrines. Among these, the important Pythagorean School asserted the principle of the substantial correspondence between numbers and reality, of which the numbers should represent a sort of epiphany. Numbers were, therefore, closely related to geometrical forms and they were not considered as equivalent terms of a succession, but as individual entities, among which there existed a plurality of marvellous relations.

For example, the Pythagoreans pointed out that the sum of consecutive uneven numbers $(1, 3, 5, 7, \ldots)$ constitutes a succession of squares $(4, 9, 16, \ldots)$, corresponding to the amplification of the sole perfectly defined shape of the geometric square, whilst the sum of even numbers produces "rectangles" of variable proportions.[8] The considerations of these properties, as elaborated by Pythagorean mathematicians, are so profound, that their depth is overlooked by people viewing them with modern eyes, who see them as elementary arithmetic results. Figure 1.6 shows how these

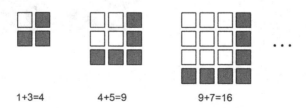

$$1+3=4 \qquad 4+5=9 \qquad 9+7=16$$

Fig. 1.6 The succession of squares of integer numbers was considered by Pythagoras to be the result of the addition of gnomonic squares that act as "containments". The gnomons are always composed of an odd number of units. According to Pythagoras, the idea of "odd" essentially referred to finiteness

---

[8] Note that the succession of squares is completed with *one* This involves the introduction of *zero* in the succession of the odd number. For the Pythagoreans zero represented the assertion of limits and the perfection of units, i.e., of the monad, whose shape is reproduced by successive applications of gnomons.

ideas were closely tied to the belief that reality consists of a combination of perfect shapes. In the example chosen, the square *monad* (the unit is the square of itself) can increase indefinitely the extension of its shape by adding an uneven number of *monads* disposed at right angles, similarly to the shape of the *gnomon*, the L-square used by architects to delimit the walls of a building. Uneven numbers were, therefore, called *gnomons*, from their ability to delimit the perfect square. For this reason, the Pythagoreans defined *limitlessness* in nature through the abstract notion of "*even*", while "*odd*" represented the limitation. It is clear that it was not an argument merely concerning numbers, but rather the quantity (μέγεϑος, *mégethos*). *Even* was, in fact, an attribute of what could be divided into two equal parts and whose dichotomy could be iterated until achieving the limit of an *odd* number, by which we are faced by the existence of an indivisible unit. Yet, removing or adding a unit, the dichotomy of an uneven quantity can continue until we find the primordial unit. Inversely, every quantity is, therefore, composed of a number of even/even or odd/odd dyads to which adding of an odd number establishes the definitive limit. In this sense infinity was considered as even, because is lacking a limit, however, the infinity was seen as a *potential* but not as an *actual* entity. Conversely, the unit, considered uneven owing to its essential indivisibility, represented an *actual* entity. In this picture, the unit (*monad*) can be as small as we want, but it is always finite and indivisible. Aristotle attacked this point in the third book of *Physics* and in *de Coelo* (L.1 c.5). We report here two main passages:

- *"But the problem of the infinite is difficult: many contradictions result whether we suppose it to exist or not to exist. If it exists, we have still to ask how it exists; as a substance or as the essential attribute of some entity? Or in neither way, yet none the less is there something which is infinite or some things which are infinitely many?*
- *It is for this reason that Plato also made the infinites two in number, because it is supposed to be possible to exceed all limits and to proceed ad infinitum in the direction both of increase and of reduction. Yet though he makes the infinites two, he does not use them. For in the numbers the infinite in the direction of reduction is not present, as the monad is the smallest; nor is the infinite in the direction of increase, for the parts number only up to the decade.* [9]*"* (Phys. III 203b; ch.4 and 206b; ch.6)

"The infinite turns out to be the contrary of what it is said to be. It is not what has nothing outside it that is infinite, but what always has something outside it. This is indicated by the fact that rings also that have no bezel are described as 'endless', because it is always possible to take a part which is outside a given part. [10] *"* (Phys. 1, III 207a; c6).

---

[9] Aristotle here means that, whenever one conceives of a number, its order of magnitude (decade) is given and, hence, the number is first expressed by a definite integer 1, 2 . . . 9 of the highest decade and then by those of the lower decades. Actually, in modern mathematics infinity is not indicated as a number, but as a symbol (∞) the use of which presupposes a complex background.

[10] The mediaeval Latin translation is conciser and clearer: " *annulos infinitos dicunt non habentes palam quoniam semper aliquid extra est accipere.*" This is only a property of straight lines and circles. Aristotle indicates the analogy between an infinite straight line and a circle. Actually straight

Finally Aristotle comes to the fundamental conclusion that forbids dealing with infinity/infinitesimal as *actual* mathematical objects (*infinitum actu non datur*).

We shall see in Chap. 4 how this conception implies *in nuce* some important arguments which have recently been taken up again by modern intuitionist theory of mathematics with its criticism of the notion of irrational numbers. It was indeed as problems increasingly arose, the solutions of which were expressed by incommensurable quantities, that cases, in which the solution was conveyed by integer numbers, were considered in greater depth. The Diophant's equations (dated back to the III century B.C., but already studied centuries before in Vedic Indian literature) having the form: $x^n + y^n = z^n$, with n being larger than 1, proposed a question that for centuries fascinated generations of mathematicians until the present day: do integer solutions exist for these equations? If yes, under which conditions and how many are they? Are they calculable?[11] It is clear that the rarity of these cases and the peculiarity of some integer numbers, that, unique among an infinity of other cases, obey important geometric relations, elicited amazement so that metaphysical or even magical properties were attributed to them. Therefore, it happened that, although they were aware of the existence and the frequency of incommensurable quantities, some mathematical schools, extending from the beginning of the first millennium B.C. until the present day, have continued to search for constituent laws of the universe only in integer numbers. Thus, Plato in *Theætetus* suggests that the surface of the five regular polyhedra associated with the five elementary substances, is constructed using a set of different triangles of incommensurable sides and that the morphogenesis of the universe is composed of an integer number of these triangles.[12]

The synthesis and organisation of geometry and mathematics, carried out by great talents, such as Euclid and Archimedes in the III century B.C., led to a consolidation of their corpus as well as to a collateral and progressive refusal of hypotheses contradictory to or theories incompatible with those previously well established. The work of Archimedes of Syracuse (287–212 B.C.) represents the apex of the development of mathematics in the classical world. His brilliant intuitions made it possible to

---

line and circle are considered as the same object in modern projective geometry: the straight line corresponds to a circle passing through a point P and having an infinite radius; whereas the circle centred in P and having an infinite radius corresponds to the straight line defined by all infinite's points (i.e., all directions in the plane).

[11] Pierre de Fermat (1601–1665) formulated one of the most famous theorems of mathematics, known as "Last Theorem of Fermat", which asserts that the above-mentioned equation does *not* have integer solutions for values of n greater than 2. The demonstration of this theorem occupied mathematicians until now and represented a source of the most important collateral results. The last theorem of Fermat was demonstrated in 1995 by using an interesting new type of number called p-adic, which will be presented in Chap. 4.

[12] Karl Popper [9], examining these ideas, also explained in *Timaeus*, concludes that Plato believed the field of continuum could be expressed as a linear combination of a countable base of irrational numbers with integer coefficients. It is not sure that this was really Plato's precise idea, but, even if not, this hypothesis clearly indicates in which direction the speculations of the Pythagorean and the following platonic schools were moving.

attack with method and clarity the problem of infinity and infinitesimal in mathematics. Using what was then called the *exhaustion* procedure [13] (a precursor of the passage to the limit of the modern infinitesimal analysis) he succeeded in solving problems that today would require the calculation of integrals. If further developed, the ideas of Archimedes perhaps would have anticipated the development of modern mathematics by a millennium. Unfortunately, the cultural and political climate of his age was not favourable to such intellectual progress. On the one hand, the devastating Peloponnesian War had started the decline of the *Magna Graecia* with the subsequent Roman conquest marking its definitive twilight. On the other hand, the policy of the Macedonian Kings had already deprived important Greek centres of power as well as their Ionian colonies, transferring their interests to Egypt and to the Middle East, thus creating a new epicentre for most scientific speculations. By the II century B.C. nobody was capable of understanding the importance of the results of the great Syracusan, whose great reputation merely survived as an echo. Of all his works, only those regarding physics and engineering were passed on to posterity, who then handed them down throughout the centuries to the generations to come.[14] Until the Western and Arabic Middle Ages, the few people who were in a position to understand the results of his mathematical analysis, were not capable of developing them further. Thanks to the work of the algebraists of the Italian Renaissance, progress was only achieved a full seventeen centuries later (Fig. 1.7).

The inability to realise what an exceptional advance in mathematics and physics the work of Archimedes represented meant that his successors were unable to channel these sciences on a unique, rigorous course and prevented them from definitively abandoning the heretical currents of scientific thought, which were thus allowed to maintain a certain prestige and authority also thanks to a continuous feeding of the parallel river beds of astrology, alchemy, numerology and other exoteric disciplines. The intellectual atmosphere of Alexandria from the III to the V century A.D. presents a paradigmatic example of how cohabitation of orthodox and heterodox scientific currents can simultaneously exist and is also mirrored in religious diatribes, which at that time were shaping the theology of early Christian religion.

This historical period is characterised by cultural exchanges over a vast geographic area which started in India, included Persia, Arabia and the Byzantine Empire, and finally extended along the North African coast until the Atlantic. It is interesting to note that the inhabitants of these regions, though having developed different intellectual heritages, were in possession of comparable technical resources. Actually, from a

---

[13] The *exhaustion* method is based on the following procedure: for example, to find the volume V of a solid whose shape, S, can be increasingly approximated by a sequence of suitable simpler shapes S'_n, whose volumes V'_n we can calculate. Then one moves on to demonstrate that, if the approximation of S by S'_n has been improved as much as needed, it is impossible that for sufficiently large n either V can be greater than V'_n or that V can be smaller than V'_n. It must, moreover, be noted that the inequalities between quantities (volumes in our case) are here founded on visual relations between different figures.

[14] The book of Archimedes, in which he explains his "*Method*" of integration, was recovered only in 1906 in Constantinople, where it was buried under the dust of twenty centuries.

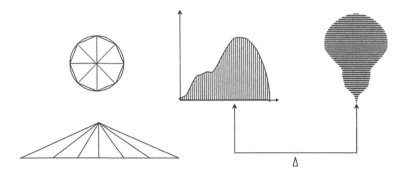

Fig. 1.7 In order to calculate complex surface areas, Archimedes developed various methods based on the decomposition of the figure into the sum of infinitesimal elements.
Two examples of his methods are given here.
On the *left-hand* side, the area of the circle of radius r is approximated by decomposing it into the sum of equal triangles; these are then aligned by deforming them, but leaving their base and their height unchanged (and hence their areas). The resulting figure is an isosceles triangle of height r, the length of whose base tends to the length of the circumference of the circle when the number of the triangles in which the circle is decomposed tends to infinity. Finally, the exact area of the circle is obtained as: $2\pi r.r/2 = \pi r^2$.
The example on the *right-hand side* illustrates a method based on physical considerations: Let us suppose that a surface, S, subtended to a curve $f(x)$, is ideally decomposed into a set of the thinnest homogenous material bars of width $1/n$ and one adjacent to the other (see the Figure); let the surface be placed on the left arm of a balance with the vertical line passing through the barycentre, B, of S. We start now moving the bars from their original position piling them vertically over the barycentre and, for every displaced bar, we lay down horizontally an equal bar on the right arm of the balance. When the process is completed, all the weight (whose value is a measure of the total area of S) is piled on the left arm and is exactly counterbalanced by the resulting stack lying on the right arm of the balance, whose area, for values of n tending to infinity is given by:
$A_S = \frac{1}{n}\left[f\left(\frac{1}{n}\right) + f\left(\frac{2}{n}\right) + \ldots + f\left(\frac{n}{n}\right)\right].$
Since Archimedes was not aware of the general concept of function, his interest was in finding the exact area of geometric figures simple enough to enable him to calculate the series in the parenthesis, thus obtaining the expression of the area.
Despite obvious limitations, the most important aspect of this procedure is that it foreshadows an integration of function $f(x)$, as defined in modern infinitesimal analysis

technical point of view, their situation remained almost stagnant for several centuries. The creative synergy that, from the VIII century B.C. onwards, had characterised the extraordinary blooming of Greek classical culture from the fertile grounds of the Near East and Egypt, had already been exhausted in the III century B.C.. In later periods, much effort had to be expended if the intellectual heritage of the past was to be retained. At the same time, schools were integrated in state bureaucracy. The compilation of important encyclopaedic works of this period, which were a substantial factor in the canonisation of scientific knowledge, was only possible within the great libraries of those times. Furthermore, scientific experiments, sometimes using equipment of great complexity, could only be carried out in court surroundings, where the finest and most specialised handcrafts were to be found. However, in this

Fig. 1.8 *Left* The so-called Antikythera's mechanism, found in a Roman shipwreck from the first-second century B.C.. It is a horological device for the analogical calculation of the planetary orbits. *Right* The photo shows a recent reconstruction by Robert J. Deroski, based on a model by Derek J. de Solla Price) (*National Archaeological Museum, Athens*)

respect, one should note that a concept of applied science, in the modern sense, was totally absent from all ancient historical cultures. Instruments, devices and machines of great complexity and precision, comparable to those of modern horology, were mainly constructed only for demonstrative purposes or in order to incite awe and wonder.

For instance, although the mechanical power of vapour was well known, nobody ever thought to put it to practical use. This might be attributed to a lack, in a certain sense, of imagination or creativity, typical of conservative societies of those times. If we consider, on the one hand, the abundant supply of slave labour which was deemed sufficient for the execution of the hard work while, on the other hand, the absolute submission of the masses to the authority of the state meant that both conditions could not yield what today we would call a "demand-based market", rising from the bottom of society. The economy itself, based on metallic currency, did not have the necessary instruments of credit to support industrial initiatives that could grow beyond the demand of the restricted dominant class.

For this reason, whilst cultural exchanges expanded to encompass more and more geographic areas, the technological platform remained almost unchanged for centuries. Therefore, the time need to effect an exchange of new ideas and to implement technical innovations became proportionally longer (Fig. 1.8).

1.1.4 Astronomy Called Mother of Physics

Since antiquity, the relationship between physics and astronomy, as defined from a modern point of view, was characterised by a substantial divergence in their objectives and their methods of investigation. The physics of the celestial bodies, consisting in the investigation of their origin and nature, belonged to the field of philosophy, whilst the determination of astral motions was considered to be its own field. As long as we know, astronomers performed measurements that were, for the most part, handed down from one generation to the next, and which were then regularly corrected and improved. The prevailing preoccupation of the time was to calculate the periods of revolution of the planets and stars, which could sometimes be rather lengthy, in order to predict their position as a function of time. The hour, the day, the lunar month, the solar year and other cycles of longer duration were defined for religious and divinatory purposes or, simply, in order to establish the dates of important private and social events. For obvious reasons, the measurement and, above all, the conservation and transmission of astronomical data, could only be guaranteed within social groups that were safe from instabilities and vicissitudes, groups, which could only exist under court protections or, particularly, within sacerdotal classes which arose around important temples.

Conversely, in antiquity, the scope of philosophy was wider and not easily defined, since it covered a whole field of speculation. Throughout the different ages and civilisations, the philosopher's image assumed different features, but was generally characterised by his impact on society, as reflected in schools and currents of thought. Cosmogonic explanations and the laws of nature have always been at the centre of philosophical speculation, an activity which was often not in line with established mythical and religious beliefs. While freedom of thought characterised the development of different philosophical currents, it sometimes gave rise to conflicting ideas. At the same time, however, it made possible the organisation of empirical knowledge in a theoretical perspective where the open critic could exert a selective and corroborating function. Thus, astronomical data were ordered within a comprehensive vision of the cosmos. The measurements were, obviously, obtained in a system of reference where the observer imagined himself as the centre of a universe, in which the moon, sun, planets and stars rotated around him. These experimental data represented a point of departure for more or less elaborate cosmological models. Confirmation of these models, though not rigorous from a modern point of view, constituted a first attempt at formulating scientific interpretations. Within these various conjectures we can find amazing intuitive conclusions, such as, for instance, the theory of the vortices of matter, developed by Anaximander, which seems to anticipate modern notions of this topic,[15] where singular points are scattered in the universe enabling the cosmic

---

[15] Anaximander of Miletus (610–546 B.C), a philosopher-naturalist educated in the school of Thales, a fellow citizen of his, proposed a cosmological theory, according to which from the limitless One (ἄπειρον, ápeiron), the principle of Creation (ἀρχή, arché) is the centre of a vortex (περίχρεσις, períchresis) that, in dilating itself, creates at its periphery smaller vortices and thus differentiates all existing things through a separation of contrary elements. The unique quality of all beings is the fruit of the embezzlement of one of two contrary terms (*injustice*); this, however, will be recomposed from the *justice* in the original One. Therefore, infinity of universes are born and die continuously. This concept presents amazing similarities to some modern cosmological models.
It is necessary to note that what remains of the work of Anaximander are only a few fragments quoted and commented on by Aristotle, but the interpretation of these extant passages is contested [4]. What, however, is pertinent here is the echo of his theories, which resounded in subsequent generations. Even today, Ilya Prigogine (1917–2003) and René Thom (1923–2002), two great experts in the theory of structures of complex thermodynamic systems, had a particular predilection for the

matter to expand or collapse. Nature ($\Phi \acute{v} \sigma \iota \varsigma$, *Physis*) was the object of speculation for philosophers, who, though interested in astronomical measurements, were seldom familiar with experimental techniques or the calculation methods of astronomers; yet they were perfectly capable of understanding the meaning of the data and of adapting them to suit their more general purposes. For this reason, primitive physical models, though based on experimental observations, were not formulated using rigorous argumentations, nor were they intended to precisely reproduce phenomena, but rather to qualitatively define their causes. In this way, although geography was the more accessible discipline, it too presented several remarkable problems: for instance, the visible earth, the $\grave{o}\iota\kappa o\upsilon\mu\acute{e}\nu\eta$, was imagined as encircled by the Ocean's waters and, consequently, how the earth could remain suspended in space was a question which needed to be answered. From this perspective, a heliocentric vision of the universe appeared, obviously, difficult to explain and was considered only by very few naturalists, among whom was Aristarchus of Samos (310–230 B.C.), who reinforced its validity intuitively on the basis of his calculations of the dimensions of the sun, which showed that this star was much larger than the earth. Yet his argument was not persuasive enough to prevail over geocentric models which, within the limits of accuracy of astronomical and geographic observations, appeared perfectly reasonable.

In the IV century B.C., Aristotle developed a comprehensive and coherent interpretation of Nature and supplied reasonable solutions to a great number of problems and questions concerning the nature of the universe, its laws and its first causes. His theory was based on the hypothesis that matter consisted of a mixture of five primordial elements, respectively called, from the "heavier " to the " lighter ", *earth, water, air, fire* and *aether*. In this vision, the centre of our planet, made exclusively of earth, is the place towards which the heavy elements gravitate, a theory which explains the equilibrium position of our planet in the centre of the universe. Conversely, the lighter elements move upwards and vertically. The heavenly bodies, made of light matter, rotate around the earth on concentric spheres composed of aether, the thinnest element which permeates the whole cosmos.

One of the more important consequences of the Aristotelian model is that, since it does not allow for privileged directions in physical space around the centre of the earth, the universe tends to be perfectly isotropic; in particular, the motions of the celestial bodies, made of pure light matter, must necessarily be periodic and circular or composed of a combination of circular motions. On the basis of this physical hypothesis, Aristotle assumes the earth to be immovable in the centre of the firmament's vault. In chapter XII of the second book of "*de Coelo*" he examines various alternative models, among which the heliocentric one which he reports to be very popular among Italian intellectual circles (*circa Italiam*), where the Pythagorean doctrine was very influential. According to this view, there is a fire (called the "Prison of Zeus") in the centre of the Universe and the earth moves on a circular orbit around this fire. The earth must be spherical and rotate around its polar axis, producing alternately day and the night in the hemisphere where we live and in the antipodal one

---

ideas of Anaximander.

(called ἀντίχϑων, *antichton*). However, Aristotle says that the only astronomical argument in favour of this hypothesis is that it accounts for the greater frequency of lunar eclipses with respect to solar ones, the other arguments being of a mystical character (he also cites similar opinions related in Plato's *Timaeus*, where he follows Pythagorean doctrines). Aristotle has, therefore, no hesitation in refuting the heliocentric model, by simply appealing to common sense. Aristotle's assumption of the earth as the fixed centre of the universe had, therefore, no ideological motivations, as has often and erroneously been claimed, even nowadays. Rather, he put forth a reasonable hypothesis, on which all astronomers and the great majority of philosophers were in agreement. The weakness of the Aristotelian model instead resides in its total lack of a suitable concept of dynamics of motion. Thus, although he had correctly devised the existence of gravitation, he did not succeed in connecting the speed of the bodies with a notion of applied force. Therefore, his model supposed that movement has its first cause in celestial spheres, which rotate around the earth, imparting different speeds to planets, the sun and the moon and to objects on the earth which are compelled into merging with the harmonic motion of the celestial spheres.

Yet, the astronomical observations in the IV century B.C. were already complete and precise enough to demand models of planetary motions much more complex than simple circular orbits of constant velocity. In particular, the planets, whose orbits had been determined from measurements carried out and transmitted by Egyptian and Babylonian astronomers during more than a millennium, presented some important "*anomalies*". The most important ones regarded the observation of regressive phases in the revolution of the planets, during which they seemed to invert the direction of their motion on the orbit around the earth. This effect, in the light of modern astronomy, is easily explicable by considering the simultaneous revolution of the earth and planets around the sun with varying orbits and speeds. In an heliocentric reference system, admitting that the orbits are circular and concentric means one can calculate the orbit of one planet, as seen from the earth, using simple trigonometric functions. An example is shown in Fig. 1.9, where the approximate orbit of Jupiter is drawn, as seen from the earth. Obviously, a change of reference system, from a geocentric to the heliocentric view point, transforms planetary orbits into much simpler geometric curves. This mathematical procedure, however, was unknown in antiquity. The regressive phases of the planetary motions were then called "*epicycles*" (literally translated as "overhanging" cycles) and interpreted as the tumbling of a small sphere, on which the planet is fixed, on to a larger sphere (called "*deferent*"), defined by the centre of former one. The heavens were, therefore, perceived to be a system of rotating spheres made of aether, which yielded and maintained planetary motions, by a mechanism analogous to that of gear machinery.

On the other hand, on the basis of astronomic measurements one could conclude that the periods of the revolution orbits of planets indicated harmonic relations [16]

---

[16] The Pythagorean School investigated harmonic relations in music by comparing sound intervals with the length of the cither strings producing them. The main intervals of the *octave*, *fifth* and *fourth* intervals were elements of harmony of sounds and related to the fractions 1/2, 3/2 and 3/4.

Fig. 1.9 The orbit of a planet,
P, around the earth, T, calculated based on the hypothesis
that the orbits of P and T
around the sun, S, are circular.
It can be seen that the orbit
of P passes through phases
of regression (called epicycles), the number of which
increasing with the period
of revolution of the planet P
around the sun.
The motion of P, seen from
the earth, corresponds to that
of a point fixed on a small
wheel, which revolves on a
circle with its radius being the
smallest distance between P
and T

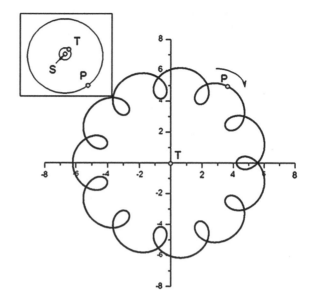

which incited admiration and wonder, and, finally, led to an interpretation of the
heavenly motions in a grandiose scenario where mathematical harmony was thought
to govern the whole Universe.

However, as astronomical measurements became more precise, the number of
small, but unquestionable "anomalies" in the motion of the planets with respect to
ideal models also increased. In order to be able to reproduce these experimental data
in a reasonable manner, additional "gears" had to be added, producing rotations on
spheres which were eccentric with respect to the earth.

One, in particular, concerned the movement of the sun. Hipparchus of Nicaea
(190–120 B.C.), the greatest astronomer of antiquity, had constructed an advanced
instrument to measure with high precision solstices and equinoxes, and, from the
position of these four points, he observed that the sun covered the ecliptic path with
a non-uniform angular velocity, lower in summer than in winter. Moreover, he was
able to ascertain that his own measurements of the longitude of some stars presented systematic differences from those carried out 150 years earlier by Timocharis
(320–260 B.C.), an astronomer of Alexandria who had published a catalogue of
stars. Hipparchus was able to estimate with sufficient precision the difference of the
co-ordinates of some stars from Spica, the brightest star of the Virgo constellation,
and he calculated a systematic deviation of his data from those of Timocharis of two
degrees with respect to the equinox, concluding that the position of the equinoxes

---

Ancient astronomy recognised five relationships based on the twelve divisions of the zodiac. Ptolomaeus taught that their significance came from an analogy with the ratios of the musical scale.
The conjunction corresponds to unison; the opposition divides the circle in a 1:2 ratio (*octave*); the
sextile (5:6) corresponds to a minor *third*; the square (3:4) corresponds to a perfect *fourth*; and the
trine (2:3) to a perfect *fifth*.

Fig. 1.10 The orbit of the
sun (represented by the sign
*) seen from the earth, T,
appears to be eccentric (in
this design the eccentricity is
exaggerated in order to render
the effect more visible). One
can interpret this orbit as a
perturbed circular motion,
produced geometrically by
an epicycle moving with
equal speed and a direction
contrary to that of the apparent
revolution of the sun around
the earth

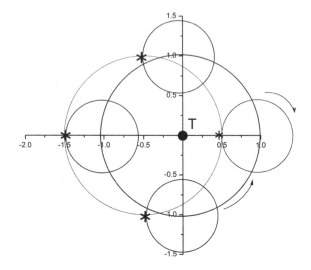

in the zodiac was moving not less than one degree every hundred years: it was the first measurement of the precession of the equinoxes. Modern measurements have yielded one degree every 72 years.

Based on these data and taking the earth as the origin of the co-ordinates, if one wanted to retain the hypothesis of the uniform circular motion of the sun, its orbit would have to be assumed as eccentric. This observation implied a crisis in the theory of the epicycles and forced astronomers to introduce more complex geometric models into their calculations consisting of an increasing number of eccentric movements. Thus the astronomers displaced the centre of the circular orbit from the earth, to a *Punctum Aequans*, which compensated for the effect of the eccentricity. Later on, in the II century B.C., Theon of Smyrna demonstrated using trigonometric methods that the eccentricity of the solar orbit could be reproduced by assuming epicycles of equal and contrary speed to that of revolution of the sun, in which case the resulting orbit turned out to be a circle whose centre's eccentricity was equal to the radius of the epicycle (Fig. 1.10).

It was in the late Hellenistic period that Claudius Ptolomaeus of Alexandria (83–161 A.D.) began his admirable work of reordering the astronomical data, and thereby constructed a system that remained in force for 1500 years. Though convinced that it was possible to calculate planetary orbits using a single geometric model, Ptolomaeus never tried to justify his model with physical or cosmological theories. He thought that his task was simply to reproduce and predict the positions of planets and stars. For this reason, he asserted that his model merely represented a mathematical method based on hypotheses, whose validity was only a function of the quality of the agreement of the numerical results with experimental observations. The description of complex planetary motions, elaborated by Ptolomaeus, were doubtless of such precision and quality that they appeared even more reliable than astronomical measurements of that time, so that in the following centuries experimental data were sometimes corrected to fit the predictions calculated.

Obviously, ancient, as well as modern, astronomers were obliged to analyse measurements in their observatories by starting from a reference system in which their eye or the focus of their telescope represented the origin of the co-ordinates used. This allows for a simpler and approximate mathematical treatment, but the more accuracy is required, the more complicated matters become by using a geocentric hypothesis. As stated earlier, planetary orbits in this system of reference have a complex shape, which depends on the radius and the orbital speed of the planet (Fig. 1.9) and, in a first approximation, motion could be interpreted as if the planet in question moved along on a sequence of *epicycles*, whose centres move on a *deferent* circle. This model, already proposed by Aristotle, was improved by Ptolomaeus in order to account for new, more precise astronomical measurements. Therefore, he moved the *Aequans* sphere to a slightly eccentric position and introduced four other spheres in order to adjust motion to compensate for errors caused by the simplest model of the circular orbits. All these spheres moved in the manner of a horological mechanism. Predictions were so accurate that even by the time of Galileo Galilei the Copernican model did not provide substantial improvements in precision. Actually, it was only with the calculation of elliptic orbits by Johannes Kepler (1571–1630) that the heliocentric model finally yielded greater precision and a concomitant definitive formal superiority.

The point of view expressed by Ptolomaeus reflects a modern phenomenological approach, where physical effects are calculated numerically by means of (to a certain extent) arbitrary mathematical procedures, and, at the same time, represents an attempt to reproduce experimental data as precisely as possible, and to progressively improve the reliability and the accuracy of the extrapolations. Ptolomaeus always referred to what he called *hypotheses* (his greatest work is entitled "*Hypotheses Planetarum*"), recognising, at the same time, that diverse mathematical procedures, such as, for instance, that of epicycles and of eccentric orbits, could be both valid. It was not a matter of crude approximations of imprecise data, for already by the III century B.C. Hipparcus had collected data from Chaldean and Egyptian astronomers and had published a catalogue containing the positions of 850 stars, totalling 1022 by the first century A.D.(all visible from Alexandria). The method of calculating these data, published by Ptolomaeus in a book that has been handed down to us to us in an Arabic translation entitled "*Almagest*" is considered his most famous work together with his "*Practical Tables*", whose scope surpassed by far the practical needs of all Byzantine and Arab astronomers in the centuries to come so that they only needed to update them. Even today, these data were used by the great astronomer Edmond Halley (Fig. 1.11).

Many modern critics tend to underestimate the work of Ptolomaeus, viewing it as the most influential carrier of a geocentric error that adversely affected astronomy for centuries. In reality, the Ptolemaic system was founded on a legitimate experimental basis and on methodological premises which are still reasonable by today's standards. In his day, and even later, there was no reason to prefer a heliocentric system of reference that appeared completely unnatural from every perspective.

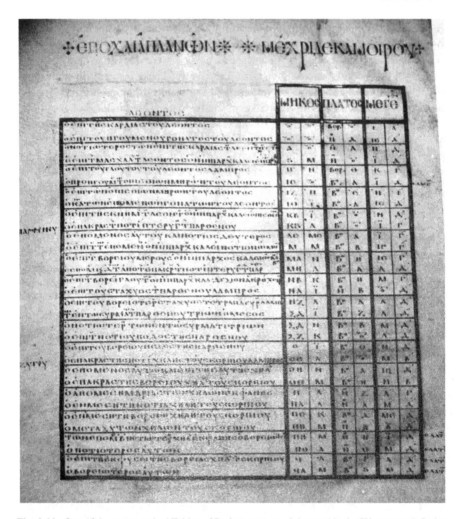

Fig. 1.11 One of the astronomical Tables of Ptolomaeus, re-elaborated in the IX century A.D. in a Byzantine context, where the co-ordinates of the stars of the constellation of Leo are detailed. The three columns on the right-hand side of the Table contain their longitude, latitude and magnitude. (*Florence, Biblioteca Laurenziana, Plutei 28-26, Sheet 124*)

The discriminating argument in favour of heliocentrism was supplied by the theory of universal gravitation, and that 1500 years later.

1.2 The Dispersion of Classical Science

In the V century A.D., after the definitive fall of the Roman Empire in the West, owing to pressure from barbarian peoples, the Byzantine world took up the Hellenistic cultural heritage in the form of a variety of philosophical and religious currents, many of which contained scientific doctrines. Stoic, Cynical and Epicurean philosophers disputed not only ethical problems, but also issues regarding Nature, *Physis*, considered as a product of the necessity or causality, chaos or rationality. In this muddle of doctrines, the Neo-Platonic and Aristotelian schools were facing off in a contest which was to implicate the future development of science.

The neo-platonic natural philosophers, referring to Pythagorean doctrines, asserted that the number was the substance of the universe and the origin of the shapes of all beings was in relations between numbers. Fixed to a static vision of Nature, they investigated symmetries, concatenations and simple, but hidden, relationships and proportions between physical quantities. From this point of view, Neo-Platonic schools represented a continuation of geometric/mathematical thought of the classical age. Mathematical formalism remained, however, too rudimentary and insufficient to proceed towards new important discoveries. These were believed to be found by searching the sky and matter for proof of ideal laws of affinity and repulsion, of cyclical recurrences and astral influences. It is a fact that, at least from our modern point of view, this method could not yield any real opportunities for scientific progress. Nevertheless, paradoxically, it was the very Neo-Platonic physicists, with their abstract tenets, who were the most inclined towards experimental verification through astrology and the ancient practice of alchemy which enjoyed a revival in the Greco-Roman world.

The stoic philosophical and scientific culture of the first century B.C. is typically represented by the work of the well-known figure of Posidonius of Apameia, a most esteemed friend of Cicero. Born in Syria and educated in Athens, of stoic formation but of platonic tendencies, rationalist and mystic, often well—though in some cases superficially—versed in all the doctrines of his time, he conducted extensive astronomical, geographical and biological investigations, in particular, on the nature of plants and animals. Astrology was, however, his preferred science. Cosmopolitan and an indefatigable traveller, he even found a way to study the high tides of the Atlantic during a sojourn at Cadiz and demonstrated that they were caused by the moon's periodic motion. These observations convinced him of the influence of the motion of planets and stars on all phenomena occurring on the earth, including the behaviour of men. Posidonius, whom St. Augustine calls "*fatalium siderum assertor*", asserted that, at the end of a defined time period, called the Great Year, all the planets will occupy the same positions which they held at the creation of the cosmos, and the universe will be consumed in a divine fire ($\dot{\epsilon}\kappa\pi\acute{\upsilon}\rho\omega\sigma\iota\varsigma$, *ekpirosis*), and then it shall regenerate itself ($\pi\acute{\alpha}\lambda\iota\nu\ \gamma\acute{\epsilon}\nu\epsilon\sigma\iota\varsigma$, *palin genesis*). These ideas constituted benchmarks of the late stoic philosophy which was widespread in the Roman Empire [7] and which survived the fall of the empire so as to even occur in the modern age in the form of recurrent millenaristic doctrines.

Fig. 1.12 Figure of *Apollo Solaris–Helios* on a Pompeian fresco. Although Greco-Roman astronomy was geocentric, the sun was considered to be the centre of the primordial force that governs all motions in the cosmos. In this figure, the Solar God is represented holding in his left hand the sphere of the firmament, on which the ecliptic and the celestial Equator are traced.
With his right hand he cracks a whip to urge and direct the horses of its chariot, dictating their course. Astronomers knew that the axis around which the firmament rotates was not fixed and that its periodic movement was the cause of the precession of the equinoxes.
Stoic natural philosophy developed its own mythology based on an astronomical model, which incorporated most data of that time

The model of the Universe that prevailed in late classical antiquity and which was then transmitted to various heterodox currents of thought of the European Middle Ages was anything but a rudimentary picture. It was, however, a model in which geocentrism with all its consequences was only justified for ideological purposes: for souls, after descending down onto the earth from celestial spheres, had to return to the heavens via dangerous astral paths. Yet, the astronomical position of the earth was not considered to be as important as one may believe. The movement of stars is, in fact, dependent on the heaven's polar axis around which the firmament revolves. One knew that this axis oscillates regularly (the precession of the equinoxes) giving rise to complex motions whose mystic significance was only revealed to a few initiates. Indeed solar cults grafted their own beliefs on to this very vision of stoic astrology, in particular the cult of Mithra which peaked in the II–III century A.D. [13] . In these religions, the sun represented the creating force that marshals the motion of stars and the destiny of men. The initiation rituals were bound to complex interpretations of the motion and the position of the stars, through which was opened a way for the salvation of souls (Fig. 1.12).

During the European Middle Ages various doctrines arose which attempted to reconcile these cosmological visions with the philosophy and theology of the Christian religion. Nevertheless, although they incorporated almost all the scientific knowledge of that time, their cyclical or static character made it difficult to develop new fruitful ideas and, therefore, they always remained relegated to marginal currents of Western culture, where, however, they have survived and prospered until now.

On the other hand, Aristotelian science emphasised the effects of general laws of the world evolution that operate in nature. The Aristotelians didn't foster any interest in a definition of the static condition that represents nothing but one of the infinite stages of an evolutionary process that extends from the *potency* to the *act* of all beings. Some Aristotelian schools preached quite explicitly against "ἀκρίβεια, *akribeia*", precision in observation and measurement, since the object of their study was the universal laws, to which nature with its elements was subordinate. Something important lies behind this position. In fact too detailed knowledge may be somehow dissipative, in some cases even destructive. A perspicacious mind must follow a fixed path to which he is confined, a path which implies economy in thinking, but does in the end hinder pursuing more advanced objectives. A thinker must be able to change his perspective, to see less or more, to decide when he has to neglect or to consider further details.

For Aristotelian physicists, natural processes were far too complex to be formulated, described and predicted in detail by mathematical methods. In modern terms, the Aristotelian model of nature could be called *analogical* in contraposition to Platonism's *numerical* model, whereby the former was totally lacking a foundation in adequate calculation and the latter in indispensable, abstract mathematical formalism.

The ship carrying scientific Hellenistic-Roman culture was wrecked by these contrasting currents of thought, which it was incapable to master. All heretical currents did nonetheless set the course for feebly divergent directions which for centuries accompanied the navigational path of science. What failed to appear in the late Greek-Roman world was the establishment of one strong orthodox current, to which all efforts for scientific development could have been tied together. Instead, centrifugal forces prevailed and led to the fall and disintegration of the entire Western classical culture. It is true that the majority of the dominant class was largely formed by stoic doctrines, which supplied a solid ethical basis for the formation of strong characters, but stoic cosmology and physics were too eclectic and fanciful: Pythagorean ideas were crossed with astral theologies and pantheistic visions of Nature, which was conceived as a living organism. Scientific progress in this intellectual atmosphere was almost impossible.

Starting from the III century A.D. the spread of Neo-Platonism, a product of mixing platonic philosophy and oriental doctrines, delivered the coup de grace to scientific thought. While, in this historical context, the work of Plotinus (204–270 A.D.) still constitutes a superb philosophical system, to which was indebted metaphysics and newborn Christian theology, his disciples, however, turned towards elusive Gnostic doctrines from Asia.[17] Hermetic cults extended in Europe and became pervasive in all great towns and attracted all social classes, but, particularly, those of the military class as well as merchants. These religions tried to maintain a scientific basis

---

[17] One may recall that the master of the great Plotinus was Ammonius Sacca, a philosopher of Indian origin, perhaps, as asserted by Jean Daniélou [69], a Buddhist monk, who naturally tended toward mystical and magical practices, which would become the carriers of a pernicious infection that affected Western thought for centuries. In metaphorical terms, science returned from Athens to Babylon.

of Pythagorean signature. For instance, we have seen before that Proclus, an acute Gnostic philosopher of the V century, wrote, in addition to his elements of theology, excellent commentaries on Euclid's *Elements* as well as treatises of trigonometry. This at least had the merit of connecting Neo-Platonic philosophy to the gene of mathematics that would accompany it along its millenarian way through the culture of Occident.

On the opposite end, in the Middle East, Hellenistic culture approached its definitive twilight. Already in the II century A.D. in the Persian Empire the barbarian Parthic dynasty of the Arsacids had begun a campaign of ethnical cleaning, suppressing the centres of Hellenism (the cosmopolitan old capital, Seleucia, was reduced to rubble). The subsequent dynasty of the Sassanids continued the anti-Hellenic policy, favouring the birth of a literature in Palhavi language and eliminating the use of Greek in highly civilised regions where for centuries it had represented the vehicular language of culture. In the VI century A.D. a series of events occurred, which marked a definitive cultural fissure between the Eastern and the Western Mediterranean world. The warning signs first appeared in Sassanid's Persia that was itself facing a social and religious revolution, Mazdakism, a Gnostic schism of Manichean origin and syncretistic character, but based on an ethics of a communist nature. The schism, which at first enjoyed the favour of the king, who discerned in this movement a way to take down the power of the feudal nobility, expanded quickly to include all of Persia. But a reaction was quick to follow; repression was violent and led to an inflexible restoration of Zoroastrianism, upon which the power of the feudal dominant caste was founded. The victims of this repression included the numerous heretical sects as well as the Christian communities of Iran and Mesopotamia, already torn by the Nestorian schism.

On the other side, the Eastern Roman Empire was also rocked by continuous religious clashes, behind which was hovering the spectre of the dissolution of Byzantine power. In this climate Justinian pursued a policy of reordering and restoration with iron determination. After passing laws prohibiting pagan cults and the plethora of heresies that pullulated in all the provinces of the Empire and which had engendered various uprisings and repressions, Justinian finally closed the Neo-Platonic school of Athens that represented the last stronghold of pagan culture by issuing an edict to this effect in 529. [18] Its members went in exile to Ctesiphon and Carrhae, in modern Iraq, where they stayed for a certain time, before dispersing themselves. What influence they exerted on their exile milieu has not been transmitted to us, but it is a fact that the works of Hellenistic philosophers were soon translated into Persian,

---

[18] In order to transform Constantinople into the new cultural centre of the Roman Empire, Theodosius II founded there a university with 31 chairs. In 866 A.D. the school was reorganised and transferred to the Magnaura Palace by Bardas the learned and munificent Caesar, uncle of Michael III. The University of Constantinople consisted of four faculties: Grammar, Physics, Mathematics and Medicine. Excellent teachers were hired and generously remunerated by the State. The lessons and courses, held in Greek as well as in Latin, were free.

While this process was taking place in Constantinople, Athens had decayed to the rank of a peripheral provincial town. However, pagan culture was still very much alive there, so that Justinian's edict suppressing the Athens School was not merely symbolic, but rich in historical consequences.

and their contents were absorbed by the subsequent Muslim culture. With regard to this influence, it must be noted that, paradoxically, the greater part of the works of the Neo-Platonic philosophers consisted of erudite and profound commentaries of the books of Aristotle, with these treatises resuming the lemma's, and integrating them with comments and, often, important amplifications. In this way, the doctrine of Aristotle was transmitted to Middle Eastern culture through the mediation of Plato to return to Europe, after five centuries, in the form of translations from the Arabic.

In this historical context, the doctrine directly founded on the Aristotelian tradition, and the only one reasonably organised and constructed on *Organon*'s logic, continued to reject mathematical method, relegating it to the function of astronomical measurements or engineering applications. However, the works of Aristotle circulated only in vanishing circles of erudite specialists whose main concern was capillary extension and the organisation of the original ideas.

The elaboration of the Aristotelian-Ptolemaic astronomic system in the II century A.D. is part of a new tendency to organise (and sometimes even to manipulate) experimental data, in order to arrange them according to a pre-established logical-philosophical outline. This sort of scholarly activity, which today we would deem a-scientific, demanded, however, an intellectual effort that should not minimised or depreciated, since it in the end had the positive effect of introducing the use of logics as a rigorous criterion for reasoning.

At the same time, the progress of calculation techniques was unfortunately accompanied by an increasing impoverishment in scientific speculation. It is interesting to note how excellent engineers, such as Heron or Philo of Byzantium, dedicated themselves to solving practical problems by means of complex numerical and geometric procedures, and sometimes constructed very complicated machines while still demonstrating, however, an expeditious depreciation of theoretical rigour.[19]

Thus the intellectual *humus* of Hellenism, was depleted of erudition, and did not find a substitute in Latin civilisation, where the interests of the elite classes were essentially absorbed first by founding, and then by defending a new political order. With the decline and final fall of Rome, the Eastern Roman Empire remained the only caretaker of the Greek and Hellenistic cultural tradition. But continuous territorial losses—to the north from the Bulgarians and the Avars, to the east from the Persians and to the south from the Arabs—had irremediably reduced numerically the sheer numbers of the cultured class. In this dramatic situation, the intellectual vanguard died out and an enormous effort was expended to conserve the patrimony of knowledge that had culminated in the tenth century with the compilation of Suda, an encyclopaedia with more than 30,000 entries on every branch of human knowledge.

Even if Byzantine civilisation reached its apogee in the VI century and fell only nine centuries later, its variegated society was, throughout this whole time, incapable of analysing critically and elaborating on the vast heritage of knowledge scattered among tens of thousands of manuscripts kept in its libraries. The circulation of

---

[19] *"Philosophers are searching for peace, but are engineers to secure peace with the launching precision of their catapults"* (Philo of Byzantium , *"Poliorcetica"*).

classical works and of the immense corpus of Hellenistic literature, accessible only to intellectual elites, kept on shrinking in size, and in the Western provinces of Italy and Africa came to a standstill, to reappear after seven centuries of permanence and elaboration in the Islamic world.

The dispersion of classical science was not only a consequence of the political fall of the Roman Empire of the West and of the slow decay of that of the East. The main reason lay in the difficulty—if not in the impossibility—of preserving the great mass of knowledge and information acquired in the fecund previous centuries and, at the same time, of fostering further substantial advancements by selecting the most important matters, discarding the useless burden of encyclopaedic erudition, renewing basic approaches of scientific analysis and changing the points of view of speculation. Finally, the increase in number of philosophical schools had the effect of weakening, rather than corroborating, scientific knowledge, turning instead their interests towards a superficial eclecticism.

1.2.1 Islam Between Hellenism and Middle Ages

Up to now we have briefly examined some important stages in the history of science in the Mediterranean area, without considering the great cultures of Persia, India and China, while these cultures have undoubtedly exerted important influences on our classical science, it is very difficult to trace the long paths and the interpolations they passed through.

It was only with Islamic expansion that a bridge was built between the East and the West Mediterranean worlds, even if this occurred in a background of permanent conflict.

During the five centuries in which Islam flourished, the political axis of this religion established a sort of cultural short circuit between Asia and Europe. The currents that flowed westwards finally exhausted their ancient sources, but the precious fluids were not lost on the way. In fact, they substantially contributed to the birth of modern science.

The development of Islamic civilisation between the VII and the XII centuries, represents a historical counterpart to the decline of western Europe. Politically, the territory over which Islam extended by means of the Arabic conquests and religious conversions always remained a conglomerate of ethnic groups governed by local potentates. Even the Abbasid caliphate, after a short period of splendour, decayed into a state of anarchy comparable to that of the late Roman Empire. However, some events were crucial in fostering the blossoming of new cultures, which are generally called "Islamic", but which must be examined in their own right.

The axis of the first Arabic conquests (when Ali assumed the position of caliph) extended, looking from east to west, over a territory that stretched from Corasan to Numidia, and covered areas which were of vital importance to Byzantine culture, such as Syria and Egypt and large parts of the Persian Empire until 656 A.D. After the Sassanids, this area experienced political decay but could still serve as a powerful catalyst, geographically open to exchanges with central Asia and India. In the countries of Asia Minor and in great cosmopolitan towns on trading roads not far from the borders of the Empire, like Palmyra, Arsela or such as Hatra, Dura Europos and Seleucia-Ctesiphon, which were destroyed by the Sassanids, Greek had for centuries been the lingua franca of the educated class, and manuscripts in Greek had already been in circulation since the III century B.C..

Because of the conquest, the Arabic language prevailed, but also for theological reasons, where all conquered peoples converted to Islam. Consequently, Arabic soon became what Latin was in the West: an international vehicular language lexically and grammatically well-constructed and, at the same time, able to absorb concepts pertaining to different cultures.

In the newly Islamic territories immense efforts were made to translate in Arabic all the main Greek and Persian philosophical and scientific works. Even today this monumental achievement should merit our admiration. In fact, when we consider translations of books of the calibre of the treatises of Euclid, of Aristotle and Ptolomaeus, we often cannot sufficiently appreciate the conditions necessary for this work, namely a perfect grasp of the original language and an outstanding competence on the part of the translator in mastering the subjects treated. Moreover, in most cases the translator's mother tongue was not the language spoken by Mohammed and the primitive Bedouin tribes of Arabia, which had to be enriched by a new terminology and adequate definitions. That this immense achievement in translation and assimilation could be carried out in less than two centuries is absolutely amazing, especially when one considers how slow and expensive the writing of manuscripts was.

The majority of Greek scientific manuscripts were translated in the philosophical school of Baghdad. This city had been built *ex-novo*, not far from ancient Babylon, by the Abbasid caliphs after the defeat of the Omayyad dynasty, whose capital was Damascus, in Syria. Though constructed on the Tigris, about 600 Km upstream of the mouth of the river, the city was for all intents and purposes the most important port of the Persian Gulf and, in the IX century, possessed some 30,000 mercantile boats that could berth directly at the wharves of its bazaars. Geographically, Baghdad was oriented towards the north of Persia and was separated from Mecca, the religious centre of Islam, by 1400 Km of desert. During the period of the Abbasid caliphs Baghdad became a centre of importance comparable to Alexandria at the apex of its splendour. Its contacts with Persia created an influx of peoples from and towards the east, in a territory that reached India and the borders of China. The munificence of the caliphs in Iraq had attracted the cream of a cultured class of a broad and variegated area. Scientific knowledge was at first furnished by Hellenistic and Persian cultures, but new constructions were not slow to appear in this new atmosphere, characterised by fervent activity. Astronomy made great progress due to the observations and measurements performed in the new observatory at Baghdad. Mathematics also profited from the assumption of a system of numerical notation already in use in India, which then opened the way to the foundation of algebra. The theory of conic sections of Apollonius of Perga (262 B.C–190 B.C.) was applied by brilliant geometers in order to solve algebraic equations. As for physics, Aristotelian models were scrutinised and discussed and, in many cases, new explanations and theories were developed, whose merits could only properly be appreciated in the modern age (see, for example, Fig. 1.13 with comments below).

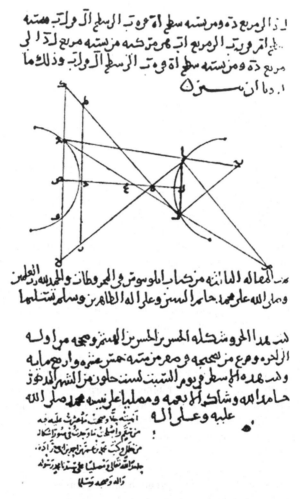

Fig. 1.13 An autograph written in 1024 by Alhacen, an eclectic naturalist, born in Basra and lived in Cairo. The drawing and the comments illustrate the properties of luminous sources and their images produced by spherical mirrors. In his main work entitled *The Great Optics*, Alhacen developed a sophisticated theory of the reflection and refraction of light. His main hypothesis was that light consisted in a wave that propagates at a finite speed varying with the nature of the transparent medium. In order to appreciate the sheer brilliance of his intuition one should remember that when Alhacen lived (and even later on) one believed that the image of objects was the result of the eye emitting rays which, in some way, probed them. Although his ideas anticipated modern theories of light, Alhacen never used any mathematical procedures in his models (his book doesn't contain a single number), deducing the property of refraction through the most convoluted geometric constructions and demonstrations. The mathematical theory of the refraction of light was definitively formulated by Cartesius only six centuries later

The expansion of Islam towards the East, after the conquest of Persia, meant that, for geographic reasons, the corridor of the "silk road" had to be traversed [20] which from Media finally led to the immense plains of Turkmenistan and, further, to China. India, protected by the impassable mountains of the Hindu Kush, remained, therefore, safe from Muslim conquest until the Mogul invasion of the XVI century. Therefore, Islamic culture in the Orient was formed in a crucible of nomadic populations of Arian stock who, for millennia were migrating in the ravaged steppes of central Asia, pursued by Mongolic peoples. These immense regions at the edge of the Persian Empire, corresponding to the ancient satrapies of Bactria and Sogdia, were weakly tied to Sassanid's Persia. After the Arab conquest, a new Iranian dynasty of the Samanids created a vast local reign with its capital of Bukhara, and which was only formally subject to the caliphate of Bagdad, and which, in the period of maximum expansion, included Northeast Persia, Afghanistan and Uzbekistan. The apogee of this civilisation was reached in the centuries following the Arab conquest, under the new Turkish dynasties of the Ghaznavids and Selgiuchids. This reign constituted the oriental pole of Islam. Filtered through ancient Persian religion and transplanted on the *humus* of Buddhist cultures, the at that time still immature Arab religion underwent a deep transformation, absorbing simultaneously aspects of philosophy, science and arts from northern Persia and India. It is not a coincidence that the most famous Sufi mystic, Gharib Nawaz, or a wizard alchemist like Abir ibn Hayyan as well as philosophers of the calibre of Avicenna and Al-Gahzali, who discussed important aspects of Aristotelian doctrines with a school in Bagdad, all flourished in Corasan. This was a remote region halfway between the centre of Persia and the Indus valley and was a country steeped in ethnic encounters, where one could hear Persian, Turkish, Kurdish and Afghan spoken, and where local Mazdaic beliefs were mixed with Avesto-Vedic doctrines. Therefore, Islamic religion developed its roots towards the East, absorbing, at their primary source, Gnostic elements that would have deeply influenced it, as had previously occurred in the case of Neo-Platonic philosophy.

---

[20] In the East, the Persian Empire was separated from Asia by the deserts of Kavir and Lut which extend from north to south for approximately 1000 km. The coastal road to Asia along the gulf of Oman crosses the desert-like, practically uninhabited and impracticable territories of Gedrosia. In the past, Persian armies had never succeeded in reaching the Indus valley via this road (Arrianus reports how Alexander, who decided to cross it with half of his army upon returning from his war campaign at the Indus mouth, risked perishing in those sandy solitudes, alternately battered by barren winds and terrible monsoonal hurricanes). The only easy access to India was to the north, following a long path that started from the Caspian Sea and, along the southern versant of the Elburz chain, until Herat; at that point a difficult mountain road towards the south-east led to Kandahar and, through the pass of Bolan, to arrive in western Pakistan. The main road to India, however, proceeded from Herat towards the north of Afghanistan and, following the border of Turkmenistan, reached the upper course of the Amu Darya (the ancient Oxus) up to Balkh, after which it turned south, towards the valley of Kabul and, finally, through the Khyber Pass, reached Peshawar, at the upper Indus water way. The spread of Islam to the East took place following this tortuous path, on which splendid towns flourished at that time, such as Marv, Termez, Bukhara, Balkh and, further east, Samarcand.

The reverse direction of Islamic expansion started from Hellenised Damascus which was ruled by the Omayyads and followed the North African shoreline, through Egypt, Libya, the ancient *Africa Proconsularis* and Mauritania—these last three regions were strongly Latinised—and, owing to the Visigoth invasions, almost all of Iberia was conquered up to its northern boundary. In these regions the Omayyad emirs and caliphs governed a variety of peoples, all deeply imbued with classical culture. On the other hand, the Omayyad's dynasty maintained a political system of late Greco-Latin classical structure, where many civil servants of high rank were Christians of Byzantine origin. However, the fall of the Omayyads produced a progressive estrangement of Muslim society from Mediterranean models: the binomial *civis-civitas*, that had characterised the Greco-Latin world for more than a millennium, finally disappeared in Islamic states that assumed the typical Asiatic form of theocratic monarchies.

There is no doubt that western Muslim culture was initially permeated by ideas originating in Greek philosophy, but oriental Islam followed a divergent course. The definitive defeat of the Omayyads in 750 was provoked by uprisings in Iraq, which were, however, supported by the recently converted populations of Persia. These events displaced the pole of Islam to the east, imbuing it with cultural elements typical of Asia. To understand these changes, it is enough to consider the case of Averroes, who commented on the books of Aristotle with scientific rigour and confirmed the rationality of the Islamic faith. Whilst in Cordoba, in his commentaries Averroes attacked the doctrines of al-Ghazali, who far away in Corasan preached the incompatibility of the Muslim religion with the philosophy of the great Stagyrite.

But by the XII century Muslim orthodoxy had long since taken the road towards the East: the philosophical school of Cordoba was declared heretical and Averroes ended his days imprisoned in exile and his works were destroyed by censors. Interest in Aristotle in Moorish Spain had also extended to include the flourishing Judaic culture, which culminated in the work of Moses ben Maimon (1135–1204). But the period of receptiveness to other ideas was over even before the Spanish *Reconquista*, with the intransigence of the subsequent Berber dynasty of the Almohads.

In most treatises of history, the floruit of the Arab-Islamic civilisation in the Mediterranean area is introduced as a process of absorption and maturation of the Hellenistic-Byzantine cultural patrimony in the fecund atmosphere of the heirs to the great civilisations of the Middle East, from Persia to Egypt, while its counterpart is identified with the sterility of the decadent Byzantine civilisation and the barbarisation of the Latin-Christian world. This sweeping judgement does not take into account the fact that objects of comparison are variegated and changed in the course of the centuries during which Islamic culture, soon after its birth, evolved and developed new traits. It is, however, incontestable that in this atmosphere science underwent a process of reorganisation and expansion of its field, whose remaining documentation is at present unfortunately very deficient and riddled with lacunae.

When the Flemish historian George Sarton (1884–1956), began his monumental History of Science [8] which was dedicated to this period, he realised how unexpectedly vast the field was and how difficult it would be to render a complete account of it.[21] The field was in fact almost unexplored. In order to even guess at its extent, we enumerate here the various fields finally considered by Sarton: mathematics, astronomy, physics, chemistry, music, geography, natural history, medicine, historiography, right, sociology and philology. When translations of these works from and into the languages of the Mediterranean and Middle East regions (Arabic, Persian, Syrian, Hebrew, Greek and Latin) began to circulate by the XIII century, an important process of dissemination and selection of all this knowledge was set in motion, involving and affecting the entire European culture.

It is difficult to comprehend this process if one omits the religious and philosophical background of Europe and of the Near East in the period from the Early Middle Ages to the end of the XIII century.

The formation of the theological doctrines of Christianity and Islam exerted a most important influence on the development not only of ethics but also of philosophy and science. Although the origins of the two religions are separated by six centuries, both had to face magical doctrines, which, as far as one could remember, appeared, disappeared and reappeared in the guise of various religious sects in all countries from Egypt to India. Even classical Greek culture was not immune to these doctrines which were propagated in esoteric confraternities of mystic cults. The message of salvation which was found in the newly ascendant Christianity was echoed in these doctrines. For this reason, the founding Fathers of the Church were attracted by the possibility of inserting Christian ethics in these eschatological visions, that seemed to complement the dry evangelic texts and to provide an answer to questions which seemed to not be explicitly answered in those texts. The most significant issue was concerned with the origin of Evil. The explanation found in Gnostic doctrines was based on an irreducible dualism: There are two principles, Good and Evil that are concretised in the cosmos as hierarchies of spiritual beings associated to celestial bodies. Matter, and hence the human body, is the lowest manifestation of the work of creation by a malevolent deity, who used it to imprison spirits. Man is impotent in a cosmic struggle between Good and Evil, and his spirit can only hope to rid itself of its body and return to the high celestial spheres from which it had fallen onto the earth.

How much Gnosticism penetrated the Christianity of the first centuries A.D. can be judged by considering patristic literature, especially that of the Apostolic Fathers. The most significant deviations were experienced by Christian communities of High Egypt (Coptos), of Syria and Mesopotamia.[22] The war against Gnosis turned out to be extremely difficult (the boundaries of orthodoxy to heresy were often vague and sometimes undefined) and hence was never concluded but is still being waged today. Gnosis was publicly banned from the Church, but it continued to exist like a subterranean current nourished by a distant source, and bursting forth as heretical movements assuming different forms and accentuations from time to time throughout

---

[21] Sarton decided to learn Arabic and then undertook a series of lengthy travels to the Middle East in order to rediscover and translate original manuscripts.

[22] What Mohammed knew of the Christian religion he had learnt from gnostic texts, which circulated in Christian communities of Nefud.

centuries, such as, just to name a few: Manichaeans, Marcionists, Cathars, Rosacrucians and Geovists.

Gnosticism found its place in late Hellenistic-Roman culture, by permeating Neo-Platonic philosophical schools, and most of the intellectual elite could identify themselves with the various manifestations of this assimilation. Yet, the first Christian philosophers, who had assimilated Neo-Platonism were, therefore, anxious to preserve their ideology from excessive Gnostic deviations, as, for instance, demonstrated by St. Augustine in his doctrine. He recognised the greatest danger to be in magical practices, with their alchemic manipulations and in astrology, which intrinsically refuted the freedom of man. In his lively debates with the Manicheans, Augustine emphasised the rationality of faith and the necessity to develop introspection and a study of the activity of the spirit. This choice, which implied a lack of interest in investigating physical nature, on the one hand, resulted in seven centuries of stagnation of science, but, on the other, immunised it from the monstrous reveries that, as history has shown, in the long run have finally succeeded in paralising scientific progress in Eastern cultures.

In its first phase of expansion, Islam did not have such concerns since, essentially, its theological content was much less susceptible to worrisome Gnostic aberrations than Christian beliefs were. However, its theological base was fundamentally in agreement with Gnostic doctrines when the arbitrariness of divine creation and the total subjugation of man was argued in an universe, in which rationality is merely a tool.[23] Gnostic religions are not *per se* contrary to science, but science is considered to be a means to discovering mysterious aspects of nature, to be able to predict astral influences and to manipulate the hidden forces by using formulae that recall the power of the number and of the word. On the other hand, an appeal for agreement between science and theology does not reappear in any Gnostic religion. The statements of science are looked on with indifference, not requiring an answer on the part of religion.

As we can see today from a subsequent history of science, from its very nascence Islam took a diametrically opposite position to that of Christianity, which for centuries has faced dramatic proofs, but has still persisted in committing itself even to scarcely defensible theses in order to maintain agreement between faith and reason and between theology and science.[24]

---

[23] The eventual evolution of Islam has been marked by an increase in the influence of Gnostic-Manichean ideas, which flowed into the political movement of Mahdism and into modern Islamic radicalism. The ominous aspects of this evolution have been thoroughly examined by Laurent Murawiec [95], who showed how these dark undercurrents are also present in European culture, where, however, they found barriers which impeded fantastic intellectual constructions and pseudo-scientific knowledge from becoming influential in Western society.

[24] On this subject it is interesting to note how Protestantism, which is nearer to gnostic Manichaeism than Catholicism, has encountered less difficulty than the latter in avoiding conflicts involving scientific theories apparently in contradiction to theological pronouncements.

Chapter 2
Birth and Floruit of Science in the Occident

2.1 Zero Year

A stage of novel developments in scientific knowledge germinated in Western Europe into a new historical context despite the fact that, after the fall of the Roman Empire and a subsequent hegemony of the Germanic peoples, conditions were created which led to almost total social and political disarray. The great towns, with their centres of scientific production situated in schools and public libraries, were lethally struck by a landslide of the entire civil infrastructure. Their decay led to a cultural regression unprecedented in history not only of Latin, but also of Mediterranean civilisations. In Europe a process of (re)acquisition of scientific knowledge had to begin from a year zero that we could fix to the middle of the V century A.D. and which lasted a little less than a millennium. This process illustrated the exponential course of the acquisition of knowledge through contacts with more advanced cultures in conjunction with their own elaborations and discoveries.[1]

Having to face a demographic and financial crisis of extreme gravity, the greatest merit of the few remaining heirs of Roman culture is that they re-organised school instruction by basing it on a concatenation of well established disciplines and avoiding a restoration of the anarchic eclecticism that characterised late Latin culture.

The subdivision and links, due to Martianus Capella (V century A.D.) of the *artes liberales* in the *curricula* of *trivium* (Grammar, Dialectic, Rhetoric)[2] and

---

[1] Nowadays the term *exponential* is often used to indicate a process which is very fast and increases much more than linearly (with time, in our case). In reality, an *exponential* increase with regard to time is initially very slow and only after a period of incubation, which is defined by the multiplication factor of the variable used in the exponent, does the process rapidly diverge. An analysis of this factor provides important information and insight into growth mechanisms.

[2] The arts of *trivium* are intended to prepare the ground for social communication. In this context, Rhetoric has the function of conveying through speech the logical arguments elaborated in Grammar. Aristotle claims that any message runs over the bridge of rhetoric, supported by the three pillars of logic (*logos*), habit (*ethos*) and passion (*pathos*). Rhetoric was first meant to be the art of political and forensic eloquence and also a canon for literary work. The schemes of Rhetoric are, however, abstract and quite independent of the contents of the message. For this reason, it was seen as an

quadrivium (Arithmetic, Geometry, Astronomy and Music) divided literary from scientific fields, subordinating all of them to philosophy, considered as the apex of knowledge. In the subsequent centuries, this discipline was in turn subdivided, following a hierarchical order, into ethics, physics and logic/rhetoric. Later, after higher instruction was re-evaluated in the Middle Ages—first in the *studia generalia* and then in the town's universities—studies of *trivium* and *quadrivium* served as an introduction to the *laurea* in jurisprudence and, above all, in theology, and were included in the faculty of arts, the first stage of the university *curriculum* to group students from the ages of fourteen to twenty (Fig. 2.1). The contents of instruction were initially based on texts which had escaped the destruction of public and private libraries in the Empire of the Occident. The constant decline of Byzantine power in Italy had, in fact, greatly reduced cultural exchanges of Europe with the Empire of the Orient, which still was in possession of the entire classical patrimony of knowledge and still produced brilliant scientific personalities of the Neo-Platonic school such as, for instance, Proclus Diadochus (412–485 A.D.).

On the other hand, the newborn universities consisted of a sort of *consortium* between teachers and students, and were free from secular and ecclesiastical authority, with the exception of theological matters. This freedom, which has been jealously defended and conserved in all universities till modern times, had in the Early Middle Ages one drawback: namely, as in their own tradition, together with rigorous classical science, universities were also allowed to engage in the most disparate notions and theories, which were sometimes completely fantastical. These sorts of eccentric endeavours never ceased to negatively affect several marginal, but not uninfluential, areas of Western culture in the millennium to come.

In fact, when translations of Arabic and Hispano-Arabic texts began to circulate in Europe from the XI century onwards, one found among them, together with valuable Greek and Hellenistic scientific works, a great number of Arab elaborations on astrology and alchemy texts stemming from "heretical" schools where Pythagorean doctrines were deeply infused with much older Egyptian and oriental traditions.

Nevertheless the work of Aristotle continued to serve as the main point of reference for European scholars throughout the entire Middle Ages. Its authority remained undisputed even when rivalries between Dominican and the Franciscan schools of European universities created an irreducible split between Aristotelian metaphysics and the newly born philosophical Empiricism. Mathematics, which had always played a secondary role in the Aristotelian system, was especially cultivated in the Franciscan school where the archbishop of Canterbury Thomas of

---

instrument of magniloquence and artificial ornamentation in the Romantic Age so that Rhetoric was degraded to an arid and empty discipline by the nineteenth century; consequently, its teaching in schools was suppressed (it survived only in France until 1880). Actually, this was a great error of judgement. Today, when messages are carried by advanced vehicles of increasing complexity, the study of Rhetoric is seen to be necessary and is thus gaining lost ground in the analysis of communication, including the scientific one.

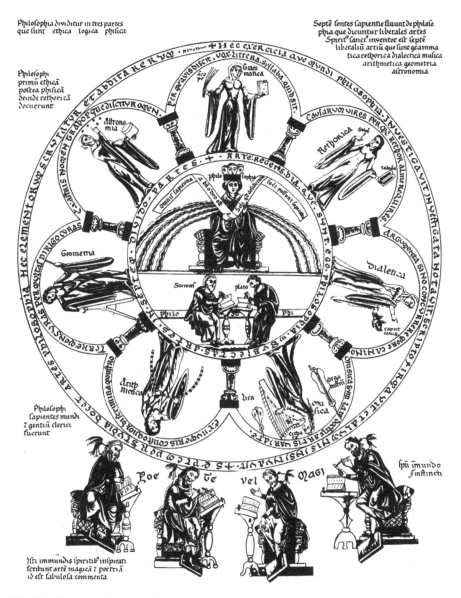

Fig. 2.1 Illustration from the *Hortus Deliciarum*, by Herrad, Abbess of Landsberg (XII century). The Liberal Arts are represented as radiating from Philosophy, on whose head rests Ethics, Physics and Logic which forms its crown. In their respective cartouches we can read that Rhetoric is the art of putting into effect the "*causarum vires*", i.e., the forces of arguments, which support the bridge between logic and persuasion in social communication, whilst Dialectics is holding in her hand a dog's head since her duty is to "*concurrere more canino*", that is to say, to have a good nose and use her intelligence to search for truth and discover error. Wizards and poets are depicted at the bottom as authors of falsehoods and "*fabulosa commenta*", instilled in their mind by a filthy spirit in the guise of a crow

Bradvardine (1290–1349)[3] took up once again, from a modern point of view, the theory of *continuum*, analysing the concepts of infinite and transfinite.

Mathematical symbolism, however, almost completely arbitrary, remained at a rudimental level. The abacus remained the principal instrument of calculation, even if it was enhanced by brilliant methods to execute complex operations and solve simple mathematical problems. The system of numeration used was generally the Roman one, founded on a bi-quinary base. When the Italian mathematician Leonardo Fibonacci (1170–1250) in his *"Liber Abaci"*, not only proposed an Indian-Arab decimal system in which digits from 0 to 9 were positioned but was even able to demonstrate its advantages with numerous examples, he obtained immediate praise and acceptance from his contemporaries in the first half of the XIII century, but the *abacists* nonetheless continued to prevail until the late Renaissance.

Meanwhile, the Abbots of Cluny, among whom Peter the Venerable, prominently figures, undertook the vast enterprise of translating Arabic works into Latin. Although his main purpose was to supply arguments used in the religious controversy with Islam, philosophical and scientific works were translated as well. In 1143 the Cluniac archbishop, Raymond of Toledo, published a complete translation of the books of Arithmetic by Muhammad ibn Musa al-Khwarismi, who had lived in Baghdad three hundred years before. In these works, various algebraic applications were clearly formulated, some of which originated from Indian works dating back to the II century A.D.. The reaction of mathematicians was very positive, with the result that they even began to treat arithmetic problems in Europe using the method of al-Khwarismi, corrupt in *"algorithms"*. However algebraic method was finally established as common practice only three centuries later, thanks to the work of valuable algebraists, in particular of Luca Pacioli (1445–1517), who introduced a great number of ingenious techniques of calculation.

It may appear inconceivable to us that such important and fertile ideas could have needed centuries in order to become established in Western scientific culture. Yet there is a profound reason for this: Algebra tends to focus on the operation itself, assigning a merely symbolic character to the number in question. In algebra there are *no* privileged numbers; a given number doesn't have any intrinsic properties within an algebraic frame except for those defined by abstract, occasional relations (equations, functions, *etc.*) to other numbers.

On the other hand, however, most mathematicians of antiquity retained a Pythagorean geometric conception of numbers, according to which to every real being corresponds a number and a relation between particular numbers.[4] After more than a millennium, Pythagorean doctrines were still enjoying a great prestige, preserving the most radical and lapidary formulation given by Philolaus of Croton (V century B.C.):

---

[3] Thomas was educated at Merton College and, later, lectured in Oxford. His main works entitled *Arithmetica Speculativa* and *Geometria Speculativa* were only printed in France fifty years after his death, in the cultural climate of the Renaissance, but they serve to illustrate the state of mathematical speculation during the high Middle Ages.

[4] Note that until the nineteenth century the Mathematician was called Geometer and, in certain academic circles, this appellation remained in use until a few decades ago.

All that can be known has a number and without numbers nothing can be known or understood.

The multitudinous visions of the Pythagorean theory of numbers, extending to include geometry, natural sciences, astronomy and music, undoubtedly possessed a glamour and charm that were enhanced by its age-old tradition. Thus, the *quadrivium*'s disciplines of geometry, arithmetic, astronomy and music, were also steeped in and affected by the persuasion that a correspondence between relations of numbers and a morphogenesis of nature existed, and by a belief in mysterious laws unifying celestial motions and harmony of sounds.

In fact, the properties of natural numbers continued to arouse the interest of mathematicians, so that, from the Middle Ages until the late Renaissance, mathematics and esoteric numerology were closely interwoven. On the one hand, the magical and cabbalistic aspects of numerology did not appear to contradict the scientific rigour of mathematics. At the same, on the other, the search for correspondences in natural phenomena of various kinds with numerical "privileged" relationships between numbers led to fascinating and inexplicable aspects, which were themselves not devoid of scientific interest.

Let us examine an example taken from the work of Fibonacci.

While studying the properties of numerical successions, he investigated his famous sequence, reproduced here below:

F_0	F_1	F_2	F_3	F_4	F_5	F_6	F_7	F_8	F_9	F_{10}	F_{11}	F_{12}	...	F_n
0	1	1	2	3	5	8	13	21	34	55	89	144	...	$F_{n-1} + F_{n-2}$

The sequence, constructed using the simplest rule, appears in the *Liber Abaci* as a model representing the increase in rabbit population, beginning with one pair and assuming certain hypotheses concerning their fertility. So far there is nothing exceptionally interesting. However, surprising features arose when one began to examine in detail the relations between the numbers of the sequence. First of all, it has been observed that the sequence of adjacent squares of side length F_n (see Fig. 2.2) defines a segmented geometrical curve, which, for increasing values of n, tends to a logarithmic spiral, a curve that possesses such amazing properties that it was called *spira mirabilis*.

The Spira Mirabilis

Let us examine the analytical properties of this curve.

The properties of the logarithmic spiral are intriguing, although its analytical expression is very simple:

$$r = a \exp(b\theta)$$

or, in Cartesian co-ordinates:

$$x = r \cos(\theta) = a \exp(b\theta) \cos(\theta)$$
$$y = r \sin(\theta) = a \exp(b\theta) \sin(\theta)$$

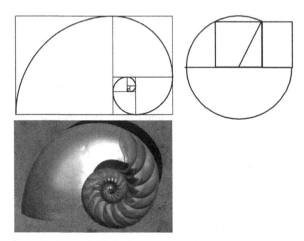

Fig. 2.2 *Top left* the spiral of Fibonacci constructed with the *squares* having as side lengths the first nine terms of his numerical succession. *Right* a construction of the golden section starting from a square inscribed in a *semicircle*; the golden section, the *sectio aurea*, corresponds to the proportion of the sides of the greater rectangle. *Bottom left* the shell of Nautilus

where r is the radial position of a point P on the curve and θ the rotation angle of r around the origin O. One can easily see that multiplying the parameter a by a factor $F = \exp(2\pi nb)$ is equivalent to adding (or subtracting, if b is negative) a multiple of 2π to the angle θ, so that the curve remains unchanged.[5] The shape of the logarithmic spiral has several remarkable aspects: it can easily be seen that the curve is *self-similar*, i.e., arbitrary scaling of its parameters produces only a rotation of the curve around the origin O and, starting with any value of $a = a_0$, if we increase a continuously, the spiral turns around O, passing through the same positions each time that a assumes one of the values of the sequence $a_n = a_0 \exp(2\pi nb)$ for $n = 1, 2 \ldots, \infty$. Looking at its *differential* properties we realise that if we move a point P along the spiral, its radial velocity is b times the velocity component perpendicular to the radius r. Furthermore, if we move P inwards along the spiral, the point turns indefinitively around the origin O approaching it for $\theta \rightarrow -\infty$; however, no matter how close P is initially to O, the distance P has to cover to reach O is always *finite*.

The logarithmic spiral is typically one of the monstrous products of the *continuum* hypothesis, but is a fascinating curve both from a mathematical point of view and for the implications connected with its use in describing certain phenomena in physics. Actually, in addition to the shape of certain molluscs or plants, it is well known that a number of physical processes, from galaxy formation to atmospheric cyclones and fluid vortices, lead to morphological features approaching the logarithmic spiral. Its shape is, therefore, connected to some general morphogenetic mechanisms.

---

[5] When $b = 0$ the logarithmic spiral becomes a circle and for $b \rightarrow \infty$ it becomes a straight line.

Let us consider, for instance, the axial section of a Nautilus shell. The growth of a living mollusc must obey a precise rule that enables the internal surface of the shell to coincide with the external surface generated in the previous cycle, like a rope of increasing diameter when perfectly rolled up.

This occurs when the parameter a of the external boundary spiral is $F = \exp(2\pi nb)$ times larger than that of the internal boundary spiral. We have seen above that this factor tells us at what rate the peripheral living cell must multiply in the radial direction with respect to the internal one. But how does the peripheral cell know how to set the factor F? The answer obviously has to do with a synchronisation process which must be established in the nucleation stage of the shell. This may be referred to as a spiral's embryo, consisting of a kind of "dipole", $P_0 - P_1$, and of an abstract rotation centre, O, situated, in the order, on the straight line $O - P_0 - P_1$. The *distance* $[P_0 - P_1]$ establishes the starting thickness of the mollusc and the *ratio* $[O - P_1/]$ $[O - P_0]$ defines the spiral's parameter, b, i.e.:

$$[O - P_1]/[O - P_0] = \exp(2\pi b),$$

that is to say:

$$b = 1/(2\pi)\log([O - P_1]/[O - P_0])$$

a parameter a is then defined by the absolute value of the distance $[O-P_1]$ or $[O-P_0]$. Thus the parameters a and b are determined by some initial features of the embryo. From the physical point of view we are facing a process where an intrinsic clock is defined based on an abstract periodic cyclic process (one full rotation of the trigonometric circle) to "count" the speed of growth of the non-periodic cyclic structure of the spiral. It is amazing that these processes, ruled by rather complex equations, do finally produce a curve, which is also described by the simple Fibonacci's recursive formula operating on integer numbers, an exercise a child can make.

Moreover, the resulting Fibonacci spiral grows outward by a ratio $\phi = (1 + 5^{1/2})/2 = 1.618\ldots$, for every angle $1/2$ of rotation, corresponding to a value of $b = 0.306\ldots$in the equation of the corresponding logarithmic spiral. The number ϕ, called the "golden section (*sectio aurea*)", represents an interesting irrational number, which can be obtained from the Fibonacci numbers sequence, since it corresponds to the quotient of the sides of the external rectangle indicated in Fig. 2.2, for n tending to infinity. The properties of the *sectio aurea*, which we cannot discuss here, are so remarkable that mathematicians have speculated on them for more than 2500 years, finding connections of ϕ with important and stunning geometric and mathematical relations. It is also worth noting that the golden ratio ϕ has been used, more or less consciously, by ancient and modern architects and artists to obtain a perfect proportion between length and height of a rectangular frame, a proportion to which some people attribute harmonic and even magical properties.

In the end, a history of multiform properties of the Fibonacci sequence would fill several volumes. It is sufficient to mention at this point that many university groups of mathematicians dedicated to its cult still exist all over the world.

We have dwelt on the description of this particular case because it represents one of the multitudinous effects, which, on the one hand, can be described by continuous functions resulting from complex mathematical procedures and, on the other, can be described as well, using arbitrary limits of precision, by simple recursive operations on integer numbers.

There is no doubt that the discrete approach seems to be more realistic when certain physical phenomena are described. For instance, the nucleation of a Nautilus shell is reasonably defined by the metric parameters attributed to the shell embryo. But the curve resulting from the Fibonacci sequence is initially only a very rough approximation of the ideal logarithmic spiral. On the other hand, if we want to attribute a starting point to this latter spiral we must operate a cut-off of a certain minimal size, which may be physically legitimate, and then proceed by applying a growth function which continuously modifies the shape of the embryo according to a pre-existent definition, which, however, may *not* be legitimate from a physical point of view.

In the case examined, by means of some not insignificant reasoning, one can demonstrate the convergence of the two respective approaches of continuum and discrete, but for more complex recursive operation rules this might be very difficult if not impossible. Then the significance of the recursive operations may remain hidden because we are incapable of understanding their implications in any detail.

The persistent attachment of ancient mathematicians to the study of integer numbers (an attachment that still persists today) is quite understandable, since it is founded on manifold reasons, not only concerning esoteric aspects, but also involving fundamental mathematical properties. It appeared, in fact, that forms and processes, which, at first sight, seemed very complicated, could be explained as the result of the simplest iterating relations between integer numbers, related to space symmetries and to periodicity or regularity of geometrical shapes. Naturally, the mysterious aspect of these properties was perceived to be a stimulus to penetrating and to understanding their nature and implications.

What is particularly fascinating about integer numbers? In set theory they represent an ordered set with respect to the operation of *addition* and to its *inverse* of *subtraction*, where zero is the unit (Fig. 2.3).[6] In this set, the operation of *multiplication* can be reduced to a sequence of additions, with 1 representing the unit, but it is obvious that the inverse operation of *division* cannot be generally applied to the field of integers, even though there are infinite pairs of integer numbers for which division can be applied.

The most interesting aspect of integers is related to the existence of prime numbers, i.e., of numbers that can only be divided by the unit and by themselves. It has been demonstrated that there is an infinity of prime numbers, but, obviously, the higher their value is, the more difficult their identification is. There are extended tables of prime numbers, and powerful computers are currently used by international teams in order discover the largest of them. Today the last one, found in 2008, is composed

---

[6] The unit is the element, I, which, operating on a general element E of the set, results in the same element. (Thus, $E + 0 = E$ for addition, $E.1 = E$ for multiplication)

Fig. 2.3 Illustration from the "*Margarita Philosophica*" (Freiburg in Breisgau, 1503) written by the Carthusian Gregor Reichs (1467–1525), representing an algebraist and an abacist in a contest of calculation speed. In the cartouches we read that the allegorical figure of Arithmetic judges the two "*typi*" of which the former is referred to as the school of Severinus Boethius and the latter as that of Pythagoras. The algebraist gains the upper hand over the abacist *Bottom left* a manuscript of the contemporaneous mathematician Brother Luca Pacioli (1446–1517), in which an algebraic method of multiplication is presented which is still in use today

of approximately 12 million figures and occupies 7 MB of memory in a computer. It might seem senseless to concern oneself with such large numbers, but interest in these numbers has recently grown by leaps and bounds since the factorisation of an integer has become part of fundamental procedures, from applied mathematics to cybernetics and physics. Perhaps the most important problem of modern mathematics is that concerning the definition of the function that defines the density of prime numbers in an integer field. This function appears to be very puzzling since there are intervals in which the frequency of prime numbers is high and others of equal amplitude in which not a single prime number appears. The most illustrious mathematicians, from Gauss to Riemann have attempted to find this function, thus obtaining most interesting results, but the problem of expressing the density of prime numbers remains unsolved and is presently bound to a demonstration of the famous "*Riemann Hypothesis*",[7] an activity which has kept—and is still keeping—generations of mathematicians busy.

---

[7] One has demonstrated that the density of prime numbers can be related to the zeros of the simple complex function: $\varsigma(s) = \sum_{1}^{\infty} \frac{1}{n^s}$ (the *zeta-function*), where s is a real or a complex number. The *Riemann hypothesis* asserts the all non-trivial zeros have a real part equal to 1/2 for any s with a real part greater than 1. Up to now all (finite) numerical calculations have confirmed this hypothesis, but
(Footnote 7 continued)
this confirmation does not constitute a valid proof, all the more since there are many cases where properties deduced from results of numerical calculations have finally been shown to be *false*.

What is interesting and significant is that some of the problems and enigmas regarding integer numbers, which were formulated centuries ago in the Middle Ages and Renaissance—and in certain cases in even more remote times—still remain unsolved despite tremendous efforts and serve to spur on the progress of modern mathematics.

2.2 Sowing the Seeds of Science Renaissance

Classical Greek logic, defined as the analysis of the structure of propositions, had become increasingly important since the V century B.C., when it appeared to be necessary to distinguish between the truth of premises and the validity of the conclusions reached by reasoning. Independent of the various schools and currents, philosophic speculation was fundamentally based on a spoken language, because it was thought this was well-suited to express all the cognitive mechanisms of the mind. Aristotle was the first to organise logic in some treatises, which were later collected in one book entitled "*Organon*". In his work, Aristotle catalogues and discusses the categories of the expressions (*substance, amount, quality, time, space*, *etc.*), he examines the connective terms (*and, or, not, if, only if, etc.*) and defines the syllogism as a verbal phrase in which, once some premises have been assumed, something necessarily follows from their being what they are. In addition, Aristotle considered their typology on a formal plan, independent of the contents of the terms considered.

In the classical world it was commonly believed that a logical language is the only instrument of knowledge that science possesses, no matter what the hypotheses on the nature of the real world are. Physical laws and their concatenation are, therefore, expressed by syllogisms whose conclusions are binding.

Here we must consider in a fundamental way the existence of physical laws, which represent the stone upon which the whole building of science rests.

According to the Aristotle's classical definition, a "*lex naturae*" is a *constant and uniform mode all physical agents maintain in operating.*

Obviously, one should first ask oneself whether these laws really do exist and whether they are necessarily as described above. The answer we can give today is the same as that elaborated in antiquity: he who denies physical laws must also reject the veracity of our senses, against the universal persuasion of all men that the senses cannot be essentially erroneous (*consensus humani generis vel sensus communis*).

One might add that, in the absence of physical laws, life, consciousness and knowledge would be impossible. Any further argument concerning the essence and necessity of physicals laws belongs to metaphysics and, specifically, to the question of the existence of some kind of design or finality in the universe. We do not wish to discuss this contested point here, but merely emphasise that if we reject the hypothesis of a universal design in nature, we encounter a number of additional problems in the light of modern cosmology. First of all, the laws of nature, expressed both qualitatively and quantitatively by mathematical formulae, contain several universal *constants* (for instance, gravitational constant, electron charge, *etc.*). We don't know who or what fixed their values and what they mean, but we can measure them and we do generally believe that they were established at the time of the Big Bang and never changed during the expansion of the universe. But we cannot demonstrate that it was effectively so; actually, there are physicists who think that they were and do change with time.[8] On the other hand, current cosmological models tell us that even a small change in certain physical constants may produce variations in the universe under which life would be absolutely impossible. We will return to this topic in subsequent chapter. At that point, however, we should stress the fundamental point that whatever the time interval encompassing the past and future of Man in the story of the universe may be, his very presence provides the only affirmative argument for the existence of physical laws and of the consequent validity of knowledge. Beyond this limit there is only obscurity.

In addition, geometry and mathematics are expressed by language and are consistent with the laws of logic. Geometric forms and numbers are seen as universal truths that can be indicated and specified by language, even though, at times, reasoning can be based on processes of visual imagination or on applying calculations. The issues of truth and self-consistency and logical rigour of the exact scientific disciplines has never been explicitly addressed, but accepting this tenet has until now constituted the starting point of any scientific speculation in our western culture.

In the centuries following the Hellenistic period, diverse philosophical schools further developed logic in several directions. The Latin world picked up where the Greek treatises left off and Severinus Boethius (VI century A.D.) was the last author to write a treatise on logic before the definitive sunset of the classical culture—we mention this author because his works were the only one to survive the barbaric destructions and to have been passed on into the hands of European *clerici* in the High Middle Ages, who developed almost *ex novo* the first elements of formal logic.

It was, however, only in the XIII century that the complete works of Aristotle reappeared in Europe through Byzantine texts and partial translations from the Arabic of the commentaries of Avicenna and Averroes. At this time Scholastic Logic was born in Europe, whose development had a most important role to play in the two centuries to follow.

The search for logical rigour was conditioned and promoted by intense theological and philosophical speculation that saw diverse schools opposed to each other and served to produce a constant critical analysis of the respective arguments. A second favourable factor was the universal use of the Latin language, more rationally constructed than Greek; though becoming lexically richer owing to a number of neologisms, mediaeval Latin maintained a rigorous and clear lexical, grammatical and syntactical structure. The work of grammarians of the Roman imperial age, which had been resumed in the Palatine Schools and later on at universities, served to settle important semantic issues and to supply clues for a classification of the various modes of reasoning. Around 1250, Peter Hispanic, future

---

[8] One might object that these changes can also be regarded as unknown physical laws. But nothing can be said about them since their discovery and experimental confirmation is definitively beyond our reach.

Pope John XXI, published a compendium of logic, which remained in use over a long period of time, and in which he listed 15 types of syllogisms that were then reordered into four categories of modern formal logic. Their use not only facilitated how argumentation in rhetorical applications was formulated and understood, but supplied the necessary means in order to find a suitable method for demonstrating theorems[9] starting from given premises.

A central issue, however, consisted in deciding whether a syllogism was a binding argument or a conditional proposition. The answer involved the semantic relation of the terms in the context of a proposition, in which these terms could be reported as objective truths (*quid rei*) or as definitions (*quid nominis*) to which no real object need correspond. Therefore, one was faced, on formal grounds, with the difference (not always obvious) between what it is true by virtue of reasoning and what is true by virtue of the intentions of the author. This distinction demanded the precision of language to be sharpened, but, at the same time, involved logic in the diatribe between *realistic* and *nominalist* schools that characterised the crisis of Scholasticism in the Low Middle Ages and impaired the development of science until the Renaissance. This crisis could only be solved when the development of mathematics, and, in particular, of algebra permitted its accessing the formalism of logic.

2.2.1 The Criteria of Truth in Science

The progress of mathematics in the XVI century and, in particular, the simplicity and transparency of algebraic formalism made it possible to reconsider the relation of this science to physics. For the first time one began to see in mathematics not just a mere instrument of calculation of physical quantities, but a rigorous and, at the same time, much more flexible language than one could have imagined from its previous applications, in particular in analytical geometry. We have seen that physics had always kept a certain distance with regard to quantitative measurements and a substantial independence from a mathematical treatment of phenomena in classical antiquity. On the one hand, wherever physics was supported by an ample and accurate experimental platform (as, for instance, in astronomy), mathematics was seen as an instrument suitable to organise and reproduce (*fitting*) available data, whilst its models had a strictly phenomenological character in the sense that they were aimed not at representing real laws and entities, but rather at predicting the properties of the phenomena under investigation with the highest possible precision. On the other hand, the Aristotelian physics, which predominated at the time, proposed supplying qualitative or semi-quantitative answers to questions regarding the natural phenomena taken in the complex whole of their interactions as a primary goal.

The so-called Copernican revolution represents the occasion on which these two points of view clashed together to result in a complete transformation.

---

[9] We remind the reader that a theorem is the last proposition resulting from an axiom followed by a chain of applications of inferences.

All began in the first half of the XVI century, when Nicolaus Copernicus (1473–1543) showed in his treatise *de Revolutionibus Orbium Coelestium* that current astronomical observations could be interpreted using an heliocentric model, in which the earth rotates, like the other planets, around the sun on an orbit that he (still) thought circular. Copernicus further developed his model in 1505, after having worked in Bologna with one of the best astronomers of that time, Domenico Maria Novara. The two were conscious of the importance of their conclusions which until that time were still, however, only circulating in the restricted circle of their friends. Nevertheless their new ideas soon provoked great enthusiasm, so that Johann Widmannstetter, a theologian and German humanist who was also secretary to Popes Clemens VII and Paul III, held conferences in Rome on the Copernican heliocentric theory. Even the Pope as well as numerous cardinals participated in these gatherings, with the former showing a vivid interest. Among these cardinals was the archbishop of Capua, cardinal von Schönberg who wrote the following letter to Copernicus:

Some years ago word reached me concerning your proficiency, of which everybody constantly spoke. At that time I began to have a very high regard for you, and also to congratulate our contemporaries among whom you enjoyed such great prestige. For I had learned that you had not merely mastered the discoveries of the ancient astronomers uncommonly well but had also formulated a new cosmology. In it you maintain that the earth moves; that the sun occupies the lowest, and thus the central, place in the universe; that the eighth heaven remain perpetually motionless and fixed; and that, together with the elements included in its sphere, the moon, situated between the heavens of Mars and Venus, revolves around the sun in the period of a year. I have also learned that you have written an exposition of this whole system of astronomy, and have computed the planetary motions and set them down in tables, to the greatest admiration of all. Therefore with the utmost earnestness I entreat you, most learned sir, unless I inconvenience you, to communicate this discovery of yours to scholars, and at the earliest possible moment to send me your writings on the sphere of the universe together with the tables and whatever else you have that is relevant to this subject. Moreover, I have instructed Theodoric von Reden to have everything copied in your quarters at my expense and dispatched to me. If you gratify my desire in this matter, you will see that you are dealing with a man who is zealous for your reputation and eager to do justice to so fine a talent. Farewell. Rome, 1 November 1536.[10]

The letter, written in Latin, was published in the preface to *de Revolutionibus Orbium Coelestium*, which appeared on the very day of Copernicus' death and bore, in addition, a dedication to Paul III. Contrary to the popular opinion, it did not cause cultured society to be confounded nor were high ecclesiastical circles unduly perturbed. As a matter of fact, heliocentric doctrines had, as we have seen, already circulated in classical antiquity as well as in Muslim and Christian Middle Ages. However, they were mostly founded on a philosophical or ideological basis while a mathematical heliocentric model had never been developed; for this reason, all

---

[10] Translation by Edward Rosen, The Johns Hopkins University Press, Baltimore.

astronomers up to the advent of Copernicus had still used the geocentric model of Ptolomaeus.

2.2.2 The Copernican Revolution

The decisive clash of the Copernican model with the dominant doctrines of the time took place a few decades later, when Galileo Galilei associated the Copernican model to physical considerations on the nature of motion. In fact, Galileo did not limit himself to confirming the simplicity and validity of the heliocentric theory, instead he directly attacked the Aristotelian-Ptolemaic system which he declared to be totally false. With this premise, Galileo directly moved the debate from the specific field of astronomy, which had well defined scopes and methods, to the more general field of physics. These assertions meant he had to face the united front of academic science of his time, which was supporting the all-encompassing Aristotelian system, a system which had been developed and extended during the course of nearly two millennia of scientific speculation. The diatribe that followed was sour and implacable and became ruinous as philosophers, theologians and, in the end, the Church itself were all implicated.

In the succeeding four hundred years, mountains of papers on Galileo's case have been written, and all protagonists of that period, even the most marginal ones, have been put through the critical sieve of the historian on more than one occasion. Today the matter appears incredible inasmuch as it was a subject for scientific discussion, and we would have expected the supporters of the new theses to be in possession of incontrovertible proofs of the new theory (in this particular case, an incontrovertible proof had to first demonstrate that at least one of the propositions of the opposing view was invalid). However, such a proof, based on astronomical arguments, could have been deemed to be authoritative only if submitted to competent judges, who were able to understand the matters in question and to follow the argumentation of the controversy. Instead of which, the dispute was prematurely made available to anyone who even had a vague opinion on the subject. In this way, any real competence inexorably ended by yielding to the decisive intervention of the established authority.

The second point concerns the probative value of the arguments. Let us try to settle what these arguments mean today and what was meant in Galileo's time.

A validation of the heliocentric model had already begun to take place in the early Renaissance, albeit in restricted circles. Its supporting arguments, however, did not have the force needed to undermine the geocentric hypothesis that, we must not forget it, represented only one aspect of a system of knowledge, on which all of science was based until the late Renaissance.

Now, if we consider the *kinematic* aspect of motion, and accept the observer's point position (the origin) and its axes of co-ordinates to be completely arbitrary, then every trajectory can be calculated in another system of co-ordinates arbitrarily moving with respect to the first one, if the movement of the new origin and the new axes (i.e., their translation and rotation) with respect to the first one is known. Therefore, from this point of view, a privileged system of reference does not exist.

On the contrary, if we want to deduce the trajectories of moving mass points using *dynamic* laws of motion, they turn out to be invariant only if we pass from one system of co-ordinates to another that moves in *uniform rectilinear* motion with respect to the first one. Otherwise, in order to calculate the trajectory of a material point, we must explicitly account for additional, virtual forces (e.g., centrifugal or centripetal forces). For instance, if we try to solve the differential equations of motion of the sun using a modern method, and taking a system of reference fixed on the earth, we merely have to apply Newton's laws of gravitation, to see that the calculation of the orbit of the sun (diurnal and annual), as observed from the earth, would entail a number of problems: it would not be sufficient to fix the speed and position of the sun at the origin of time, we would also have to introduce complicated forces of reaction, in the absence of which our calculation would produce absolutely false predictions, for instance, a rapid fall of the sun onto the earth.

If, on the other hand, the origin is fixed on the sun, our calculation is simplified because, in order to obtain a sufficiently realistic orbit of the earth, it would be enough to write the radial forces in the simplest form of gravitational attraction, as we know them. Applying successive corrections that account for the attraction of the other planets, the accuracy of the predicted orbit can be gradually improved, even though rather complex calculation would be needed. From a modern point of view, the choice of reference system is, therefore, mainly of a practical nature, dictated by the features of gravitational attraction and by the formulation of current dynamic equations of motion.

The difficulty in establishing a privileged reference system arises from a very central point of physics concerning the significance of inertia, which manifests itself in the acceleration of a body. Thus, for an observer B rotating together with his axes of co-ordinates with respect to an observer A, the physical laws appear to be different. For this reason Newton attributed reality to the rotation (which affects the laws) and not to the translation (which doesn't). But how can one confirm that A is *not* in some manner rotating? The question is not a simple one and will be answered only with the theory of the general relativity, where inertia is directly related to the structure of space and its metrics.

If we consider the problem from the standpoint of Galileo, we realise that in order to obtain clear physical proof of the movement of the earth from merely kinematic observations, one should have waited another two centuries, when very precise measurements of the parallax position of certain stars were performed, or when experiments like that of Foucault's pendulum could be carried out. However, when these proofs were made available, the heliocentric model had long since been accepted by all astronomers and physicists, on the basis of simple common sense inspired by Newton's laws. But at the time of Galileo common sense was inspired by Aristotelian physics and, even if Galileo possessed some kinematic proofs (the phases of Venus and the periods of apparition of Jupiter's satellites), which could have challenged the geocentric model, they were such specialised matters that their relevance could have only been recognised by a discussion among expert astronomers. Surely one could not expect to refute the entire body of physics of that time on the basis of a discussion of such limited arguments with philosophers and theologians. It is certainly true that in the history of science there have been cases of small, inexplicable anomalies, whose final explanation has produced a revolution in current physical models, but this has always been accompanied by a gradual development of novel, underlying theories.

Returning to Galileo, although he was able to reiterate and confirm the conclusions of Copernicus that a heliocentric model was more suitable for describing planetary motions, he could not demonstrate that the geocentric model could not do this just as well. On the contrary, the Copernican system, still based on the hypothesis of circular orbits, yielded in certain cases poorer results than that of Ptolemaic orbits. It was thus unavoidable that in such circumstances the controversy overflowed to engulf fields unfit to judge these matters and that the adversaries of Galileo, in conclusion, resorted to appealing to the established authority, namely to theological or pseudo-theological orthodoxy (Fig. 2.4).

One of Galileo's most implacable adversaries was Cesare Cremonini, a friend of his, who taught natural philosophy at the University of Padua. Cremonini was offended by the superficiality with which Galileo introduced his objections to Aristotelian physics in his *Dialogo sopra i due massimi sistemi del mondo*. It seemed to Cremonini that Galileo saw nothing but crude errors and deleterious aspects. That is, Galileo had written a *Dialogue* in which a certain Simplicius defended the Aristotelian position. The reasoning used for his defence appeared grotesquely dogmatic and puerile (in Italian his name sounds like simpleton). In reality, the historical Simplicius was an excellent critic of the Neo-Platonic school, who, in his unsurpassed *Commentaries* on Aristotle's physics, had dealt with epistemological topics, from which Galileo himself could have had drawn arguments to avoid the ruinous pitfalls of his subsequent odyssey.[11] Unfortunately, Galileo had the bad habit, common in his days, of ridiculing anybody who did not share his opinions, and of not seriously taking

---

[11] It is worth noting that Galileo repeatedly cited Johannes Philoponus in his early works even more frequently than Plato. Philoponus was a Christian philosopher contemporary with the pagan Simplicius. In his commentaries of Aristotle, Philoponus, starting from platonic critics, had formulated an original and revolutionary theory of impetus in dynamics, according to which the impetus imparted by a launcher finally resides completely in the projectile. This was neither trivial nor self-evident in Aristotle's analysis of the concept of change (of which motion was just an aspect), where the relation between *agens* and *patiens* (active/passive) lead to questions which were difficult to answer. The impetus theory allowed Philoponus to assert a finite age of the universe and to deny its eternity, establishing its temporal limits in addition to its spatial ones which had already, been established by Aristotle. Simplicius, though not an Aristotelian philosopher (in fact he tried to reconcile Aristotelian with Platonic doctrines) felt the theory of Philoponus had produced a true betrayal of Aristotle's ideas and attacked him repeatedly. The work of Philoponus came to us in the XI century through a translation from the Arabic and it had greatly influenced Thomas Aquinas in his understanding of Aristotelism and, subsequently, in the Dominican school. Perhaps it was for this reason that Galileo felt himself to be involved in the controversy between Philoponus and Simplicius. Thus, Galileo's interest in the theory of motion developed by Philoponus followed a heretical tradition of Aristotelian physics and attests to his great intuition, an intuition which allowed him to discern valid innovative ideas even in a context of controversies which had arisen a thousand years before.

Fig. 2.4 Roberto Bellarmino (1542–1621) (l.) and Galileo Galilei (1564–1642) (r.) supported antagonist positions on geocentric and heliocentric models of the solar system. Actually, from a certain point of view, the two theses were not essentially incompatible. Bellarmino was ready to accept the heliocentric hypothesis as a mathematical method for calculating planetary orbits (today we would say "as an arbitrary choice of the co-ordinates reference system"), resuming the justifications of Ptolomaeus, who merely based his "*Hypotheses Planetarum*" on the ability of his model to correctly reproduce planetary motions as observed from the earth. Galilei, however, insisted on declaring that it was false to assume there was an immovable earth at the centre of the solar system, thus challenging the whole *corpus* of physics of his time. Objectively, Galilei did not possess definitive experimental evidence, which could unquestionably prove his assertion, but his intuition told him that the Copernican system with its simplicity was more consistent with the natural movements of the planets than the complicated cycles and epicycles of Ptolomaeus. The decisive proof was supplied some decades later by Isaac Newton's law of gravitational attraction and by Johannes Kepler who calculated elliptical planetary orbits. The formulation of gravitational attraction is perhaps the most important intuitive theory that has ever been made in the history of physics. The acting force in the mathematical analysis of the planetary orbits calculated by Kepler agreed perfectly with the law of Newton, which thus appeared as a-priori properties of geometric objects. We don't know how Newton came to discover gravitation law, which was essentially predicative and *not* explicative. Did his magical-alchemic practices perhaps play an important role, where he was dealing with mysterious mutual attractions and repulsions of various substances? Or was it purely mathematical intuition at work? We don't know, and, when asked, Newton himself was never clear on this subject. We can only, once again, consider the recurring evidence that great discoveries in science often do emerge from the most unfathomable activity of the human mind and defy every deductive logical argument. The laws of Newton represented for two hundred years the foundation of all of physics. However, today we know that simple heliocentrism does not exactly correspond to reality, since in this model the mutual attraction of all the planets of the solar system are assumed to be negligible -which isn't always true—and, furthermore, the theory of relativity has invalidated the laws of Newton, which represent an approximate formulation of much more complex laws. Obviously, it cannot be excluded that even the theory of relativity may turn out to be imperfect in the future. Therefore, the opinions of Bellarmino, who emphasised the merely phenomenological character of the observed laws of motion, might seem to be closer to modern views. However, the advancement of science would hardly be possible without a certain dogmatism, such as, when climbing a mountain wall, the firmness of one hob nail makes it possible to drive another in at a higher position

into consideration the arguments of his adversary. Instead, he merely attempted to discredit his opponent with jokes that, as a good Pisan, he could come up with on any occasion.

We have selected here one of the many examples to illustrate this point.

An important part of his controversy with Aristotelian theory of motion concerns the fall of bodies. According to Newton's gravitational law, the force applied to a body is proportional to its mass, while the resulting acceleration is equal to the force divided by the mass. Hence acceleration is independent of mass. As a consequence, all bodies should fall onto earth from a fixed height at the same time. Galileo, although he did not know the law of attraction of masses, had rightly intuited this property.

Obviously, the effect holds true only if the experiment is conducted in a vacuum—even a child would correctly predict that a paper ball would fall more slowly than a cannon ball in the atmosphere. How motion depended on air friction was not sufficiently well understood at that time to exactly calculate the speed of a body falling in the air. Nevertheless, various experiments were performed that roughly confirmed the predictions of Galileo compared to those of Aristotelian physicists, who asserted that bodies of higher density should fall faster than those of lower density, even though they were unable to calculate by how much the speeds differed.

However, the conditions of motion examined by Aristotle were based on the hypothesis that a vacuum could not exist (his universe was filled with aether and light substances) and that a moving body had to pave the way by displacing matter.[12] He very clearly had the cause of this effect in mind, in his asserting that *all bodies in a vacuum would have the same speed* (*Physica* IV, 8, 216a, 20). Galileo certainly was familiar with this text of Aristotle, but he was careful not to mention it.[13]

A similar incomprehension occurred in another episode, in which, once again, the phenomenon of air friction was being considered. Aristotelian physicists reported that the Babylonians were able to fabricate such powerful bows that the thin layer of lead, which had been stretched on the arrow's point in order to stabilise its trajectory, had melted during the flight. Probably the fusion of the metal occurred during target penetration and not in the air. However, the effect is real and of primary importance for objects moving at high speed in a fluid. Galileo used the observation to make one of his ironic comments, asserting that if Aristotelian physicists believed that one could cook an egg in Babylon by letting it whirl in a sling, he had enough eggs and slings to demonstrate the opposite, even if, unfortunately, he was lacking the

---

[12] One may recall, incidentally, that a partial vacuum in a container was obtained for the first time in 1643 by Evangelista Torricelli (1608–1647), the successor of Galileo to the chair of mathematics in Pisa, who clarified the notion of air's density and pressure.

[13] Today this can easily be demonstrated by observing the fall of two objects of different weight in a glass bell in which a vacuum was created. This is one of the common and simplest physics experiments performed in schools. Yet, the effect seemingly contrasts so sharply with common sense, that during the first mission to the Moon, Neil Armstrong—we don't know whether on his own initiative or previously instructed—repeated the experiment by dropping a feather and a hammer at the same time, both of which, given the absence of an atmosphere, touched the ground simultaneously, at which the astronaut exclaimed "Galileo was right!". While this was set up to entertain the vast television viewing audience, some newspapers, however, reported it on the next day as an epochal experiment confirming gravitational laws.

Babylonians.[14] The humour of the comment would have been better appreciated if, after the witty comment, the problem of the conversion of momentum into heat had been seriously resumed. But here, as in other circumstances, the spirit of that time prevailed, whereby the aim of every dispute was the humiliation of one's adversary and personal prestige was placed above all. However, at this point, it is necessary to consider the following:

The genius of Galileo consisted in having reasoned by excluding indeed the very evidence that puts the phenomena of friction in the foreground, phenomena which are independent of general laws of motion and which may mask them completely. Yet this brilliant intuition troubles us deeply: we must, in fact, recognise that too precise observations may represent an obstacle to understanding the laws of nature and that the truth is often the product of a higher negligence of experimental data. Paradoxically, such a consideration was reported to us from the late Aristotelian school, as we have seen in previous sections: namely, Aristotelians warned against rigour in observations and measurements. However, if this remark is true, the criteria for settling *hic et nunc* any scientific controversy becomes labile and nebulous. Indeed, in a number of cases, the final judgement must be put off to a future, when a wider experimental context can possibly be made available, where a hypothesis can, step by step, gain or lose ground. Yet, in practice, leaving a judgement suspended may make acquiring new knowledge impossible. It is perhaps on this consideration that the first act of Galileo's drama begins.

When Galileo was summoned for the first time before the Inquisition to defend himself on the charge of heresy, Cardinal Roberto Bellarmino, advised him to shift the controversy to the specific area of astronomical calculations, where mathematical methods were completely free and separate from the uncertainties of physical theories. After all, he said, the Copernican model, until then, had not provoked any objections and was even taught at the Roman College of the Jesuits. Bellarmino pointed out that the situation was similar to the one encountered by Ptolomaeus with regard to different hypotheses concerning epicycles and the eccentricities of planetary orbits. Ptolomaeus did not try to find physical proof to solve the dilemma; he instead explicitly asserted that it would have been worthwhile further developing the two models in order to verify their respective predicative capacities. This position is clearly expressed in a passage of Simplicius [11], the best commentator of Aristotle, who lived in the VI century. A.D.:

... the astronomer does envisage, on a hypothetical basis, a certain method where he asserts under which conditions the examined phenomena can be saved. As an example: Why the planets seem to move non-uniformly? Answer: If we assume that

---

[14] The laws of aerodynamics predict that significant heating through air friction occurs in objects that move faster than 1 Km/s and is proportional to the square of speed. The speed of an arrow launched from a wooden bow is under 50 m/s. However, it is interesting to note that Ammianus Marcellinus, who served for a long time as an officer in the Roman army, reports (*Hist.* XXIII, IV) that the tip of the arrow of the *ballista* (an enormous steel crossbow set in action by two winches) sometimes started *nimio ardore scintillans*. There are phenomena which require careful and thoughtful explanations.

(a) they follow eccentric orbits or (b) that they move along epicycles, this seeming anomaly can be saved.[15]

This methodological concept is much closer to that of modern physics, but, if it had been adopted in the XVII century, it would have remained sterile for at least three centuries. Galileo, in fact, aimed at a different justification of his theory and he tried to offer conclusive physical proof without being in possession of it, a proof that should have been the basis of his mathematical model, and only of his, thus providing his theory with the unequivocal character of "truth". In modern language, the problem was to distinguish between a phenomenological model, based on empiricism, and a mechanistic one, based on the formulation and application of general physical laws. Galileo had realised the importance of this distinction, but this strategy was far too advanced for the scientific culture of his time.

In fact, the accusations made against Galileo were finally formulated in a way that didn't leave any room for criteria of scientific judgement. It is worthwhile examining them as summarised by Stillman Drake, an historian of science, curator of recent English translations of the scientific works of Galileo [10].

There were two propositions submitted in 1616 by the Inquisition for the consideration of the experts ([12], vol. XIX, pg. 321):

1st *Thesis*: That the sun is absolutely fixed, as far as its motion is concerned, at the centre of the world.

Censorship: All assert that this proposition is foolish and absurd in philosophy and *formally* heretical inasmuch as it contradicts the opinion expressed in the Holy Scriptures.

2nd *Thesis*: That the Earth is not fixed in the centre of the world, but moves in it as a whole and in a diurnal spin.

Censorship: All assert that this proposition receives the same censorship in philosophy and, with regard to the theological truth, is, at least, erroneous in the faith.

We note that both censorships mainly refer to Philosophy and assert that the propositions are "foolish and absurd" and not "false". No reference is made to astronomy, which was to be expected as this matter fell under the jurisdiction of Philosophy. However, Drake justly remarks that, if the judges had asked the opinion of astronomers, the verdict would have been undoubtedly the same, since, in every conceivable committee of astronomers of that time, the large majority would have disavowed Galileo's doctrine. Yet, if the sentence had been motivated by their opinion, the historians would have had to inculpate the astronomers. It is, in any case, at least curious that the historians have accused the theologians and not the philosophers for a decision taken against the freedom of scientific opinion in astronomy.[16] The fact is that the

---

[15] *"To save the phenomena (σώζειν τὰ φαινόμενα)"* is here set against *"to explain the phenomena (ἀποδιδόναι τὰ φαινόμενα)"*, as meant by Plato, in order to emphasise, on the one hand, the hypothetical nature of a new model compared to a previous one, without any pretensions to explain the final truth and, on the other, to establish the capacity of a new model for exactly reproducing "anomalies", which the previous model was unable to do.

[16] Galileo had a clear opinion on who was responsible for his sentence of 1616. In the *Letter to his daughter Cristina* he asserts that it was the philosophers who, with swindles and stratagems, cheated the supreme authority of the theologians.

philosophers would normally have asked the theologians for advice, trusting that the latter would have been on their side.

In the years to follow, Galileo elaborated on this argument again stating ([12], vol. VII, pg. 540) that he could have never imagined that the Church would have accepted that the Bible be used in order to confirm the Aristotelian cosmology, and rebutted his charge of wanting to introduce innovations at all costs, by asserting that it was just to allow persons totally ignorant of science, wearing the garment of judges of astronomers, to turn matters in the way they wanted, thanks to the authority bestowed on them, which involved innovations that would finally ruin the States (and here for States he meant the Church).

Evidently Galileo thought that the Church would have never turned an astronomical dispute into an article of faith, and was surprised by this shift of responsibility from theology to philosophy in interpreting the Bible, a decision that, in the following centuries, did cost the Church dearly which had committed the error of overrating the reliability of current scientific views and of underrating the rigour and dogmatism of the academic class, which was not less than that of the theologians.

In concluding this short *exposé* of *Galileo's case* some considerations inevitably arise regarding the problematic mutual relationships between science, philosophy and theology, three forms of knowledge, which operate in different, but partly overlapping, fields. Their limits of competence and influence are illusory, since they are outlined from time to time by conflict- or dissent-matters that vary continuously. While these conflicts are normally shaped by choices of a subjective character at an individual level, at a collective level, however, they tend to take deep roots and may finally generate pernicious effects of intolerance and repression.

The doctrines of science, philosophy and theology do not have a definitive order, but are in continuous evolution each following their own respective aims and methods. It is, therefore, improbable that they continually be in common agreement. In fact, each one possesses criteria to appraise their own contents, which cannot be imposed on the others without calamitous effects. In Galileo's time, physicists viewed geocentrism only as a question regarding planetary motions whilst philosophers associated this concept to a cosmic model of the evolution of nature and its influences on man. Theologians, on the other hand supported geocentrism mainly for scriptural reasons in order to place man at the centre of creation. These visions were the results of ideas, which sometimes germinated and matured through conflicts that were later solved by means of a slow process of assimilation that lasted almost two millennia. To be sure, an immediate adaptation of all disciplines to every new pertinent discovery or development and in disagreement with their contents would not even be conceivable. The most productive strategy would instead be to proceed fully conscious of the conflict at hand and to establish its resolution as the next goal. Unfortunately, those who are mistakenly convinced that they possess the universal key to knowledge are often tempted to turn these occasional conflicts into a permanent state of out and out war. The case of Galileo has made history, but has taught little to the generations to come. The *mea culpa* recited by the Catholic Church four centuries later, for whatever reason or motivation it was thought necessary to issue, has been used in certain quarters of modern scientism as grounds for a renewed attack on religion. Today the positions of the inquisitor and of the accused are inverted, but the war is still being waged.[17]

In conclusion, this digression on the Copernican revolution has proven necessary in order to point out problems that reveal important social aspects of modern science, problems which first appeared when science definitively shook off the tutelage of philosophy.

Whilst the history of philosophy has always been marked by a sort of dialectic, involving coexisting diverse and often contrasting schools, science presents itself as a monolithic system of knowledge. For this reason, it can evolve, but in no way can it tolerate any alternative system. Attacking and eliminating dissenters is its *raison d'être*. If these turn out to be correct, their valid ideas are conveniently absorbed; if the dissenters are proven to be wrong, they do not have right to exist. In other words, any discrepancy is supposed to be immediately eliminated. This is an awkward position, all the more since science admittedly renounces the ability to produce a definitive truth and is even prepared to contradict itself whenever necessary, if a world reality has been observed which is to a great extent still unknown; yet this position must be accepted because the claim that science is unique is necessarily founded on the common belief of the rationality of reality. The dilemma faced by modern society when confronted by this totalitarian tendency of science is, on one hand, to allow its free development and, on the other, to set some limits to its authority in pragmatic matters.

After Galileo, some decades were still required in order to rectify the Copernican model and almost a century was needed in order to find the physical proof of its "truth", based on the formulation of the law of universal gravity. This goal was achieved in two stages. First, Johannes Kepler (1571–1630) established the orbits of the planets with greater accuracy using mathematical techniques, thus formulating his three fundamental laws:

The orbits are ellipses of which the sun occupies one of the two *foci*.

A line joining a planet and the Sun sweeps out equal areas during equal intervals of time.

The square of the orbital period of a planet is directly proportional to the cube of the semi-major axis of its orbit.

Kepler, like Galileo, did not know the law of universal gravitation, but it is wonderful that the mathematical functions he produced describing the planet's motion contain it implicitly: it can, in fact, be demonstrated that the mathematical form of the planet orbits implies that acceleration is always directed towards the sun and is inversely proportional to the square of its distance from the planet in question.

Isaac Newton finally experienced that flash of intuition which led to him explaining this law, whose application makes the calculation of the trajectory of any heavy body in space possible. At this point, the heliocentric hypothesis appeared to have

---

[17] Of course not all scientists have maintained an antireligious stance. However, in 1992 the centenary celebrations of Galilei in Pisa, Venice and Florence were largely used for anticlerical propaganda. In Florence this was reduced to inviting some Apache Indians to protest against the astronomical observatory that Vatican City had constructed in the desert of Arizona.

a physical basis supporting it and one would have expected the diatribe with the supporters of geocentrism to be definitively settled. In reality, the Newtonian definition of the solar system was not a point of arrival, but rather of departure towards our modern conception of the solar system. The force that acts on planets is not, in fact, only due to solar attraction, since the attraction of other planets must also be accounted for. While its contribution is small, one felt the necessity to calculate these complicated corrections, when more precise astronomical measurements were obtained. Moreover, the motion of Mercury was irremediably in disagreement with Newtonian laws and the necessary corrections could only be produced in modern times by applying the theory of General Relativity. A correct calculation of the planetary orbits represents, even today, a serious mathematical problem that cannot be solved with simple heliocentric models: the most accurate corrections are only feasible numerically using powerful computers. Presently, an application of mathematical perturbation method implies by necessity the introduction of approximately 100 harmonic functions, defined empirically, in order to be able to apply the corrections demanded by modern astronomical measurements to the Keplerian orbits.

Faced with these conclusions, we may imagine the spirit of Ptolomaeus to be rejoicing: for he had followed an analogous method in his system and he had to introduce 18 additional eccentric movements in order to reproduce the observations of the planetary orbits measured in his time.

If then today we must resort to the same methodological approaches as in antiquity, what is the true significance of the Copernican revolution? Various answers can be given to this question, but the most important one is substantially tied to the definitive criterion of the formulation of physical laws in terms of mathematical language. Physical laws must be obtained from experimental data, even if these refer to conditions that cannot be perfectly reproduced, but only ideally defined; the laws, however, are perfectly valid under these conditions. Hence, physics turns out to be constituted of two buildings: an experimental one that grows incessantly and a second, theoretical one consisting of fundamental laws and mathematical relations, whose predictions must be in agreement with the experimental data, within the limits of maximum attainable accuracy of methods of observation and measurement. Thus, science turns out to be an object in a state of permanent evolution, whose soundness and legitimacy are continuously being challenged, but whose development is aiming for a mutual corroboration of our knowledge of the world of phenomena as they are perceived, and their models which are elaborated and validated by these very experiments.

This "experimental method" constitutes the base of modern science, which no longer seeks to offer universal explanations, but rather to maintain coherence between the observations of phenomena and the generalisation of their course. From Galileo onwards the progress of science, realised by *research*, has continued to follow these two paths, whose direction has been dictated at times by experimental observations and at times by theoretical predictions. The two horizons, that of mathematical implications of physical laws and that of the observed phenomena remain, a priori, disengaged, and the function of science is to make them coincide on the widest possible arc. Science finds, therefore, vital hints of progress outside of itself, whilst, paradoxically, its death would be a self-perfecting process, resulting in a *closed* system of definitively self-coherent experimental and theoretical knowledge, whose only use would be to tell us what we already know.

2.2.3 The Role of Logic

With the Copernican revolution the issue whether the instrument of logic is sufficient to understand reality remained suspended. There was, however, general agreement on one important point: logic, applied to abstract, but clearly defined objects and axioms—such as, e.g., those of mathematics—is able to supply a self-coherent description of its entire field of implications. But this thesis also needed to be demonstrated. However, in the age of the Renaissance, Humanism, with its return to Classicism and its rejection of "barbaric" mediaeval culture, was no longer able either to understand or to continue the development of formal logic.[18]

At the close of mediaeval culture, the contrast between Humanism and Scholasticism is, once again, the result of the clash between Aristotelian and the Platonic methods of investigation, each one crystallised in its own *forma mentis* without any possibility of a true conciliation. Aristotelism was essentially aimed at interpreting the reality as a system of relations between objects in perennial evolution. The variety and transitoriness of their properties could only serve to indicate that science could only consist in tracing their evolutionary paths, reconstructing the entire map of the cosmos which itself constitutes a Perfect Design, ordered by rigorous laws of causality, which themselves reflect the exact laws of logic of the mind.

The approach of Platonism is quite the opposite: objects possess eternal properties and are in mutually harmonic relations, hidden to the human senses, but which the mind can perceive in a glimpse, contemplating the harmony of ideas. The science of shapes (geometry) and of numbers (arithmetic) is essentially nothing but the discovery of their perfection and simplicity, which we perceive only in a diffused way in the observed phenomena.

Paradoxically, Platonism, in search of the deepest abstract truths, has provoked analytical attention in its speculations and, hence, research which addresses the tangible world. Conversely, Aristotelism was mainly concerned with the construction of intellectual models based on processes of synthesis and generalisation and with their endless improvement. From this perspective, the explanation of phenomena merely represented an exercise in applying these models.

Today, if we had to judge the respective merits of these two approaches, we would still be confronted with an insoluble dilemma. Considering the issue in terms of methodology, we must admit that modern science advances by following one of the two approaches, depending on what specific requirements are entailed. But in the

---

[18] Starting from the first decades of the twentieth century, the rediscovery of classical and scholastic logic—mainly due to the work of Jan Lucasiewicz (1878–1956) and his school—has supplied important clues for the study and development of modern formal logic.

past the choice methodology being based on pragmatic principles would not have been considered sufficient. Science could not have even been conceived if it had not been encompassed in a philosophical system. In this respect, we can say that the modern crisis of philosophy and the consequent disengagement of science could, on the one hand, have been beneficial to the free growth of modern science, but, on the other hand, would have left science in a labile position, where its social and cultural value would be exclusively tied to its ability to promote technological progress. We shall return to this important topic in subsequent chapters.

Coming back to the decay of Scholastic logic during the Renaissance, it suffices here to remark that this discipline has been taken up again only in the modern age, culminating in the theorems of Kurt Gödel[19] on the *completeness* of the calculation of the first order and on the *incompleteness* of all theories apt to supply proofs of the logical foundation of mathematics. We anticipate here briefly one of the aspects of the problem regarding the Non-Contradiction of the formal theory of numbers which, from the beginning of the past century, has engaged many eminent mathematicians for decades.

With the introduction of formal logic it became possible to demonstrate that, in some cases, even the most rigorous application of mathematical formalism could lead to antinomies. One of the first examples, produced by Kurt Gödel, is reported here as reformulated by Herbert Meschkowski [14]:

- Let us assume an arbitrary method to express any mathematical proposition with an integer number, using the key of a simple cipher code. The table below gives an example of one of the possible codes that establish a correspondence between numbers, alphabetical letters and mathematical symbols (top row), and integer numbers (bottom row).

1	2	...	9	0	a	b	...	v	z	=	+	-	·	(	)	→	:	if	else	etc.
1	2	...	9	11	12	13	...	37	38	39	41	42	43	44	45	46	47	48	49	etc.

Here the number 0 does not appear in the second row because its symbol is used to indicate the division between two subsequent characters in the ciphered expression. In practice, by adding a few tens of other definitions to the table, it is possible to convert any mathematical expression or proposition into an integer number. Inversely, any integer number, if deciphered, produces a sequence of numbers, letters and symbols, which, obviously, in the vast majority of cases, has no meaning. There exists, however, a subset of integer numbers, whose deciphering leads to a calculable mathematical function. Though extremely rare, there is an infinity of these numbers.

Let now define a function $g(n)$ as follows:

$$\textbf{if}\quad n \to f(n)\qquad g(n) = f(n) + 1$$
$$\textbf{else}\qquad\qquad\quad g(n) = 1$$

---

[19] Kurt Gödel (1906–1978) was with Bertrand Russell, Alfred N. Whitehead and David Hilbert one of the principal founders of mathematical logic.

where **if** and **else** are respectively meant for, *if* and *otherwise*, as used in some current computer programming languages.

The definition of $g(n)$ is clear: if the decoding of n produces an unequivocal definition of a calculable mathematical function (called $f(n)$), then the value of $g(n)$ is given by $f(n) + 1$, otherwise the value of $g(n)$ is set equal 1. Function $g(n)$, literally codified with the previous table, corresponds to the number:

$$n = N =$$
$$48028046017044023045018039017044023045041010490180440230450390010$$

Now, let us assume the above value of N to be the argument of function g. If g were a calculable function we would yield the result that:

$$g(N) = g(N) + 1,$$

which is absurd. Would it be, therefore, false if function $g(n)$ were calculable? No, since if it were *not* calculable, its value would be 1, i.e., it would be calculable.

The cause of this contradiction is a most subtle one: the definition of $g(n)$ implies a specific *decision* (Gödel calls it *Fallentscheidung*) on the computability of a function corresponding to the code cipher n. If the result of this decision were calculable, it could be made automatically and the difficulty would be overcome, but this is impossible, since judgement and decision are taken by the thinking subject on the basis of his knowledge.

It was later attempted to solve these contradictions, pathologically present in certain recursive logical processes, by using the so-called Principle of Transfinite Induction. But this principle shifted the problem on to even less solid ground, and these tentative answers remind us of somebody trying to get out of the water by pulling his own hair. Today the three famous statements formulated by Gödel are universally accepted:

1. There are insolvable problems in the theory of numbers.
2. The non-contradiction of the mathematical formal system cannot be proved with means supplied by the system itself.
3. The axioms of the theory of numbers do not contain an *implicit* definition of the concept of numerical sequence.

The idea that, once formalised, mathematical logic can be applied to all disciplines with a certainty of not encountering antinomies must be abandoned. A self-guarantee of the infallibility of our thoughts does not exist, whatever the field may be. In other words, it is not possible to gain knowledge without trusting, a priori, in *something* that justifies, *a posteriori*, its consequent implications. The investigator must be ready to use, if necessary, new instruments of logic, but he must also be aware of the possibility of failing, since every cognitive process possesses inherent limits of application. From this point of view, the laws of logic have a fundamental nexus with metaphysics, without which they lose any significance.[20]

We must, however, realise that the splendid evolution of mediaeval logic in the centuries to come only weakly affected the development of sciences. Concerning physics, as long as logic remained attached to Aristotle's teleological models, no significant progress was observed, with the exception of alchemy, which continued to progress through the Renaissance until the end of the XVIII century (see Fig. 2.5 and legend).

2.2.4 The Role of Algebra and Infinitesimal Analysis

From a modern point of view, we are in a position to understand why in the cultural climate of the Middle Ages, mathematics could hardly develop beyond the horizons of the classical antiquity, not because speculative ability was wanting but an adequate formal language was lacking. This was first supplied by algebra, but its importance

Fig. 2.5 Gottfried Wilhelm Leibnitz (1646–1716) (l.) and Isaac Newton (1643–1727) (r.) both set in motion modern infinitesimal analysis. They independently developed the same ideas and in the same period, but using different formalisms. The fight for claiming the *right* to have discovered *"calculus"* was very energetically waged and from Newton's side at least was sometimes conducted using reprehensible means, to put it mildly. However, although at that time research was carried out privately, communication among scholars was excellent. Thus, very soon the European scientific community was able to correctly appreciate the respective advantages of the two formalisms

---

[20] Some mathematicians, among whom David Hilbert, do not agree with this conclusion. They recognise, obviously, that insolvable problems exist, but they think that, having demonstrated its insolvability, a problem is thus intrinsically solved, since an answer is given to the question it poses. But what can be said if the problem regards the very foundations of mathematics?

was not recognised for a long period, and only towards the end of the XVI century, mainly thanks to the work of the French mathematician Franciscus Vieta (1540–1603), was literal calculation introduced with logical rigour, clearly defining the basic principle of homogeneity that for centuries was only indistinctly understood.[21]

Algebra—the *Ars Magna*, as Gerolamo Cardano called it—represented a powerful instrument of mechanisation of thought in the centuries to come, because it yielded an immense economy of mental effort. This in turn allowed the following generations to instead focus on more and more complex problems. If, in fact, until the first half of the XVI century algebra was mainly used to study solutions of non-linear equations, in the period to follow it supplied a conceptual and formal base for the development of functional analysis and infinitesimal calculus.

The opening of modern horizons of mathematics can be viewed as the wonderful, autumnal fruit of Humanism. With the fall of Constantinople, in addition to countless philosophical texts and to comprehensive commentaries to Aristotle, [22] the works of great mathematicians of the Classical and late Greek antiquity, such as Archimedes, Ptolomaeus, Euclid, Aristarchus of Samos, Pappus of Alexandria, Apollonius of Perga, Serenus of Antissa, Heron of Alexandria , and Eutocius of Ascalon, also reached Italy. There was an immense patrimony of ideas, of applications and comments, seeds, that for centuries had been waiting to sprout.

Two Italian astronomers and mathematicians, Francesco Maurolico (1494–1575) and Federico Commandino (1506–1575), were pioneers in this regard because their perfect acquaintance with ancient Greek allowed them to understand and study these mathematical works and to publish their Latin translations, allowing for their distribution throughout all of Europe.

---

[21] The principle of homogeneity asserts that a given symbol can signify different quantities in diverse circumstances, but these quantities must be of the same type, that is to say, they must be expressed by the same dimensional formula (e.g., seconds, litres, joules, metres/second, $/barrel, *etc.*).

[22] From the III century B.C. until VII century A.D. writing *commentaries* to works of the greatest authors of the past, mainly of Aristotle, was the surest way to ensure broad circulation. These commentaries, often very extensive, sometimes contained amplifications and important original considerations that, thanks to a vast apparatus of quotations, allowed one to comprehend in depth the various topics discussed. The commentaries to Aristotle can be divided into three groups. The first ones, which cover the Hellenistic period until the IV century A.D., were written by the direct heirs of the Aristotelian tradition. The second, by far more numerous, group dates back to the three subsequent centuries and are composed almost exclusively by Neo-Platonic philosophers, many of whom were trying to conciliate Aristotelian and Platonic doctrines. The third group was compiled in the Byzantine cultural milieu of the XI–XII centuries. Only this last group immediately reached the West, while part of the ancient commentaries had been previously absorbed by our mediaeval culture through Arabic translations. In order to get an idea of their size it suffices to say that the commentaries to Aristotle alone compiled between the III and VII centuries A.D. occupy approximately 15,000 pages in the original Greek manuscripts. They were only published for the first time in 1909 (only with the Greek text and the title *Commentaria in Aristotelem*) by Hermann Diels, the result of thirty years of labour made possible by a munificent grant of the Prussian Academy of Sciences. Till now only minimal amounts of these Commentaries have been translated into modern languages. With the study of these texts having become more widespread, the modern critic has been able to unearth ideas and cognitions which were commonly believed to be original products of the Arabic or mediaeval culture.

In the wake of the renaissance of ancient mathematics, in the 80 years between 1550 and 1634, in which *De maximis et minimis* by Pierre de Fermat (1601–1665) appeared, Italian algebraists reaped this inheritance and opened a window to the novel field of differential calculus.

The subsequent XVII century represented a period of tremendous fecundity for mathematics. For the first time, after seventeen centuries, talents bloomed in Europe who were able to further develop on the ideas of the great mathematicians of classical antiquity, as Eudoxus , Euclid and, especially, Archimedes.

It was thanks to Kepler that the problem of integration according to the method of Archimedes was taken up again, a subject he called *Archimedaea Stereometria* and which he first applied in his now famous calculation of the volume of a wine barrel (*dolius austriacus*), where he arrived at important conclusions on the maxima and minima of isoperimetric areas. Equipped with his exceptional intuition, Kepler introduced the most complicated integration methods which were different from those of Archimedes, sometimes avoiding very insidious obstacles and arriving, for the most part, at correct results. For the first time in the history of mathematics he was able to put in relation, the maxima and minima of a function with the annulment of its derivative, despite having used an approximate formulation. Yet the methods developed and applied by Kepler were too manifold and complicated to be used by other scholars. It was in fact still necessary to characterise a general method of integration, which could be clearly and rigorously applied to every problem.

Once again it was the Italian mathematicians who led the way in this regard.[23] Among them, Bonaventura Cavalieri stands out prominently, whose work represented a welding of new ideas. Cavalieri introduced the definition of *indivisibilia continuorum* as a base for the performance of integral calculus, a definition, that, already owing to its verbal formulation, gave rise to animated discussions. On the wake of Archimedes, he started from the hypothesis that a plane surface could be decomposable into a set of rectangles of infinitesimal thickness, a concept that was known since antiquity, but was never thought to be a valid instrument for demonstrating mathematical theorems. Cavalieri answered these critics by considering infinities of discrete *indivisible* areas only in order to mutually compare them, and stressing that not these indivisible objects, but rather their relationships (ratios) had any real significance when considering different parts of a figure. It was a step forward, although not a definitive one. Within its limits, this method enabled competent mathematicians, such as Evangelista Torricelli, to calculate integrals of more and more complex functions. The limit of Cavalieri's approach was, however, reflecting in the importance of the choice of integration method needed for the curves to be analysed of the specific geometric form in question. For practical purposes, it was necessary to establish a procedure for every type of curve which varied from case to case. Progress would have only been possible by starting from the concept of curve

---

[23] A centre of excellence in this period was the University of Bologna and gracing the chair of Mathematics were outstanding mathematicians one after the other, such as Piero Cataldi (1548–1626), Bonaventura Cavalieri (1598–1647), Evangelista Torricelli (1608–1647) and Pietro Mengoli (1626–1686).

as a geometric representation of a general function $y = f(x)$ and from there to a surface subtended between a curve and the axis of the abscissa. This idea made its appearance very soon indeed. Luca Valerio (1552–1618), a mathematician of the Roman College, was the first to introduce the concept of arbitrary continuous curves and to enhance algebraic formalism with the introduction of differential calculus. The decisive step was, however, taken by the great Cartesius (1596–1650) and by his contemporaneous French mathematicians, when, as a result of the development of analytical geometry, the concepts of functional derivation and squaring could be visualised and, at the same time, formulated algebraically. The essential requirement was to operate a synthesis between the geometric definition of infinitesimal difference (differential) and the property of a curve (or surface, or volume) defined by abstract mathematical functions.

At last, Isaac Newton and Gottfried Leibnitz had clearly formulated and solved this fundamental problem. It is well known that their hotly disputed claim to having made this discovery involved scientists from almost all of Europe. The current view is that the two reached the same result independently and through different ways: the imagination of Newton was strongly influenced by the equations of motion, whilst Leibnitz founded his mathematical ideas on his corpuscular theory of *monads*. Both methods, however, could lead to developing integral and differential calculus thanks to an unconstrained, at times even excessively so, application of algebraic formalism. It is interesting to note that, from a strictly mathematical point of view, their procedures were not at all rigorous nor were their definitions sufficiently precise.

The formalism of Leibnitz was eventually favoured and has been at least preferentially used up to now. According to this formalism the derivative of order n of a function $y = y(x)$ is given by the quotient of two differentials: $d^n y/d x^n$. This was an intuition of geometric character, but whose criteria of validity were only demonstrated when the limit concept was definitively elucidated in the 19th century, including rigorous definitions of continuity, passage to the limit and of multiple differentials (Fig. 2.5).

It is known that the majority of contemporaneous mathematicians at first did not take seriously the ideas of Newton and Leibnitz, and in some cases they even ridiculed the new infinitesimal analysis. They were, in fact, intelligent enough to see the weakness of some of its initial foundations, but sadly not enough to appreciate the enormous importance and inherent potential of the new approach. Thus, once more we are here confronted with the alarming consideration that even in mathematics great discoveries are first made by intuition, and are rigorously demonstrated only later, often by others and after a long time.

The word *calculus* (Latin for "pebble"), as infinitesimal analysis was thereafter called, is referred to the concrete element of an abacus, but was given a new meaning, which was no longer part of a legacy of the various philosophical theories of the *physical* indivisible unit, but rather a *concept* of an abstract numerical variable, arbitrarily large or small, but a priori not specified. Actually, algebraic formalism made it possible to define variable physical quantities and their functions, and to apply the conventional operations of addition, subtraction, multiplication and division to these objects. General symbols were introduced to precisely indicate the nature of the mathematical objects they represented and the way through which, at least in principle, these could be evaluated numerically. Without any doubt the automatism of algebraic formalism not only opened the road towards new intuitions, but also contributed to specifying basic theorems that were later demonstrated in the modern age.

During the XVII century the concept of rational function (corresponding to the quotient of two polynomials with rational terms) was studied and extended to defining transcendental functions of a real variable. Moreover, the concept of imaginary numbers, already enunciated in the XVI century by Cardano, was later developed by Euler and Gauss, who established the fundamental relations between trigonometric functions and exponential functions of complex numbers. Finally, it was the speculation of several generations of mathematicians that made it possible to consolidate the various aspects of differential calculus. Once the way was open, however, the different efforts could be co-ordinated to a greater extent so that a certain synergy increasingly began to accelerate a scientific progress that was going to change Western civilisation entirely.

2.3 Harvesting the Crop

From the end of the Middle Ages natural sciences progressed at the same pace as mathematics, while physics was previously considered to be a section of philosophy with its privileged instruments being logic and geometry (Fig. 2.6). Nobody thought that mathematics could be of great importance in the cognitive processes of nature. This was in part a consequence of the lack of interest in the practical applications of scientific knowledge and in part in the conviction that *quality* had a higher rank than *quantity* in all phenomena, which was considered an accidental property. It must, however, be considered that it would be very difficult to see something in the simple act of calculation that goes beyond basic problems of metrics and measurements if algebraic formalism is lacking. Physical laws were, therefore, expressed analogically or through geometric models.

On the other hand, the mature age of science is marked by the emancipation of mathematics from the constraints of an exclusive introspective analysis and from its metaphysical aspects that, starting from Pythagoras, had conditioned its objectives and development lines. In fact, physics began to mature as a discipline when objects in mathematics began to assume a character of generality that enabled one to relate them to real objects, no longer on the basis of the principle of *identification*, but rather, on a (weaker) one of *representation*. Hence, mathematical operations could be interpreted as descriptive (*models*) of physical processes.

The advantages derived from a precise mathematical language and from the concision of its formalism caused analogical language to progressively lose its importance in the formulation of physical laws. Actually, beginning from the end of the nineteenth century, a physics treatise could have been written using a mathematical symbolism, almost without the aid of any spoken language. On the contrary, today

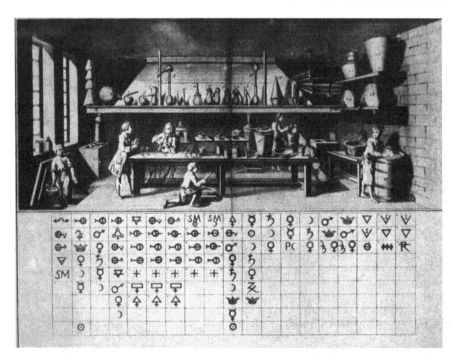

Fig. 2.6 The engraving, taken from *Table des Rapports* by Etienn-François Geoffroy (1672–1731), shows a chemical laboratory of the first half of the XVIII century. The table at the bottom contains results of experiments on the chemical affinity of various substances. The reported symbols of the "*elements*" are still those used in alchemy, based on complex magical-esoteric doctrines, a field which cast its own spell of fascination and even today doesn't cease attracting disparate followers. Notice that in the 4th column from the *left*, among other *symbols*, there is a *triangle* (indicating *fire*) with three small *circles* (the *essential oil*) on its vertices; this combination indicates the *phlogiston*, which, in those times, was believed to be the fifth fundamental element (the others were: *fire*, *air*, *water* and *earth*). The absorption and release of *phlogiston* was explained to be the cause of, respectively, endo- and exothermic reactions. In fact, the only alchemic *elements* that can be considered as such from a modern viewpoint were the metals; as for the rest, hundreds of definitions of heterogeneous substances, states and processes were defined by applying classification criteria which had been established from experiments and observations carried out during the course of more than a millennium. It is truly amazing that alchemists were able to gain valid general knowledge from such an accumulation of inorganic and organic reagents. Even more remarkable is the correspondence they found between chemical reactions and the Aristotelian laws of causality, which were believed to govern all phenomena, from reactions in the microcosm to astrological influences. It should be finally be noted that, apart from thermodynamic aspects, theoretical chemistry could only develop as a modern science only once a model of atomic bonding was devised and rendered accessible for general applications. By nature, modern chemistry was, therefore, subordinate to progress in physics for a long time. For this reason, we have not dealt here with the interesting story of its birth and growth—even though chemistry is today at its own frontiers of development

we recognise that there are concepts in modern physics whose verbalisation would represent a futile effort and, in some cases, entail enormous difficulties in comprehension. This does not mean that the progress of science proceeds exclusively on the rigid and pre-arranged tracks of a mathematical language. Rather, this language is continuously being required to evolve in step with the necessities and intuitions of the physicists using it.

2.3.1 Physics and Infinitesimal Calculus

A critical and historical milestone was reached in the second half of the nineteenth century, when infinitesimal analysis could finally be applied to a wide range of functions, whose properties had meanwhile been studied in depth. This milestone could only have been achieved after infinitesimal analysis had been consolidated based on rigorously demonstrated theorems. Differential equations turned out to be a particularly important field for the development of a modern language for natural sciences. In these equations the unknown is not expressed as a numerical variable, x, but as a function of this variable, $y(x)$.

In its most general form, a differential equation is written as a relation F between $y(x)$ and its derivatives with respect to x, where the highest n^{th} derivative determines the degree, n, of the equation, i.e.:

$$F(x, y', y'', \ldots, y^{(n)}) = 0$$

The problem is finding a function $y(x)$ that satisfies this equation and possesses a sufficient degree of generality.

In modern analysis this concept is expressed as the definition of *general integral* of the equation, which is as follows:

"The general integral of a differential equation is a function

$$y = \varphi(x, c_1, c_2, \ldots c_n)$$

of the variable x and of n constants c_k with $k = 1, 2 \ldots n$, which satisfies the equation and arbitrary initial conditions (i.e., for $x = x_0$) imposed on y and its derivatives."

Note that one does not exclude the possibility that different solutions may exist apart from φ. Here, however, the only requirement is that the function be sufficiently flexible to reproduce the arbitrary initial conditions. One knows that the determination of the general integral of a differential equation is only possible for a few, simple cases. Furthermore, its expression is often so complicated so as to be of little practical use. A common method of solving differential equations is, therefore, to directly study the properties of particular integrals in connection with the initial conditions. In addition, in mathematical physics a function y often represents a quantity that varies with space, x, and time, t, and the differential equation which has to satisfy it must contain combined partial derivatives with respect to both x and t. The solution of partial differential equations is much more problematic than that examined above, and the cases where integration methods are available are only a few handful and of degree lower than fourth.

Partial differential equations could, therefore, be used in mathematical physics only for describing simple fundamental laws or standard ideal cases, such as, for instance, the theory of motion in a potential field, some cases of dynamics of fluids, heat conduction, elasticity theory, vibrations, electromagnetic waves and, in particular, fundamental phenomena of quantum mechanics.

Only in the last decades could more complex differential equations be solved numerically with adequate approximations, owing to the use of powerful computers. But these procedures require skill and caution since an automated numerical programme may lead to treacherous conclusions while masking problems which can only be revealed by infinitesimal analysis.

The importance of differential equations in mathematical physics is not due to its practical applicability, but, rather, lies in methodological reasons inherent in the description of observed phenomena. Actually, if $y(x)$ represents a physical quantity of a system as a function of a variable x, the *derivatives* of $y(x)$ describe its *changes* as they occur in nature: mostly, changes as functions of space and changes as functions of time. Therefore, the expression of this function must implicitly contain the pertinent physical laws, which govern the evolution of the system. However, except for a very few cases,[24] the direct construction of such a function would hardly be possible without having previously defined these laws, which are expressed as *rules of variation*, i.e., as differential equations. In effect, mathematical functions, $y^n(x)$ do often exhibit a simpler dependence on space-time variables than $y(x)$.

Differential equations finally came to constitute the new backbone of modern mathematical physics. Their impact was enormous since the advantages obtained from them were such as not only to agree with a general and effective formulation of elementary physical laws, but also with an analysis and precise description of complex phenomena.

Their influence on epistemology was even greater by the end of the nineteenth century. Theorems of existence and the uniqueness of solutions of equations of motion for given initial conditions were interpreted as being mathematical proof of mechanistic determinism, already expressed at the end of the XVIII century in a famous statement by Laplace [25]: *We can consider the present state of the universe as the effect of its past and the cause of its future. A mind that at a certain time knew all the forces that act in nature and all the positions of the objects of which it is composed, if were large enough to analyse these data, could embrace with one single formula all the movements, from the greatest bodies of the universe to the smallest*

---

[24] The typical case is that of the elliptic planetary orbits found by Kepler: the second derivative of the planet's radial position versus time corresponds to the gravitational law as written in its simplest form.

[25] Pierre-Simon, marquis of Laplace (1749–1827), had studied methods of solving differential equations by applying variable substitutions (Laplace transforms) which make it possible, in a (Footnote 25 continued) number of cases, to solve them more easily. Laplace was obsessed by the idea of being able to foretell the future—hence his interest in astronomy and in the theory of probability. His reply to Napoleon, who remarked that there didn't appear to be any mention of the Creator in his books is famous: "This hypothesis was not necessary" he said, adding later on: "It explains everything, but predicts nothing".

atoms; for such a mind nothing would be uncertain and the future would be present to its eyes as the past is.

Today we know this assertion to be false from a mathematical point of view as least, but until about the end of the nineteenth century the physical phenomena investigated were satisfactorily described by mathematical models which had been solidly demonstrated. Moreover there was a general belief that once some basic postulates had been accepted, all that still remained unknown was potentially contained in formulae and their mathematical implications. This meant that they could still be discovered independent of experiments or of mathematical calculations—as, in fact, has often happened. Nevertheless shadow zones existed in science which became more obscure the closer one approached them.

The mathematical formalism which was adopted presupposes a biunivocal correspondence between physical quantities and real numbers. Thus, from a theoretical point of view, measurements of the basic quantities time, position, mass and electrical charges and of the other derived quantities, consist, once an arbitrary unit has been fixed, of *real* numbers, entailing, as a consequence, an unlimited and continuous property for all physical quantities, although any actual measure can only be expressed by a *rational* number. The difficulty inherent in procedures demanded by such ideal measurements are immediately revealed in conceiving a physical measure expressed by a real number. However, this central problem was eliminated as with peripheral issues dealing with the treatment of errors and experimental uncertainties. Once this position was accepted, the superiority of mathematical data with regard to experimental data was implicitly asserted, in the sense that the former can reproduce the latter, but not vice versa.

Once again we are confronted with the problem of *continuum* in mathematics and physics. The objections of classical antiquity, based on the grounds of rational numbers, still persist. However, mathematical instruments have now made it possible to navigate on the sea of a continuum following precise routes, without worrying about its unfathomable abysses, although nobody can imagine all the possible effects of the currents arising from these depths.

2.3.2 The Concepts of Force- and Potential-Fields

The first of the great innovations brought about by physicists of the eighteenth and nineteenth centuries is a generalised concept of force. Although vaguely intuitive, force must necessarily refer to concrete situations, in which it manifests itself as a mechanical acceleration of the object to which it is applied. Force is characterised by its intensity and direction and, therefore, is represented by a vector, F, and by its point of application; thus, in an extended body, different forces can act on various application points and the motion of this body depends on their vectorial combination. By analysing these cases a concept was developed of a force depending on spatial co-ordinates of a general point of application (x, y, z), that is to say, of a function $F = F(x, y, z)$ defined in the field of real numbers which represent the space of the object. By abstracting the materiality of the application point, function $F(x, y, z)$ was assumed to represent a "property" inherent in space and was called *force field*.

Force fields may be described by complicated functions such as, for example, those representing mechanical forces applied to a deformable body, or static stress in a complex building. However, the force fields that we know in nature have their origin in elementary interactions, of which the main ones are attraction of masses and repulsion/attraction of electrical charges (generalised in the concept of electromagnetic forces). If masses and charges are punctual, in both cases, their interaction force decreases with the square of the distances between two interacting points; i.e., the intensity of the force is given, in both respective cases, by the simple formulae $F(r) = \varepsilon q'q/r^2$ and $F(r) = gm'm/r^2$, where m are the masses and q the charges, and where ε and g represent universal physical constants; the direction of F is that of the line joining the two interacting points.

It is not easy to mentally describe the intuitive notion of applied force, which entails an action of a body onto another, to that of a force field. Moreover, behind the reassuring simplicity of the formulae, certain considerations lead us to face the difficult explanation of more fundamental issues: firstly the reason for this particular spatial dependency of elementary interactions and, secondly, the meaning of universal constants ε and g. Furthermore, apart from their mysterious nature, these constants are immensely different: the intensity of a gravitational field is almost infinitesimal if compared with an electrical field. If, for example, we consider the interaction between a proton and an electron, the electrical attraction of their charges is approximately 10^{39} times stronger than that due to the gravitational attraction of their masses. This ratio is a constant number; therefore, whatever the explanation of the two constants may be, we wonder which physical property can have determined this enormous difference. Whatever this number means, it is certain that it must have fundamentally determined the entire form and evolution of the Universe.

In addition, the spatial dependency of gravitational and electrostatic forces has most important consequences for physical laws. These fields have, in fact, an important mathematical property: they can be defined as the derivative (gradient) of a *scalar* field $U(x, y, z)$, called potential field, through the relation:

$$F = \text{grad}\,(U) = \left(\frac{\partial U}{\partial x}, \frac{\partial U}{\partial y}, \frac{\partial U}{\partial z}\right)$$

where U has the dimension of an energy and it can be easily demonstrated that in a potential field, $U(x, y, z)$, the energy spent or gained by a punctual mass (or charge) moving from points P_1 to P_2 does not depend on its intermediary trajectory, but only on the difference of the values of U in its points of departure and arrival. Therefore, we may interpret the quantity $-U(x)$ as a potential energy which the material point possesses in a position P and that the force tends to convert into a gain or loss of kinetic energy. This property has placed *potential* at the centre of mathematical formalism of differential equations of motion. For a system of N, mutually bounded, material points, these equations can be expressed in a more general and, at the same time, compact form than the well-known equations of Newton. We report them here in the simplest case where the bonds between the points are rigid:

First of all, a new function, H, is defined, called Hamiltonian function, given by the sum of the potential energy, $-U$, and the kinetic energy, T. If q_k are the space co-ordinates of the points of mass m_k of a system (consisting of N points labelled with index k), and p_k their kinetic moments ($p_k = m_k \, dq_k/dt = m_k v_k$), then the differential equations of motion are reduced to the simple and symmetrical form:

$$H = \sum_{i=1}^{N} \frac{1}{2} p_i v^i - U$$

$$\frac{dq^k}{dt} = \frac{\partial H}{\partial p_k} \qquad k = 1 \ldots N$$

$$\frac{dp_k}{dt} = -\frac{\partial H}{\partial q^k} \qquad k = 1 \ldots N$$

These relations, called *canonical equations* of motion, together with constraint conditions[26] between material points, have as their solution a complete description of the trajectories of the N bound particles, expressed by variation with time of 6N variables of the system: the 3N co-ordinates of the points plus their 3N moments (or velocities).

The structure of canonical equations shows that the Hamiltonian function possesses the property of a generalised potential, whose gradients in the space of the variables $\{q, p\}$ are respectively equal to the speed and to the acceleration of the single points. These equations have a form that can be extended to more general co-ordinates; for instance, p and q can represent statistical or thermodynamic variables. In order to emphasise the importance of canonical equations, we may note, incidentally, that the first formalism of quantum mechanics was established by starting from their expression.

To conclude these considerations on force and potential fields, it is worthwhile noting that the Hamiltonian potential energy that appears in the canonical equations of motion to replace force in Newton's equation may hardly appear to be an intuitive quantity. However, even the concept of force, which is seemingly more concrete, is after all no less abstract, while, at least formally, a clear and rigorous definition of force is indeed better provided by the derivative of potential energy.

In fact, in the motion of a particle in a potential field, $U(x)$ (we suppose it here to be one-dimensional for the sake of simplicity), energy $E = 1/2 \, mv^2 - U$ is invariant with time. Speed must, therefore, vary according to the function $v = 2(E + U)^{1/2}$ and the derivative of energy with respect to time must be equal to zero. We obtain, therefore:

---

[26] For instance, if all N points belong to a rigid body, the constraint conditions specify that their mutual distances remain constant as time elapses.

$$m\frac{dx}{dt}\frac{d^2x}{dt^2}(=mva)=\frac{dU}{dt}$$

which implies

$$F=ma=\frac{dU}{dt}\frac{dt}{dx}=\frac{dU}{dx}$$

This expression of force, F, appears to be clearer than that (only seemingly) simpler and intuitive expression of applied motive force. It states that force corresponds to "the local slope" of the potential energy function, $-U$, along which the mobile point "slides" acquiring speed, when attracted (or pushed) in the regions of the space where the value of the potential energy is lower.[27]

In problems of applied physics it is not always possible to obtain a mathematical expression of the potential. For example, it would be very difficult to calculate the mechanical interactions between rigid bodies starting from repulsive interatomic potentials. For this reason, the formalism of rational mechanics remains in great part based on a phenomenological definition of force fields, although every physical force is actually generated by a superimposition of interatomic attractions/repulsions generated by elementary potential fields.

Before closing this short exposure of the classic definition of physical force and of potential energy, some fundamental considerations are here in order, whose implications will accompany us in the following chapters.

In problems of physics, the mathematical formulation of the acting potential- (or force-) field supplies, at least in principle, the instrument necessary and sufficient to calculate every type of movement, from sidereal to atomic motions. As we are accustomed to the formalism of modern physics, this appears obvious to us and our interest is essentially focused on the functions that describe these fields and on an analysis of their properties. In the last three centuries the formalism of mathematical physics has, in fact, allowed us to cover up the essential questions regarding these concepts. In reality, nobody knows what a potential field is and why it even exists. As Newton answered people who asked him what universal gravitation is, the modern physicist must also admit that he simply does not know: he knows *how* to deal with these fields, but he must confess that he ignores *what* they are.

At this point, It is worthwhile travelling back in time twenty-five centuries and reconsidering an analysis of motion as it was set up by the first Greek naturalist philosophers.

Beginning from the VI century B.C., on the basis of the mythical or pre-scientific knowledge available, a fundamental issue was raised: namely that of the nature of change (κίνησις, *kínesis*) of all beings (ὄντα, *onta*), of which movement is only one particular aspect. This led the philosophers to face an infinite chain of causes and effects, to which they had to assign a more or less imaginary origin or starting point; however, the real problem lay in understanding the nature of the relationship between *agent* and *patient* and, consequently, the correspondence between the experience of

---

[27] We recall here that in the theory of general relativity (gravitational) potential defines the curvature of space and the lines along which the masses move.

the world of senses and its representation as elaborated by the mind. The position of pluralistic philosophers of the school at Athens in the V century B.C., in particular of Anaxagoras,[28] constitutes the first attempt at formulating this issue in scientific terms.

Developing the ideas of Empedocles and Hippocrates , Anaxagoras asserted [5] that in the beginning the whole universe was condensed to an "Intelligible One" that "was outside time" and contained the seeds of all things ($\dot{o}\mu o\iota o\mu\acute{e}\rho\varepsilon\iota\alpha\iota$, *omoioméreiai*). Manifesting themselves in the world we perceive through our intellect ($\nu o\tilde{\upsilon}\varsigma$, *nous*) the creatures, having developed from these seeds, maintained a confluence, a common breath, which even now make them co-participants one of the other, owing to their shared origin. Time is a product of our intellect which, in its weakness, is unable to entirely perceive the timeless aspect of the infinite, but needs some reference as a starting point.[29] If, therefore, the universe possesses an intelligible unity, a unity that is original only in a hierarchical and not in a temporal way, human intellect must then resign itself to its limits.

The atomistic philosophers, headed by Democritus, supported instead a more radical position, asserting that eternal atoms of various shapes move in empty space and, aggregating together, engender variety in things and their phenomena. Their motion is, however, chaotic and without scope; even the mind, our perception of the world, is itself a product of this aggregation.

Aristotle critically attacked the problem of change in the third and fifth book of *Physics*, while reviewing previous theories. He refused to acknowledge the atomistic position and their acceptance of empty space ($\kappa\varepsilon\nu\acute{o}\varsigma$, *kenós = not-being*) as a necessary condition for the variety of beings. He also criticised Anaxagoras, arguing that his conception of Nature ($\varphi\acute{\upsilon}\sigma\iota\varsigma$) is in every way similar to the atomistic view, and accuses him of resorting to the action of the mind only for questions that could otherwise not be answered.

His theory, on the contrary, becomes part of a system, where change is seen as a movement of all of nature, originating from a first motor and directed towards the *form* of all beings. Force, therefore, is the attraction of every creature towards its perfect final state. Excluding possible properties of a final state (the purest, the most stable, the most adapted, the most probable, *etc.*) we can find in this concept a strict analogy with modern views. However, at the base of the Aristotelian conception is a divine Design, due to which science ($\dot{\varepsilon}\pi\iota\sigma\tau\acute{\eta}\mu\eta$, *epistéme*) expresses the ordering of phenomena in reality. For almost twenty centuries, Aristotelism, in all its schools and interpretations, has defended the thesis that physical forces in Nature are nothing else but the realisation of logical necessities. Modern science intends to do without this thesis. A negation of any aprioristic science, from Cartesius onwards, has produced a radical reversal of the Aristotelian point of view which marked the

---

[28] Anaxagoras (500–428 B.C) was born to Clazomenae in the Ionian province, but while still a young boy he moved to Athens where he became a friend and collaborator of Pericles, in whose political ruin he too was finally brought down.

[29] This aspect of the ideas of Anaxagoras has reached us thanks to a late commentary of Aristotle's *Physics,* compiled by Simplicius of Cilicia [11].

separation of physics from metaphysics. The promotion of experimental data to the highest hierarchical rank of knowledge has freed physics from pre-ordered schemes of interpretation, but, at the same time, has deprived it of the status of possessing an absolute and incontrovertible knowledge.

2.3.3 The Motion of Continuum

The above-mentioned canonical equations of motion, which are nothing but a symmetric and comprehensive formulation of the second Newtonian law establish an important property which states that, in a given potential field, the trajectory of a material point is completely determined by its initial speed and position; that is to say, at any given time, it is always possible to exactly predict the position and the speed of a point. Since this property applies to every system of points for which the forces of mutual interaction are known, Newton's law was assumed to serve as irrefutable evidence of determinism in Nature, according to which all natural phenomena are determined by the conditions of motion at the birth of the universe. Even admitting that we would be never able to write the relevant system of equations for all existing particles, it would still remain true that these equations do exist and determine for all time the course of all past and future phenomena.

Physical determinism seemed, therefore, to be able to supply the mathematical proof needed to bolster the truth to be found in philosophical determinism so that in certain circles it was held as a commonplace sort of definitive argument.

Yet determinism is a mental abstraction corresponding in mathematics to the concept of a not always guaranteed integrability of equations of motion, but this concept has little in common with physical reality. In the following chapters we shall discuss the validity and the limits of this kind of determinism, but at this point it is convenient to consider its significance in the context of mathematical formulations of general laws of motion.

In nature all bodies are considered to be deformable, continuous systems. Some, like liquids or gases, apparently do not have a fixed shape and are continuously being shaped by the variable forces that act on them; but even solids are always subject to deformations due, for instance, to thermal expansions and contractions, elastic or plastic strains, chemical reactions, inner restructuring, and so on.

Therefore, in order to describe the motion of a continuous body in a force field special equations with suitable variable have been developed: to the co-ordinates of one material point (x, y, z), is associated a variable density function, $\rho\,(x, y, z)$, which represents the mass per unit volume around the general point (x, y, z) of the space. Hence, the motion of a deformable body is described by the variation with time of function $\rho\,(x, y, z, t)$. In its simplest form the equation of motion, due to Joseph Liouville, is written as:

$$\frac{\partial \rho}{\partial t} = \frac{\partial H}{\partial p}\frac{\partial \rho}{\partial q} - \frac{\partial H}{\partial q}\frac{\partial \rho}{\partial p}$$

where H is the Hamilton function (the generalised potential), q is the co-ordinate of the moving point (x, y, z) and p its moment ($p = mv$). Compared with canonical equations, Liouville's equation immediately appears to be more complex on a formal level and, in fact, its solution introduces remarkable problems. Of these we quote here a fundamental one:

Even when the equation is integrable, the determination of a trajectory may require a precision of the initial conditions, which is unattainable not only from a practical, but also from a theoretical point of view.

In other words, even infinitesimal variations of the starting conditions may produce completely different trajectories. Therefore, in certain cases the trajectory may pass through bifurcations points emerging into alternative trajectories, without having established which one should be chosen. We are, in fact, dealing with critical states that, in certain contexts, are called "points of catastrophe", from which the real physical system can find an exit only by leaving behind an absent ring in the causality chain of its chronological evolution. It is not worth objecting that determinism should not fail, inasmuch as the outcome of the catastrophe will always depend on forces, however weak they may be and which we do not control or even perceive. Actually, the instability of the solution is of a mathematical nature, tied to a property of continuous variables. In order to overcome this difficulty, it is necessary to postulate an arbitrary external intervention or to renounce the *continuum* hypothesis, on which, however, the laws of motion have been founded as well as the consequent thesis of physical determinism. In Fig. 2.7 an example is illustrated where even equations of motion of a simple point may predict random trajectories oscillating between circular orbits and spiral curves.

Since the second half of the XVIII century, the study of this problem has presented an inexhaustible source of discoveries and surprises for mathematicians and physicists, who have increasingly devoted more attention to the probabilistic and statistical aspects of physical laws and their respective equations. Today a physicist cannot believe in the existence, even at a hypothetical level, of the famous "Demon of Laplace": to be able to calculate the entire course of all cosmic events, past and future.

Yet, despite this denial, philosophical determinism has survived, as in the past, although no longer benefiting from any endorsement on the part of physics. Physical determinism, however, has got a footing in disciplines like biology and psychology, where it cannot be undermined by any rigorous mathematical verification.

2.3.4 The Theory of Electromagnetism

The completion of the edifice we call classical physics can be dated to 1864, when the great Scottish physicist and mathematician, James Clerk Maxwell, published his work on the novel theory of electromagnetism. Maxwell had developed his ideas starting from the experimental results of Michael Faraday and André-Marie Ampère on electromagnetic induction phenomena. His brilliant insight was to associate

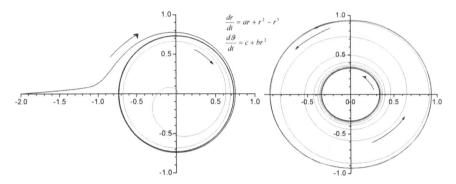

Fig. 2.7 The two diagrams show that even the motion of a punctual mass can be unforeseeable, even if governed by completely defined, integrable equations. The example illustrates the trajectories of one point in a plane, as calculated by the equations reported in the figure, where r represents the distance of the moving point from its origin, O, and ϑ the angle of rotation around an axis passing through O and perpendicular to the plane. On the *left-hand side*, the trajectory is traced for the values of the differential parameters of the attracting force as follows: $a = -0.25, b = 0.1, c = 2$. One can see that the mobile point is attracted toward the origin and, at a critical distance, it enters in a circular orbit of radius $r = 0.73$; however, this orbit is unstable and with minimal perturbation the point leaves the orbit and falls down to the origin O (the attractor point). On the right hand side the trajectory is calculated by lowering the attraction coefficient, a, to the value $a = -0.1$ and leaving b and c unchanged. In this case we can see that two orbits exist: one with radius $r = 0.33$, the other with radius $r = 0.94$. The inner one is, however unstable, and with any minimal perturbation the point jumps onto the external trajectory which turns out to be more stable. Nothing can be said, a priori, concerning the time at which the point changes its orbit. In mathematics there is an infinite number of cases of this kind where bifurcations are met. In nature, continuous systems involving gas, liquids or, generally deformable bodies are described by equations, the solutions of which often contain bifurcation points. For example, they typically entail reliability limits which cannot be avoided in current models of fluid dynamics for meteorological forecasts. On the other hand, the study of these unstable states has made it possible to make great progress in the description of criticality phenomena in thermodynamic systems through statistical mechanics. Today, even within the framework of classic physics, causality laws are not to be found in the rigorous Laplace determinism, but rather in the statistical properties of complex systems, where predictability limits are inherent in possible fluctuations

the interpretation of electromagnetic waves to the equations governing mutually dependent electric and magnetic fields. It is worth examining in detail the equations of Maxwell, because, for the first time in physics, they describe the propagation of a field of forces at finite speed in a vacuum space and connect its property to the mathematical formalism of vibration waves in fluid and elastic bodies, a subject which had been studied since the end of the XVII century.

According to Maxwell's equations, an electric field, E, and a magnetic field, B, linked by the law of mutual induction, can be expressed as follows [30]:

---

[30] The reader who is not familiar with vector differential analysis used in mathematical physics may find helpful a few elementary explanations contained in the short glossary in the Appendix. However, in order to understand the equations of Maxwell it suffices here to recall a few definitions:
- Given a vector field $F(x, y, z)$, its *rotor*, rot(F)—also written as curl(F)—is a vector perpendicular

$$\frac{1}{c}\frac{\partial E}{\partial t} = \text{rot}(B) - \frac{1}{c}j$$

$$\frac{1}{c}\frac{\partial B}{\partial t} = -\text{rot}(E)$$

$$\text{div}(E) = \rho \quad \text{div}(B) = 0$$

where c is the speed of light, j the electrical current density and ρ the density of electrical charges. These equations are clear: the increment in time of the electric field has two origins, namely the density of electrical current and the rotor of the magnetic field; on the other hand, the increment of the magnetic field has as its sole source the rotor of the electric field, because, owing to the dipole character of the magnetic "charge", there are no magnetic currents analogous to the electric ones.

One can easily see that Maxwell's equations contain the well-known electrostatic attraction/repulsion law of positive and negative charges (obtained by putting $j = 0$ and, consequently, rot(E) = 0). But of greater interest to us is the case where there are no currents ($j = 0$) and charges ($\rho = 0$) but, due to boundary conditions, the field vectors in a given point O are orthogonal and start oscillating regularly (we do not ask why for the moment). Solutions of the Maxwell equations are then E and B functions representing a *spherical wave* that from point O propagates in the space at a speed c; on the surface of this wave the vectors E and B are perpendicular, and oscillate with the same frequency as the initial oscillation in point O and have intensities that decrease with the square of the distance from O.

The equations of Maxwell immediately found their confirmation in experiments in electromagnetic optics and other domains. This result was conclusive for definitively establishing the *reality* of electromagnetic waves and *not* relegating to being a *mathematical artefact*. Actually, the importance of this result was such that Einstein, in developing the theory of special relativity, did not hesitate to modify Newton's laws in order to guarantee the invariance of Maxwell's equations with respect to uniform rectilinear motion from an observer's point of view.

Although the equations of Maxwell have defined solutions when j and ρ are fixed functions, in reality however, the density of charges and electrical currents vary

---

to F and its direction is that of the axis around which $F(x, y, z)$ rotates within an infinitesimally small volume around the point $P(x, y, z)$; the rotor module is given by the amplitude of this rotation. The somewhat complex rotor expression is reported in the glossary—The *divergence* of a vector field $F(x, y, z)$, is given by:

$$\text{div}(F) = (\partial F_x/\partial x + \partial F_y/\partial y + \partial F_z/\partial x)$$

and represents the sum of the derivatives of F in a point P (x, y, z); that is to say, it represents a measure of the average increase of F obtained by exploring the field F around a point P.—In Maxwell's equations the divergence of the electric field, div(E), is equal to the local density of electrostatic charges; while that of the magnetic field, div(B), is always equal to zero, because the magnetic elementary charges consist of dipoles of which the effects of the opposite poles cancel each other out. For this reason, magnetic and electric fields exhibit fundamentally different properties. The significances of these differential operators is explained in more detail in the glossary.

with time as they are affected by the electromagnetic field in question(remember that j is given by the product of ρ and v, where v's are the speeds of the electrical charges). Moreover, electrical charges and magnetic dipoles are assumed to be point-like entities, with the force field approaching them tending to infinity, but if these have a finite extension and move in the very same field they generate, the solution of the equations becomes much more complicated and the existence of a general integral of the equations can no longer be guaranteed.

It was Paul Dirac who developed a mathematical procedure aimed at separating the effect of the charge's self-generated field from that of the surrounding charges. The result was, however, alarming, since the solution turned out to exist only if the initial value of acceleration was given (in contrast to the deterministic conditions of mechanics, where this specification cannot be required). Furthermore, for certain values of the starting acceleration, the mathematical solution predicts an "escape" of charge particles at increasing speed—utter nonsense from a physical point of view. The consequence is that the solution is to be found by fixing the initial conditions as functions of the results; a solution which evidently raises fundamental objections. The answers to these objections is normally to assert that these phenomena must be described by taking into account quantum-mechanical effects. This, in fact, is the case in the theory of quantum electrodynamics, but even in this context the same problem reappears, although possessing a different form.

2.3.5 Energy: Attribute or Substance?

The gravitation and electromagnetic equations account in classical physics for all forms of energy exchange, where electromagnetic energy can be transformed into mechanical energy and vice versa. The principle of conservation of total energy is, however, strictly maintained. The old question regarding the significance of energy in this more complex context is taken up here once more: what is energy and what is its purpose in the universe?

Our primordial awareness of *energy* is subjective and arises from the interaction of our body with the objects surrounding it, which can be moved or stopped, constructed or destroyed through actions, which take place because there is something contained in our body which can be spent at will.

What is this *quid* that is present in Nature and man feels he possesses to some degree, and whose manifestation has roots that reach the mysterious centre of his will? Even using our modern dictionaries, when we try to look up definitions of this entity and find terms like force, power and others similar, we remain deeply perplexed.

Physics has adopted these technical terms in order to define corresponding quantities and instructs us on how to deal with them, but not on their essential nature.

The greatest abstractive effort made by man in order to define these notions on the basis of first principles date back to the Ionian naturalist philosophers of the V and IV century B.C.. It was, however, Aristotle who introduced the terms that are used in modern science, although with different meanings. Yet Aristotle did not introduce them in his books of *Physics*, but in those of *Ethics*, where he defines energy (ἐνέργεια, *enérgeia*, from ἐν = *in* and ἔργοη, *ergon* = *work*) as man's activity, both material and mental, directed for a purpose.[31] Later on, in the eighth and ninth book of *Metaphysics*, he resumes this term together with others, used today in modern physics, to describe his vision of reality: every being is born from a possibility (δύναμις, dynamis, translated into Latin with *potentia*) and its process of realisation (called ἐνέργεια, *enérgeia* = *energy*), through the acting change (κίνησις, *kinesis*), guides the being to its perfect form (ἐντελέχεια, *entélecheia*, from ἐν, en = *in* and τέλος, *telos* = *end, aim*). In the history of science the first three terms were fortunate and passed through different philosophical schools, often changing their significance, to finally be adopted by modern physics: the latent energy of an object in a force field is now called *potential* energy, while the energy associated with the motion is called *kinetic* energy. The fourth term, the entelechy, fell into disuse when Aristotelian physics was categorically abandoned in the XVII century and was finally even banned as it implied a concept of finalism in Nature.

In the grand Aristotelian system the terms *dynamis*, *kinesis* and *energy* did not primarily refer to the theory of motion, but, more generally, to any change in which movement is only one aspect. Actually, all beings change from their potential state to their perfect form toward which they are attracted and that energy is the motor of this process. It is, therefore, understandable that Aristotle had a greater regard for energy in biological and mental processes (growth and learning), in which the concept of entelechy, appeared more self-evident.

In the Renaissance, until the XVII century, the Aristotelian term *energy* remained in use in Europe to indicate the force and effectiveness of speech and of words as well as in physiology to refer to the force of muscles and nerves, especially that pertaining to virility: a meaning that was still hinting at a background of Aristotelian doctrine.

However, when Newtonian laws of motion were expressed in contrast to Aristotelian physics, Leibnitz proposed a new term for energy: *vis viva*, living force, to indicate the kinetic energy associated with all types of motion, a quantity that, when rigid bodies collided, could be transmitted from one to the other, but whose total value remained unchanged. The definition of *vis viva* was contradicted by Newton, who proposed a law of conservation of impulse or *momentum* and, once more, a long debate followed between the two. Today we know that the two laws are both true and complementary, but the law of conservation of energy formulated by Leibnitz was destined to develop in a more comprehensive way. The term *vis viva* soon fell out of favour but even today represents a semantic conjunction between the mathematical formulation of modern physics and the analogical model of the Aristotelian φύσις.

---

[31] The Aristotelian manifold significance of energy and force-field was acutely analysed by Pavel Florenski where he claims that both mechanical and psychological forces and the changes they produce are related to the same, general definition of space's geometry [100].

After the energy associated with all elementary force fields that, supposedly, exist in nature was formulated, one felt that physics had fully matured. Nevertheless, in spite of this reassuring feeling, one had to admit that the concepts of force or potential field and that of a moving body (mass, charge or any other particle) could not be separated since fields and movement of their sources are fundamentally entangled in all real phenomena.

In fact one faced, albeit in a new form, the old dilemma of the relation between *agens* and *patiens,* which had intrigued Aristotelian physicists and which is expressed by the question: what is transmitted to an object set in motion, and by whom? The actions viewed at distance between masses, charges and nuclear particles rendered the answer more problematic. For instance, even a simple definition of mass can refer to three different contexts: inertial mass, active and passive gravitational mass, which are assumed to be equal. Whether there is an underlying substance common to them all we do not know.

However, in the intriguing context of these questions regarding the physical theory of changes and mutations the concept of energy became of primary importance. Because of its general valence, energy became the universal exchange currency for all interactions, and questions on the nature of physical objects finally focused on the significance of this quantity.

Curiously, the term energy was not in the standard vocabulary of the past centuries; it was indeed one of the last physical quantities to be clearly defined towards the first half of the nineteenth century, in spite of being a quantity with fundamental properties. Let us briefly examine its basic definition in rational mechanics and then consider its manifold meanings in order to see, by following their historical development, how energy has become part of fundamental physical quantities and how its importance has constantly grown in step with the progress of modern physics. To this end, let thus us start with the laws of inertia.

The first fundamental law of inertia as formulated by Galileo (and taken up again by Newton), asserts that:

In an absolute vacuum, any body that possesses speed v, maintains it invariant of any time elapsed.

With speed being a vector defined by a direction and an intensity, it follows that the trajectory of this body is a straight line, along which the distances covered are proportional to the time elapsed (*uniform rectilinear motion*).

This law, which to a modern reader might seem trivial, is anything but an obvious statement. Indeed, we can assert with some assurance that an experiment that validates this law has never been made and never will be; on the contrary, common sense tells us that all bodies in motion sooner or later change their speed for one reason or another, since an infinite vacuum is not physical space, but a pure abstraction. Moreover, what a straight line is cannot be explained without assuming some postulates of geometry, also founded on pure intuition.

But matters become more intricate, when we ask by whom and how speed is to be measured, inasmuch as any observer can be considered as himself moving somehow. The answer is contained in the second law, called the Law of Galilean relativity:

It is impossible to define a privileged observer at rest in the Universe. Nevertheless all physical laws of motion are valid and invariant for any observer moving with uniform rectilinear motion with respect to the observer who established the laws of motion; such an observer is, however, not in a position to establish his own speed with respect to an immobile system of reference in the universe.

Even here it is necessary to make two fundamental observations: (1) the law states that the observer, the *subject* of the observation, must be himself an *object* of the space defined by the postulates of geometry and (2) the observer not only must possess an instrument to measure time, i.e., a real object that, for instance, moves with a perfectly cyclical or periodic motion, but this instrument must be synchronous with those of any other observer, who moves at any uniform rectilinear speed at any distance from the first observer.

Let now consider the second Newtonian law, expressed by the equation $F = ma$, which connects the acceleration, a, of a body of mass m with the force applied, F. Based on our perception of force, it can be intuitively understood that a reasonable definition of the work or "fatigue" spent in moving the body is represented by the product of F times the displacement, l, produced.[32]

It is important here to observe that both force and displacement are vectors, i.e., quantities defined by magnitude and direction. Work is then given by their scalar product, that is to say, by the product of the force component in the direction of the displacement (or vice versa) and is, therefore, calculable only in relation to a precise trajectory of motion.

We see how this relation is expanded, by assuming for the sake of simplicity an arbitrary rectilinear motion on the axis of co-ordinate x, starting from $x = 0$.

Newton's law for a point of mass m, initially at rest, can be written in differential form and one-dimensionally as:

$$F = ma = m\frac{dv}{dt}$$

Let suppose that the speed of the point, v, increases with the force applied, then the work produced must be calculated as the integral:

$$L = \int_0^l m\frac{dv}{dt}dx = \int_0^v mv\,dv = \frac{1}{2}mv^2$$

The work done, L, is, therefore, directly proportional to the square of the speed imparted to the mass at the end of the displacement; the product $1/2\,mv^2$ is the familiar kinetic energy.

---

[32] As well as for every physical quantity, the definition of work as $L = Fl$ is a substantially arbitrary one and is based on the thought that if, for example, the mass to be moved doubles, the work needed doubles accordingly, and the same holds true for displacement. However, only the particular properties of this quantity, defined here heuristically confirm the real value of this intuitive thought.

In the following step we consider that force F is, in general, produced by a potential energy field $V = -U$ (which has the dimension of energy) and is expressed as $F = -dV/dx$. We can then write:

$$L = \int_0^l -\frac{dV}{dx}dx = \int_0^l m\frac{dv}{dt}dx = \int_{v_0}^v mvdv$$

which results in:

$$L = -(V - V_0) = \frac{1}{2}mv^2 - \frac{1}{2}mv_0^2,$$

that is to say:

$$E = V + \frac{1}{2}mv^2 = V_0 + \frac{1}{2}mv_0^2 = \text{costant}$$

where index zero refers to the initial conditions. This equation represents the simplest formulation of the law of energy conservation, which asserts that the sum of kinetic and potential energy of a moving material point is constant in a force field, independent of the trajectory it follows between its departure and arrival points and the time needed to cover it. The sum of kinetic and potential energy can be interpreted as the total energy, E, of the point of mass m, that it possesses and maintains as long as it moves within the potential field U.

We also see that these conclusions are not bound to the form of the potential. This cannot be totally arbitrary, but the only condition to be fulfilled is that the force is always parallel to the potential gradient. These kinds of field are called *conservative*. Not all fields are conservative, (for instance magnetic fields are not). Conservative fields are those that satisfy some analytical properties, but only from a physical point of view. Experiments alone can tell us whether an observed field is conservative or not.[33]

An important feature must also be mentioned: the conservation law stipulates that the total energy E of a mass remains constant, but it does not tell us how to calculate its value. In fact, in agreement with the law of inertia, the speed of a mass can be calculated with respect to any point of reference moving with a uniform rectilinear motion; moreover, the force field does not change if an arbitrary constant is added to the potential U. Strictly speaking only the *increment* of E is produced when possible forces *external* to the field U act on the mass m.

From the law of energy conservation, a physical process can be imagined, in which, for example, the potential energy of a body is spontaneously converted into kinetic energy (e.g., a falling body). It is, however, impossible to define the real conditions under which the potential energy of a material point attains its absolute minimum. One could imagine that for repulsive forces, like those between two charges of equal sign, the minimum potential corresponds to positions of the two charges at

---

[33] For example, it is impossible to represent magnetic forces as gradients of a scalar potential. If a potential is to be defined for magnetism as well, it must be of a vectorial character.

an infinite distance; for attractive forces (for example, between two masses) the minimum corresponds to their zero interdistance. From a mathematical point of view, these conditions are clear, but from a physical point of view they entail substantial interpretative problems as it would be difficult to achieve the conditions in real space at an infinite as well at a null distance. For this reason, even in this very basic case the intrinsic total energy of a body is undefined.

The tendency of every body initially at rest in a potential field is to start moving towards positions in the space where the potential energy is lower. The conversion of potential energy into kinetic energy is *not* always spontaneous and may require the participation of forces external to the field. For instance, the potential energy of the water contained in a mountain basin can be converted into kinetic energy only by conveying it through a siphon up to the saddle point. The energy necessary to initiate the process is then completely returned when the water falls. These processes are called "activated", in the sense that the system must first be raised from a metastable state to an unstable one in order to spontaneously reach a more stable state than its initial one.

Many nuclear or chemical reactions consist of activated processes. After all, it is certainly conceivable that matter in the universe with its atomic, nuclear and sub-nuclear components could correspond to a metastable system, which can be entirely converted and dissipated into electromagnetic energy.

Energy is evidently a quantity of cosmological significance, a quantity that we have heuristically discovered, but whose nature remains a mystery. A mathematical demonstration of the invariance of energy may appear obvious, even trivial. Actually, in the case examined above, it is nothing other than a mathematical consequence of Newton's law; however, this simple law contains concepts such as force and mass, concepts which arise from some sort of insight or from an abstraction of rudimental sensory experiences, whose implications leave us at a loss. Richard Feynman, despite possessing the finest imaginative abilities, confessed his bewilderment to his students in the face of this law:

There is a fact or, if you want, a law that governs physical phenomena. It is exact, without exceptions and is called "law of energy conservation". It establishes that a certain quantity that we call energy does not change in the manifold changes to which nature is subordinate. We deal with a most abstract idea, a mathematical principle, not a mechanism or something concrete …. It is a strange fact that we can calculate a certain number and that, if we recalculate it after having observed the mutations that nature produces through all its tricks, this number is still the same [74].

In modern parlance energy has assumed a general meaning, sometimes expressed in terms of vague units, but having a common denominator: when we speak about energy we always imply an appraisal of an economic nature, as we instinctively feel that one deals with something that can be acquired or sold, be horded or spent, be changed from one currency to another, but never created.

On the other hand, the concept of energy appears so natural that its use was extended to the most disparate fields: in addition to thermal, kinetic, electromagnetic, chemical and nuclear energy, we rightly speak of intellectual, mental, decisional, behavioural energy and so on.

Among all quantities defined in modern physics, energy is that, which more than any thing else evokes a perception of reality as being something in continuous evolution. In the Aristotelian conception, motion is defined as work (ἐνέργεια, *enérgeia*), through which all beings attain complete development (ἐντελέχεια, *entelécheia*) of their potential, that is to say, the attainment of their ideal position and form in the universe. This process should gradually move the entire universe towards a final order. According to Aristotle, change is, therefore, not the product of inordinate agitation of atoms, as Democritus asserted, but a course established by the laws of nature. Energy is something that defines the effort of all beings to achieve their just end (τέλος, telos). This concept contains *in nuce* our notion of energy, but is contained within a comprehensive theory. However, its physical definition is too general and abstract to be of use quantitatively even though it puts the focus on the nature of motion.

Only Newton's laws made a modern definition of energy possible, for three fundamental notions of physical dynamics: speed, mass and force are connected. But there are two important aspects to be considered:

- First, as we have seen before, the *single* energy conservation equation allows us to calculate the absolute value (or module) of the speed of a body as a function of time, while the calculation of its trajectory requires solving *all* equations of motion.
- Secondly, it can be demonstrated that the natural motion of a system in a potential field is that, which, among all possible trajectories between two points of the space, produces the minimum difference between the mean value of kinetic and potential energy.[34] This property is generalised in statistical mechanics as the Principle of Energy Equipartition, which asserts the tendency of all spontaneous changes in a multi-particle system to equally distribute the total energy among all accessible forms (translation, rotation, vibration, ionisation, *etc.*). These two properties assume a fundamental importance in systems of particles subject to chaotic motions, in which the calculation of trajectories does not have any relevance or practical use. We shall see in the following chapter how thermodynamics represents the development and crowning achievement of this notion of energy.

Presently, energy has been analysed in the frame of all known force fields while its principle of conservation was extended to all complex aggregate systems defined in chemistry and thermodynamics. In classical physics, the calculation of the state of the most complex and largest system is always related to the constancy of the sum of its masses[35] and of its total energy.

---

[34] The difference between kinetic and potential energy is required to be at a *minimum*, but not necessarily *null*, because their respective values are determined except for an arbitrary additive constant.

[35] We may recall at this point that the concept of mass is applied in three diverse contexts: that of inertial mass, that of gravitational attraction mass and that of passive gravitational mass. In modern physics the three masses are considered equivalent.

With the advent of the theory of relativity, the principle of equivalence of mass and energy has made only the law of energy conservation the very base of our universe and of all physical reality.[36] According to modern cosmological models, the evolution of the universe consists of a sequence of multifarious transformations of some kind of primordial energy into radiation and matter, whose creation and annihilation took place as space expanded. Therefore, space is substantially defined by the existence of energy.

In this scenario, matter corresponds to a "state" characterised by the highest density of energy (the mass at rest of a proton corresponds to an energy of approximately 938 MeV). In order to appreciate this figure it should be considered that the energy of a proton in the potential field of an atomic nucleus is of the order of magnitude of some MeV while the bonding energy of atoms in liquids and solids is only a few eV. Moreover, we know that every chemical compound can tolerate a limited thermal agitation corresponding to kinetic energies of atoms not higher than some tenth of a eV (1 eV corresponds to a temperature of 11600 K). On the other hand, chemical compound formation is only possible if the reagent's atoms are subject to random motion, i.e., if they possess a finite kinetic energy sufficient to statistically overcome possible saddle points of the existing interatomic potentials.

Therefore, in the evolution of the universe it seems that various bodies (elementary particles, nuclei, atoms, molecules and aggregates of every dimension) are characterised by the intensity of the energetic *exchanges* that mutually take place among them.

In this perspective we may consider the crucial aspect of the conditions of our planet as established by the distribution of its total energy in all possible forms. The first condition is that of a thermal quasi-steady state where the average temperature of its surface had stabilised to a range of 280–295 K more than two billion years ago. This condition depends in the first place on the equilibrium between the energy that the earth disperses in space, mainly in the form of electromagnetic radiation (mostly in the infrared spectrum), and that it receives from the sun (mostly in the visible spectrum) or produced in its core by nuclear reactions. Much more complex are the effects due to geochemistry and, in particular, to the chemical reactions of carbon, whose energy aspects define the composition and the property of the important thin layer we call our biosphere. Finally, the chemical reactions governing the growth of living beings are based on the reduction and fixation of carbon that mostly took the form of stable carbon dioxide in the primeval atmosphere of our earth. If we consider our planet as a hypothetically close system, these reactions produce a net entropy *decrease* and, consequently, do not occur spontaneously. It is thus necessary to activate them by supplying energy from outside. We know that the most important reaction of carbon fixation takes place in the growth of vegetal plants by capturing (within small temperature intervals) energetic photons coming from solar radiation. Since macromolecules of living organisms have a limited life

---

[36] We briefly note here that negative masses and energies have already been introduced in certain cases in the theory of elementary particles. However, even these models maintain the principle of energy conservation.

span, at the end of which they oxidise and decompose into CO_2, in order to maintain equilibrium, photosynthesis reactions must not be interrupted or even reduced in speed; their persistence in time is in fact guaranteed by the continuous absorption of energy of electromagnetic radiation of visible frequencies, that, at the end of a life-death-decomposition cycle (often of a very long duration), is converted into thermal radiation and dissipated in space.

In conclusion, one can easily understand that all beings, and in particular all living organisms, are assigned a small and precarious space in the universe, situated in small regions at the borders of zones of the highest energy densities. These beings are nevertheless bound to strictly defined energy exchanges, whose order of magnitude, in comparison with the intrinsic energies of matter almost vanish. Therefore, energy represents the substance (matter), the support (environment) and the nutrient (growth) of all living beings in this particular scenario.

2.4 Separating the Wheat from the Chaff

If we consider the *corpus* of theoretical physics at the beginning of the twentieth century, we realise that—for the first time after Aristotle's work—a complete and coherent system was proposed, founded on a small number of elementary physical laws and on an underlying mathematical apparatus and operational system, whose extension and performance largely surpassed their field of application. To be able to appreciate this jump in quality, it is enough to give a comparative glance at the physics books of the end of the XVIII century. Except for astronomy—to that time the physical science for excellence—these texts consist of collections of problems of disparate nature to which follow an analysis and an explanation, sometimes incomplete or referring to other texts equally nebulous. The experiments were planned and executed without a precise programme and often were used just in order to entertain the good society where—it must be however said—there were many amateurs provided with an excellent scientific culture and not little talent (Fig. 2.8).

2.4.1 Life and Death of Determinism

In spite of these limits, it was just in the second half of the XVIII century that science began to exert a direct action on the culture of that age. The persuasion that Newton's laws were sufficient to explain and to predict all motions in the universe gave a new impetus to the philosophical doctrines based on mechanical determinism. Philosophers and men of science were convinced that, sooner or later, the physical laws would have explained everything, including the systems apparently unforeseeable as the living beings that seemed to be subject to a particular criterion of ordering.

Therefore, determinism became the official doctrine of the "illuminated" social class, and whoever did not embrace this faith was taxed with obscurantism. The

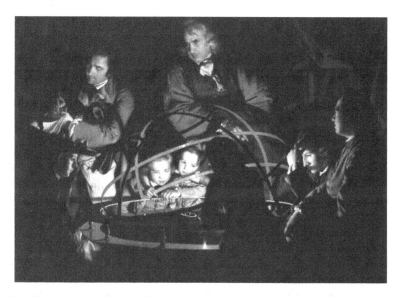

Fig. 2.8 The painting by Joseph Wright illustrates a lesson of astronomy in front of a small planetarium, in a pleasant homely atmosphere of XVIII century's England. The interest for such lessons and their high scientific level should excite our admiration. (*Derby Museum and Art Gallery*)

philosophical implication of determinism appeared in an optimistic vision of the knowledge and of its benefits. Trusting in the capacity of human reason to decipher the design of the evolution of the cosmos, one considered the progress as conquest of the reason against the darkness of the ignorance that was finally destined to succumb. Man, like a cog-wheel in a clock, has obligatorily to turn in accord with the natural harmony, a condition from which ignorance and superstition tend to divert him. Yet, the general enthusiasm concealed the shadow that his doctrine projected. Only one century later one became aware of the dark side of this kind of doctrines. Arthur Holly Compton (1892–1962), an American scientist and philosopher, Nobel Prize winner for physics, expressed a warning against the psychological implications of the scientific determinism [30]:

The fundamental question of morality, whether man is a free agent, is not only a problem which regards the religion, but is also a subject that regards the research activity in science. If the atoms of our body follow immutable physical laws, like those of the motion of the planets, why should we take pains? Which difference makes every effort, if our actions are predetermined by physical laws?

Compton spoke of determinism in nature as of an incubus. A deterministic system is completely self-contained, without any space for participations from outside. The very progress of the knowledge would be a mechanical process, in which the free creativity is completely absent. In this picture, our thoughts, feelings and efforts are mere illusions or, at most, by-products, epiphenomena of physical events that do not have any influence on reality.

The popularity the determinism enjoyed for almost two centuries is due to the reassuring simplicity of its premises. Firstly, in front of the apparent complexity of the real world, one believed that the inflexible action of the physical laws constitutes a transparent, indefatigable process in the evolution of the universe, more than the plans of the Providence or the events of Chance do. Moreover, determinism should have made it possible, at least in principle, to predict the course of phenomena and events.

Was this belief justified in the context of the scientific knowledge of that time? The answer is affirmative if we consider the generations that lived in the cultural climate of the late 1700s, until the post-Napoleonic epoch. But already towards the half of the 1800s in physics the mathematical construction of determinism began to waver when the two pillars of its carrying arch, the infinitesimal analysis and the theory of the physical measures, exhibited, in certain applications, clear signs of yielding.

Differential equations of motion had always been successfully solved for rather simple cases; however, when one begun to examine more complex systems, one noticed, as previously said, that, sometimes, the solution is split in two completely different trajectories without offering a criterion to decide which of the two is the right one. In some branches of physics, as, for instance, in dynamics of fluids, these cases are very common. As an example, the transition between a laminar and a turbulent flow takes place through a point of bifurcation mostly with an unforeseeable outcome. The deepened study of the differential equations showed that only for enough simple cases it was possible to demonstrate the existence of a unique solution. For the overwhelming majority of the cases this was not possible or sometimes the solution displayed instability points. Today innumerable cases of this sort have been studied, regarding the most various applications (some of which are examined in the following chapter).

Finally, the determinism of the physical motion appeared as an intellectual construction, in which mathematical laws are applied to idealised simple phenomena and one postulates the possibility to calculate the effects of all the relevant forces at any time and position.

It must be considered that the development of mechanical statistics, which began in the end of the 1800s, starting from the application limits of the Newtonian mechanics in systems containing a great number of particles, led to the definition of the concept of statistical equilibrium and fluctuations of the system's properties. The evolution of complex physical systems appeared, therefore, in a more interesting perspective, where determinism and chance alternate in guiding the course of the events.

The second failure of the justification of the physical determinism regards the perfection (in the sense of correspondence between reality and model) of the objects as they are and as they can be observed. No motion occurs exactly as predicted by the mathematical equations. Even Newton thought that the description of the solar system was imperfect and that one day it would have been destroyed in some way. Obviously he could not precisely say why and how, but as a good mathematician he knew that even a smallest perturbation term in the motion equations could carry to unforeseeable developments (Fig. 2.9).

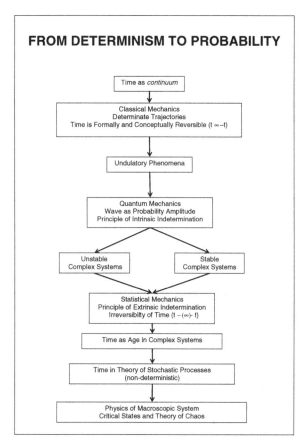

Fig. 2.9 In the XVIII century, the Newtonian mechanics produced a widely spread persuasion that all motions in the universe are determined by the initial conditions and by the differential equations of the physical laws. Yet, already in the nineteenth century, one realised that it is impossible to mathematically predict the behaviour of some complex macroscopic systems. Furthermore, quantum mechanics did finally establish that all atomic and sub-atomic processes are not compatible with the determinism as formulated in classic mechanics. Today, complex systems are analysed and described with statistical methods, accepting that the measurements of their properties can only be expressed as averages around which fluctuations can take place with amplitudes and frequencies depending on the state of the system. The state of the system may evolve with time, but this definition of time is not the same as that figuring in the equations of motion of the particles that constitute the system, but rather represents what we may call *the age* of the system, that is a parameter depending on the system's equilibrium conditions or on the degree of stability of possible non-equilibrium states. The study of these fluctuations represents one of the most interesting chapters of modern physics, where the idea is more and more making its way that the evolution of the universe is ruled by the convergence of general laws with accidental, random events

Also some experimentalists raised doubts on the perfection of the predictive ability to the physical laws. Charles S. Peirce (1839–1914), a distinguished personality proficient in logic and mathematics, as well as versed in philosophy and semiotics wrote in 1892:

Who is behind the scenes of the experiments knows that the most precise measurements, even of simplest quantities like masses and lengths, are far from possessing the precision of a bank account, and also the physical constants can be paired with the measurements of curtains and carpets.

His opinion was that, in the context of certain phenomena, this imprecision can invalidate the mechanistic analysis and compel us to consider the results from a statistical point of view. Peirce was sourly attacked in the positivistic circles, which, until the development of quantum mechanics, remained attached to the ideas of the mechanistic determinism.

After the shattering of its mathematical foundations, the physical determinism survived only as ideology. However, if it is not always possible to predict the course of the phenomena, what can imply such a principle except that the world is such as it must be—and nothing else? An assertion totally lacking scientific content, but quite paradoxical from the psychological point of view. Nevertheless, whilst today the determinism in physics is completely declined, its principles, transplanted in the ground of philosophy resist, in spite of all contra-arguments, so that Karl Popper began the second of its Compton Memorial Lectures (1965) at the University of Washington [31] with the following provocative statement:

The philosophical determinism and, in particular, the psychological one, does not have even the old tip of diamond it had in physics. If in Newtonian physics the indeterminacy of the properties of a system can be attributed to our ignorance, psychology does not have even this justification...If the physical determinism was nothing but a day-dream, that of the psychologists has never been more than a castle in air: it was nothing but the utopian dream of attaining a level of equality with physics, with its mathematical methods and its powerful applications—and, perhaps, also with the scope to reach the superiority on other sciences, with the intent of moulding men and society.

During the last two centuries, the Western civilisation has searched in the method of scientific investigation the means to manage the asset of other activities, from economy to sociology, from psychology to theology. Today there is no branch of knowledge that does not strive to embrace a "scientific order" interpreted as based on deterministic models, which should warrant indisputable validity to their conclusions, but which often are only a substitute for the lack in concrete knowledge. It is unfortunately probable that in the future certain doctrines will remain tied to the old deterministic modules, which, outside their original context, will entail the risk of rendering them definitively sterile and useless.

The laws of classical physics are intimately tied to a mathematical formulation that has made it possible to describe many aspects of reality, but that has inexorably encountered definitive limits in the complexity of the systems that are to be investigated. Modern physics has found itself wedged between the two frontier walls of the properties of the extremely simple and extremely complex systems. On the one hand, the development of quantum mechanics has radically changed, at least from the epistemological point of view, the conception of our relation with reality, on the other, reflection on the nature of the statistical phenomena has opened a new perspective that limits, but at the same time strengthens, our methods to analyse the phenomena when these are unfolded in increasingly complex forms.

2.4.2   The Quanta in Physics

The organisation and the mutual contacts among the various branches of classical physics, which was finally established in the second half of the nineteenth century. was of fundamental importance in establishing frontiers and links of physics. From this base it was possible to chart a course into *terra incognita* where experimental physics was growing and flourishing. Agreement between theory and the experimental results of that time was initially almost complete and the few discrepancies were generally due to errors of procedure or to incorrect interpretation of the data. However, this agreement became strained and finally and irremediably came to an end due to new experimentation in the field of optical spectroscopy and solid state physics, where a defined class of phenomena seemed to invalidate existing models.

Yet, as fate would have it, the decisive challenge to classical physics came from statistical thermodynamics, a most general branch, where attempts were being made to calculate the statistical effects of mechanical equations for systems consisting of a great number of objects. It was felt moreover that this application of statistical thermodynamics represented an exact science.

One of the main results of this study was that, in a closed system of particles approaching equilibrium, the total energy is homogeneously partitioned in all accessible forms and manifestations, called degrees of freedom (for example, kinetic energy is distributed in all possible types of movements of the particles, the potential energy in gravitational, electrostatic, magnetic energy, and so on). The energy equipartition principle is of fundamental importance and is validated by the properties of the equation of state of a variety of thermodynamic systems, from simple gases to liquids, solids and plasmas.

Physicists of the nineteenth century had already noticed worrisome deviations from this principle in solids at very low temperatures. But at the beginning of the past century indisputable proof for this inconsistency in laws of classical physics resulted from the application of Maxwell equations, i.e., from this recently published and yet almost immediately prominent theoretical model. When, in fact, attempts were made to apply the energy equipartition principle to a system of electromagnetic waves confined in a three-dimensional, perfectly isolated cavity kept at temperature T (called "black body"), one realised that classical thermodynamics predicted that the cavity would emit electromagnetic waves of intensities increasing proportionally to their frequency and, hence, that the total energy irradiated from the cavity diverged; a prediction which is evidently impossible. This failure could not be attributed to lack of precision, but was plainly and undoubtedly a faulty prediction.

It took the splendid intuition of Max Planck and the genius of Albert Einstein to abandon the set of models that definitively led to erroneous predictions. The key to freeing physics from this situation was hidden in the theory of mechanical statistics that Ludwig Boltzmann had been developing for twenty years. Planck, however, did not like (at that time) either Boltzmann or mechanical statistics. When, finally, he saw no other recourse but to use this key, despite having vainly tried many others, he made it—we took this decision, to quote—"out of despair"; so, after his success, he still remained convinced for some years that his achievement was a matter of mathematical expediency, a merely convenient formal manoeuvre .

Planck's idea was to assume that energy exchanges between radiation and the black-body walls took place through discrete radiation *quanta* whose energy was the product of their frequency, v, with a universal constant, h, (the famous Planck constant): $E = hv$. By re-calculating equilibrium conditions of the cavity based on this hypothesis, Planck obtained an expression for electromagnetic energy irradiated from a black-body of volume V at frequencies comprised in an infinitesimal interval between v e $v + dv$:

$$E_v dv = \frac{8\pi h}{c^3} V \frac{v^3 dv}{e^{hv/kT} - 1}$$

where k is a constant (the Boltzmann constant) and c the speed of light. At low frequencies this function increases approximately with the third power of v, but, after reaching a maximum, decreases towards zero as v tends to infinity. The essential difficulty of the classical model appeared thus to have been overcome; moreover, this solution yielded excellent agreement with spectroscopic experimental data. In this way, a "formal expedient" became the cornerstone of this new edifice called *Quantum Physics*. Yet, for a certain time, Planck's hypothesis was almost unanimously rejected by the scientific community as being a physical law. It was generally believed that sooner or later one would have found a classical explanation for what was considered to be a fortunate formal attempt.

What followed is well known. An avalanche of new experimental results (photoelectric effect, Compton effect, interference phenomena indicating a duality in behaviour of waves and elementary particles, etc.) demonstrated an irreducible incompatibility with the classical physics models; at the same time, the Planck hypothesis and its consequent implications appeared perfectly in line with those results. But all these coincidences appeared as shining light on a few distinct objects in a dark atmosphere, so that they suddenly displayed new contours, which had merely been overlooked.

At this point an important observation must be made. Quantum theory started from physical phenomena—the emission and absorption of energy quanta—which lie beyond our direct perception. Correspondingly, concepts were introduced that are foreign to the world we experience. Therefore, their significance remained confused until mathematical formalism was completely developed. Actually, it is important to remember that, in the beginning, quantum physics formalism was not unique and the various approaches first had to be compared with one another in order to be interpreted in physical terms. This makes an historical introduction of new ideas rather difficult since the theoretical developments appeared plausible but were not conclusive.

What was extraordinary about new quantistic models was the biunivocal similarity between radiation quanta, photons, and elementary particles. Both possess characteristics that, in classical physics, could at the same time be classified as undulatory and corpuscular. That a photon behaves like a particle was already difficult enough to conceive, but that a particle could behave as a wave exceeded every possible expectation. The only way to ensure further progress seemed to be to trust the mathematical formalism supporting these hypotheses and consider further predictions of the new theories and their agreement with experimental results. These confirmations were in fact quick to appear from even the most disparate fields of physics. Therefore, one began to analyse a wide gamut of effects of undulatory mechanics that soon consolidated in a general theory of quantum physics.

Only five years after the discovery of *quanta*, Einstein published his first article on special relativity in 1905, and in the following decade he completed the construction of the theory of general relativity integrating and encompassing quantum mechanics. In less than ten years theoretical physics was completely rewritten.

Quantum mechanics is based on the classical mathematical formalism of undulatory mechanics, but, while the latter deals with the propagation of the periodic motion of particles or electromagnetic waves, the wave function of *quanta* results in a broad interpretation margin. On the other hand, from its beginnings, quantum physics was developed from various points of view by emphasising different particular phenomena. Consequently, mathematical methods were devised using a variety of formalisms chosen to facilitate the interpretation of certain phenomena or the calculation of specific effects. The plurality of all these successful approaches, with their respective merits and limits, is an indirect indication of the impossibility of resorting to a single physical principle that, at the same time, singles out the limits of the validity of classical physics and demonstrates the necessity of undulatory formalism for both radiation and particles.

The basic difficulty consists in having to reconcile the firmly intuitive concept of material point in space (e.g., an electron of mass m that moves with uniform rectilinear motion) with dispersion phenomena which manifest themselves whenever we attempt to measure their trajectory (e.g., the diffraction of electrons produced by microscopic diaphragms or gratings). Indeed, these phenomena are typical and explicable when vibration waves are propagated in various media, and the first attempts to explain them at a mathematical level date back to the XVII century.

When one thinks of a mathematical expression of a wave, we immediately think of a periodic function of space and time describing a sinusoidal motion of particles of matter. Yet, normally, a real wave may assume any shape and its mathematical description can be quite complex. A theorem exists, however, which asserts that a periodic function $f(x)$ (let us take x as a one-dimensional variable, but the theorem is valid for every dimension), of period L and of any form, can always be written as a linear combination of infinite sinusoidal waves:

$$f(x) = \sum_{n=0}^{\infty} \left[a_n \cos \left(\frac{n \pi x}{L} \right) + b_n \sin \left(\frac{n \pi x}{L} \right) \right]$$

The coefficients a_n and b_n are numeric constants, which can be easily calculated from $f(x)$. This theorem can be generalised by demonstrating that a function, $\varphi(x)$, although non-periodic, can be written in the form:

$$\varphi(x) = \int_0^{\infty} \Phi(u) \cos(ux) \, du$$

where, in order to simplify the formulae, $\varphi(x)$ is here supposed to be symmetrical with respect to x; in general cases, an analogous integral is added in the term on right-hand side, with the cosine replaced by the sine. Function $\Phi(u)$ represents a transform of $\varphi(x)$ and is given by:

$$\Phi(u) = \int_0^{\infty} \varphi(u) \cos(ux) \, dx$$

Now, let us imagine a quantistic particle of finite dimensions, b, and of density equal 1, whose centre is found in position $x = 0$. The square of the wave function represents the space density of the particle and $\varphi(x)$ is, therefore, defined as:

$$\varphi(x) = 1 \text{ for } x \leq b \text{ and } \varphi(x) = 0 \text{ for any other value of } x$$

One sees immediately that the transformed function $\Phi(u)$ is given by:

$$\Phi(u) = \frac{\sin(ub)}{u}$$

Then the function representing our particle is:

$$\varphi(x) = \int_0^{\infty} \frac{\sin(ub)}{u} \cos(ux) \, du$$

Let us now try to make sense of this formula: the simplest function $\varphi(x)$, equal to 1 inside the particle and equal to zero outside, becomes the sum of infinite standing waves of type $\cos(ux)$ of wavelength $1/u$, and oscillating amplitude, with maxima and minima decreasing in height with u. The diagram of Fig. 2.10 l.h.s. shows the value of $\varphi(x)$ obtained by taking $b = 1$ and integrating it until $u = 500$, that is to say up to values of wavelength $1/u$ much smaller than the dimension, b, of the particle. Function $\varphi(x)$ is reproduced sufficiently well inside the particle ($x < b$)

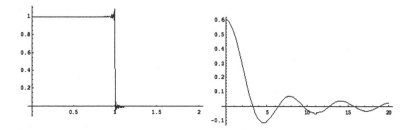

Fig. 2.10   On the *left* Value of the wave function of a quantum particle $\varphi = \varphi(x)$ as calculated as the sum of a "package" of sinusoidal waves with decreasing period, that extend from an infinite wavelength ($u = 0$) down to a wavelength equal to 1/500 of the particle's diameter ($u = 500$). The resulting wave function predicts a space confinement of the particle. On the *right* the same wave function, but consisting of a package of sinusoidal waves with wavelengths that go from infinity ($u = 0$) to the magnitude of the diameter of the particle ($u = 1$). In this case the particle is not precisely confined in space

and in the surrounding empty space ($x > b$), however, for values of x close to b, small oscillations are observed that even extend to empty space. If the integral is calculated until $u = \infty$ these perturbations disappear and $\varphi(x)$ assumes the shape of a perfect step. Let us then see what happens if we arrest the sum at $u = 1$. The result is plotted in Fig. 2.10 r.h.s.: $\varphi(x)$ in the centre of the particle ($x = 0$) has the value 0.6 and is oscillating elsewhere, even far off in empty space. The apparent bulk density is therefore distributed in the space with a sequence of maxima and minima. The figure corresponds to what is called diffraction spectrum, a well-known effect in optics when a light beam passes through a small opening. But, while the effect can be explained in this case by interference of electromagnetic vibrations out of phase, in the case of a particle the effect is produced by a vibration whose nature remains unknown.

Let us finally consider once again the physical significance of the mathematical terms introduced, imagining that the particle moves at a speed v and that the waves propagate with it. Without resorting to a rigorous treatment, it is nevertheless evident that the wavelength $1/u$ is inversely proportional to v and, therefore, if the particle really consists of a packet of waves, our first calculation has yielded a good prediction of its position by assuming a packet of waves with a large spectrum of possible speeds. If we had to calculate the *average speed* of the particle and its standard deviation, i.e., its uncertainty, we would obtain a high value for this latter quantity, that would diverge if the calculation were extended to include, at its limit, infinite speeds. In the second case, where we have limited the values of speed in a much narrower interval (note that a limitation would necessarily be present in any experimental measurement of the particle's trajectory), the *average position* of the particle is no longer well-defined and the value of its standard deviation is elevated. What we see here, in simplified terms, is the famous Heisenberg indeterminacy principle which holds that the product of the measurement's precisions of two independent physical quantities is constant: if one increases, the other decreases proportionally.

This simple example illustrates the conceptual difficulty that an undulatory hypothesis introduces in quantum physics. In and of themselves, the mathematical relations examined, which describe the shape of a packet of waves, reflect properties of continuous functions that exist independently of quantum mechanics. An interesting aspect is that these properties are intrinsic to a space of greater generality, called *function space*. In this space a Cartesian-Euclidean point (x) is replaced by a general function $\varphi(x)$ and the axes of co-ordinates are represented by a group of functions (in our case simple trigonometric functions), called *base*, whose linear combinations are able to reproduce any function $\varphi(x)$ in the function space.

Here we may observe a strict analogy with vector space inasmuch as function $\varphi(x)$ can be interpreted as a generalised vector, which can be represented as the sum (the integral) of its components (the scalar products defined by the integrals of $\varphi(x) \cdot \cos(ux)$) in the directions of generalised versors represented by the functions $\cos(ux)$. Thus, while *operations* (sum, multiplication, *etc*) are defined which are applicable to its *points* in a Cartesian vector space, *operators* (derivation, integration, etc) applicable to its *functions* are defined in a function space.

Quantum physics is thus established in a space where functions are general substitutes for Cartesian vectors. In this space, called Hilbert space,[37] all the objects and phenomena of physics can be described with a much more powerful and flexible formalism than that inherent in the Euclidean space.

Hilbert space geometry is well-established and provides the most powerful formalism needed for any given wave function calculated from differential operators, which describe its variation produced by physical agents.

Wave formalism made it possible to overcome fundamental difficulties faced by classical physics, but the cost was a substantial increase in the complexity of its mathematical language and abandoning any geometric reference to describe physical reality. We are dealing, in fact, with a description of phenomena based on objects completely uprooted from a common perception of physical space. Matter, which in classical mechanics is considered as an *object* which has a defined position and shape in a Cartesian space, becomes a *state* in wave mechanics, whose properties vary with its definition (where the word assumes here its etymologic significance of *confinement*) caused by any kind of physical measurements.

One might imagine, once this difficulty had been accepted, that mathematical formalism of wave functions and their differential operators in Hilbert space could be trusted. Unfortunately, there are still numerous problems in quantum physics which cannot be rigorously solved in this space, but require additional approximations and hypotheses. For this reason modern quantistic theory is proposed as a model of reality, without any aspirations to represent a definitive physical order of laws.

Yet it must be emphasised that the quantistic revolution erupted just at the moment when the scientific community was profoundly convinced that classical physics had

---

[37] Simply defined, Hilbert space is the ensemble of functions, whose square integrated over the field of their variable, is finite. Analogous to vector space, the operations of sum and product can be defined for functions as well as differential operators, which transform a function into an other belonging to the same space. Generally the functions of Hilbertian space are defined in the complex field with their required condition being that their *norm* be finished.

reached a stage of perfection and depth.[38] The deep conviction that the intellectual class of the end of the 1800s possessed the absolute truth of science is nicely illustrated by an autobiographic page from philosopher and epistemologist Alfred N. Whitehead [15], in which he describes the blow he had to endure when he was forced to recognise that the new theories demonstrated that classical physics, like all human creations, was not infallible:

...I had a good classical education, and when I went up to Cambridge early in the 1880s, my mathematical training continued under good teachers. Now nearly everything was supposed to be known about physics that could be known, except a few spots, such as electro-magnetic phenomena—which remained (or so was thought) to be co-ordinated with Newtonian principles. But, for the rest, physics was supposed to be nearly a closed subject ... By the middle of the 1890s there were a few tremors, a slight shiver as of all not being quite sure, but no one sensed what was coming. By 1900 the Newtonian physics were demolished, done for! It had a profound effect on me: I have been fooled once and I'll be damned if I'll be fooled again. Einstein is supposed to have made an epochal discovery. I am respectful and interested, but also sceptical. There is no more reason that Einstein's relativity is anything final than Newton's "Principia". The danger is dogmatic thought: it plays the devil with religion and science is not immune from it.

Moreover, significant technological and industrial developments of that time had infected the academic world with a spirit of fervent optimism and confidence in the progress yielded by the applications of physics and chemistry in several fields. This world felt uneasy about theories of quantum mechanics and relativity, which were to be taken as the new pillars of the edifice of science, but which were not of the sort to provoke great enthusiasm.

Furthermore, the theory of quanta was developed in the years between the two wars in different schools, each one developing its own formalism. This rendered its comprehension and interpretation more difficult. The most important approaches were, respectively, those of Werner Heisenberg and Erwin Schrödinger with a third one developed by Paul Dirac unifying the two previous ones. In all cases, the models were subject to various interpretations, some of which were in dramatic contrast to the premises of classical physics. For instance, the formalism of Heisenberg, based on matrix calculation, asserted the non-commutability of mathematical operators that determine position and momentum of a particle, with the consequence that the product of the precision of their simultaneous measurements results in a non-zero constant. A premise that implies a general principle of indetermination of any physical state as described by classical mechanics.

On the other hand, Schrödinger, who based his hypothesis on mathematical wave-formalism, focused on the results that questioned the classical principle of localisation of a particle. Moreover, wave functions representing different interacting

---

[38] When Max Planck, a young and brilliant student, intended to matriculate at the Faculty of Physics of the University of Munich in 1874, Philipp von Jolly, Dean of the Faculty, tried at that time to dissuade him with the argument that "almost everything had been already discovered, and nothing remained but to fill up some holes". Ironically enough, Planck replied that he was not interested in discovering new laws, but rather in understanding their existing foundations.

particles, resulted in a mutual involvement of their properties (so-called *entangle-ment*) by virtue of which they should *instantaneously* influence each other, indepen-dent of their interdistance; this represents a serious attack on conventional notions of sequential causality. This material was so incendiary as to disturb a philosopher's sleep.[39] This atmosphere was further inflamed by Einstein himself, who, though having contributed to the formulation of quantum theory—which was not at all in disagreement with the theory of relativity—was nevertheless convinced that the the-ory of quanta was a mere mathematical artifice or, at least, an incomplete model, and he never ceased to criticise its pseudo-ontological aspects. Even in 1935 he ridiculed entanglement effects in a famous article [16] calling them "quantistic nonsense" or "magical manipulations at distance".

Therefore, it should not come as a surprise that quantum physics remained con-fined to academic circles until the 1950s where it was cultivated by a minority of specialists. There were also objective reasons for this limited interest. Firstly, the experimental data with which theoretical predictions could be compared consisted almost exclusively of measurements of optical spectroscopy. Second, the founders of quantum theory were not specialists in mathematics, and for this reason the for-mal apparatus they used was not optimally consolidated and, sometimes could not stand up to rigorous examination.[40] Furthermore, the instruments available for ana-lytical and numerical calculation were limited and applicable to solving differential equations only for simple systems.

Quantum physics could only develop and be strengthened later on in more exten-sive surroundings and when much more powerful equipment was available than was possible in pre-war universities and research centres. What happened in the years to come not only created more favourable conditions for experiments, but also radically changed the structure of scientific research.

2.4.3 The Theory of Relativity and the Fall of Euclidean Space

The theory of quanta produced a serious earthquake which affected all laws concern-ing the atomic structure of matter. However, the construction of quantum mechanics was first pursued in the context of the old Euclidean space. The concepts of potential- and force-fields were taken from classical physics, where space and time co-ordinates were defined in a Cartesian system of reference in which Galilei's principle of relativ-ity held sway. Yet the theory of electromagnetic waves had been making theoretical

---

[39] It is well-known that quantum physics was publicly banned from Soviet universities since it was thought to be in contradiction with dialectic materialism. Niels Bohr and the school of Copenhagen had supported a probabilistic interpretation of quantum mechanics and bore the main brunt of the attack of the Communist Party which basically represented the Soviet Academy of Sciences. Even a few European and American physicists with communist sympathies joined in this campaign.

[40] The properties of the space of wave-functions, in which quantistic operators must be applied, were thoroughly studied and defined by David Hilbert. However, his fundamental treatise [20] became a reference text only twenty years after its publication.

physicists uncomfortable for a long time since they felt it necessary to find a support for these waves. Thus they ended by conjecturing a hypothetical substance, called *aether*, by which all of space should be permeated—not unlike what Aristotle had imagined. This hypothesis, which was later revealed to be false, triggered a sequence of discoveries which finally led to the complete fall of classical mechanics, the very core of physics.

This time the epicentre of the earthquake was located in the field of astrophysics, where attempts were made to find an experimental confirmation of the existence and effects of aether. If, in fact, aether was permeating space like a gas, its relative motion had to be somehow perceived on sidereal bodies of our galaxy as a sort of "wind". In particular, in our solar system, the earth, turning around the sun with a speed of approximately 30 km/s, should be found in its annual cycles both "windward" and "leeward" with respect to any direction of the aether's relative movement, with a difference of speed of the order of 60 km/s with respect to the speed of light emitted by the sun. To study this phenomenon, an American physicist, Albert Abraham Michelson (1852–1931) developed a precise way of measuring the variation of the speed of light and began to estimate the expected effects of the aether wind. If the light waves were really transmitted by the aether, their speed of propagation should depend on the speed of the observer with respect to the source of light, producing a phase shift in the spectral lines. However, from the very beginning, interferometric measurements showed that this effect did not appear at all. The result was disappointing since one had initially hoped to be able to exploit this property of aether for important astronomical applications. In fact, some laboratories spent considerable amounts of money to construct more precise interferometers than that used by Michelson, but the more accurate the measurements were, the stronger was the confirmation that the speed of light be independent of that of the observer. The first consequence was, finally, to reject the hypothesis of aether that appeared completely superfluous. Galilei's principle of relativity was finally examined in some depth, as it, apparently, was invalid for the propagation of light.

Michelson, blessed with keen acumen, started to see a possibility for explaining the results concerning the invariance of the speed of light in the recent work of a Dutch physicist, Hendrik Lorentz (1853–1928), who had analysed the effects of co-ordinate transformations on physical laws.[41] He had indeed found a correct method for interpreting the effect, but his cautious attempt was swept away by a cyclone that struck physics when Albert Einstein published his theory of special relativity. This event took place in 1905—later called *annus mirabilis*—and the young physicist, then almost unknown, proposed to make *tabula rasa* of the principles of mechanics.

Einstein used the results of Michelson on the invariance of the speed of light as a postulate and did not hesitate to repudiate Newton's laws which contradicted it. It is not fully clear how he arrived at this audacious step. Very probably he also started with the results of Lorentz which showed that if the speed of light is to be maintained

---

[41] Lorentz, on the other hand, had developed the theory of electrical and optical phenomena in moving systems that provided the foundation for Einstein's special relativity. However, he was convinced that Michelson's result could be explained by the time contraction predicted by his

invariant for any observers in uniform rectilinear motion, it was necessary to assign to each of them, apart from their own spatial co-ordinates, their own time as well. In fact, once this hypothesis was made, the invariance of *physical laws* was still guaranteed for transformations of co-ordinates between two reference systems with different rectilinear uniform motion, however, quantities such as lengths/distances, masses and time intervals, always considered as invariant, became functions of the observer's speed.

While the novel quantum theory had expressed the theoretical impossibility for an observer to analyse phenomena without irreversibly perturbing them, the theory of relativity placed the observer at the centre of his world of perceptions, but confined him to a kind of island where any absolute determination of his position in space and time was excluded.

The impact on the structure of the edifice we call physics was to shatter it, for the main pillars represented by current models of mechanics had to be replaced. While the four-dimensional space-time of relativity still maintained some mathematical properties of Euclidean space, its geometric-intuitive character was completely lost when the speed of an object approaches the impassable limit of the speed of light: distances undergo contractions, times are dilated and mass increases. The increase in mass is proportional to kinetic energy: $T = 1/2\,mv^2$ and from this property one deduces the relation of equivalence between mass and energy and, in particular, the correspondence of the rest mass, m_0, of a body to its intrinsic energy $E_0 = m_0 c^2$. Thus kinetic energy and mass appear to be two manifestations of the same, mysterious substance whose variable forms are bound to their local density. This represented an extraordinary contradiction of the traditional vision of mass as an objective reality, which was preserved in all physical processes in which it participates, according to the fundamental mass- conservation law which holds that *mass can neither be created nor destroyed.*

But there was more: In relativistic space two events contemporary for an observer A are not for another observer B who moves with respect to A, unless the two events occur exactly in the same place—which is not easy to imagine (a geometric representation of these effects of special relativity is briefly introduced in the Appendix). It is obvious that the kernel of the paradox resides in the relativity of time. For centuries,

---

(Footnote 41 continued)
transforms. Lorentz, however, developed his theory on the aether hypothesis where the speed of light, V, is referred to a system absolutely at rest. Actually, he wrote in a paper of 1899: *Michelson's experiment should always give a negative result ... because the correspondence between the two motions we have examined (the light beams emitted by sources S_0 at rest and S moving at velocity v.) is such that, if in S_0 we had a certain distribution of interference-bands, we should have in S a similar distribution, ... provided, however, that in S the time of vibration be k ε times as great as in S_0,* where:

$$k = \frac{1}{\sqrt{\left(1 - \frac{v^2}{V^2}\right)}}$$

Here he means that Michelson 's effect is due to the compensation of the Lorentz time-contraction $1/k$ through a constant ε which Lorentz found to be *an indeterminate coefficient differing from unity by a quantity of the order of 1/k [102].*

time was considered to be a continuous variable in physics and one had forgotten or neglected, at least in current formalism, the problem, raised by Aristotle that time being a product of the mind's activity, must be considered as overlapping an *observed* movement. The proof of the absolute objectivity of time was, therefore, inextricably tied to possible *instantaneous* communication between different observers. That this cannot effectively be realised is clear, but for physicists it was enough that it could be reasonably imagined. On the contrary, Einstein stated that a physical theory of motion could not be absolutely valid if this Gordian knot had not been cut.

In the decade that followed his development of *special relativity* (whose results mainly concern kinematics), Einstein studied the effects of the transformations of space-time co-ordinates for observers moving with a general, not necessarily rectilinear uniform, motion. His work represented an attempt to study the trajectories in space as defined by universal gravitation. The result was the theory of *general relativity*, published in 1916. The mathematical form of the co-ordinate's transformations between two observers is in this case much more complex and, in addition it includes the expression of the attraction of the masses present in the universe. Einstein, starting from the proposition that the laws of motion have an invariant character with respect to any moving observer, was able to formulate [42] the property of space-time (*chronotope*) in which the motion takes place.

The problem of the significance of inertia and of centrifugal forces, which all physicists had inherited since the time of Newton, was now directly attacked by Einstein. The question was from where do these forces derive their origin? The answer at the time was that they originated in an absolute physical space with its fixed metric structure. Now, assigning reality to the metric structure of space, as Einstein did, entails that this must change by interacting with matter, a property the Euclidean metric was not capable to provide.

The fundamental idea of Einstein's general relativity consists in his definition of a more general metrical structure of world-space with a quadratic differential form of the co-ordinates x:

$$ds^2 = \sum_{i,k} g_{ik} dx_i dx_k$$

instead of the Euclidean length invariant:

$$ds^2 = dx_1^2 + dx_2^2 + dx_3^2,$$

and of its formally equivalent expression in *special relativity*:

$$ds^2 = -c^2 dt^2 + dx_1^2 + dx_2^2 + dx_3^2$$

where c is the speed of light.

---

[42] Thanks mainly to the work on tensor analysis developed at that time by the Italian mathematician Gregorio Ricci Curbastro (1853–1925).

In *general relativity*, the tensor g_{ik}, which varies with the space-time co-ordinates, defines the metric of a curved space and can be regarded as analogous to a gravitation potential field depending on the matter distribution in the universe. This tensor defines an abstract non-Euclidean geometry, which is known as an object in mathematics, quite independent of physics; but general relativity asserts that this tensor is real, in the sense that its actual present state is the product of the past states which characterise the evolution of the universe. A metrical field tensor g_{ik} can actually be constructed with some additional hypotheses. Indeed there is no difficulty in adapting physical laws to this generalisation.

We quote here three results that are particularly relevant for our purposes:

- First, the speed of light, V, is *not constant* in the chronotope of *general relativity* but depends on the intensity of the normalised gravitational potential, U, of the point in question, namely:

$$V = c \sqrt{1 - \frac{2U}{c^2}},$$

  where c indicates the speed of light in an absolutely empty space, whose value appears here as a universal constant. This means that, when light enters zones of high mass density, and hence of strong gravitational potential, U, its velocity slows down and, at the asymptotic limit where $2U = c^2$, light stops propagating. This prediction by the theory of general relativity was soon confirmed experimentally by several phenomena (among which the existence of black holes, around which light is trapped because of their enormous density of mass). One should note that, paradoxically, this result of general relativity contradicts both the revolutionary Michelson discovery and the original Einstein assumption on the constancy of the speed of light, from which he had deducted the theory of special relativity—although one can immediately see that the latter represents a particular case ($U = 0$) of the former.
- The second example concerns the dynamics of a point of mass m. In Einstein's chronotope, defined by a given distribution of interacting masses, the "inertial" trajectory of a material point corresponds to a curve (called *geodetic*), which is defined a priori by the metric of space, i.e., by the local curvature produced in a metrical field. In this scenario we are no longer compelled to admit actions at distance between the masses (actions in which Einstein had never believed) and the two concepts of inertial mass and gravitational mass turn out to be equivalent. Practically, the concept of force becomes a definition without any direct reference to the "cause of motion", while gravitational potential defines the real features of curved space, whose metric and topological features are inherent in the tensor g_{ik} which determines the motion of all the masses in the universe.
- How far does this explanation go towards satisfying our common sense aspiration for an intuitive clarity of reality? Certainly in such a scenario we definitively lose our *primordial* feeling for an essential correspondence between our senses and mental elaboration and the world, which might be condensed in the assertion that if we belong to the world, the world belongs to us. In Einstein's chronotope

the world of our experience is nothing but an infinitesimal space element where time and distances are determined by the surrounding universe, which is *a-priori* precluded from the direct experience of man.

- Finally, whereas in special relativity each point of the chronotope is the origin of the two *separate* cones of *absolute past* and *absolute future*, according to general relativity the two cones may, at least in principle, overlap. This means that today we could conceivably experience events, which are in part determined by future actions. Obviously these paradox conditions are given by features of the metrical field which are far different from those of the region of the world where we live. But they remain a challenge for cosmological models and represent a philosophical warning whenever time is referred to perceived phenomena or designated as cosmic time.

Historically, Einstein's theory of relativity was promptly well received in Academic circles and soon became a fundamental part of what was taught at all universities. However, the significance and the interpretation of its new axioms and hypotheses remained difficult to understand for some decades. In this context a German physicist, Hermann Weyl (1885–1955), started a systematic analysis and synthesis of the mathematical formulation of relativity. His splendid book, entitled *Space-Time-Matter* [90], first published in 1918, contains a detailed and comprehensive description of mathematical objects and methods used in relativity theory. His book begins with basic tensor analysis and examines the properties of Riemann Geometry and the metrics of curved spaces, within a scenario where the old concepts of space, time and matter had been swept away in a deeper vision of reality. Weyl realised that the new notions entailed a series of questions, which the (at that time dominant) positivistic philosophy could not answer. Nevertheless he expected that new philosophical currents might shed some light on these problems. Practically, there was no alternative but to dogmatically follow mathematical formalism, trusting in the difficult but composite harmony of Riemann's geometry of non-Euclidean spaces.

Yet, in an abstract relativistic space-time, the physical content of the world is conceptually fixed in all its parts by means of numbers, and Weyl emphasises the risk of such a condition with a significant comment:

All beginnings are obscure. Inasmuch as the mathematician operates with his conceptions along strict and formal lines, he, above all, must be reminded from time to time that the origins of things lie in greater depths than those to which his methods enable him to descend. Beyond the knowledge gained from the individual sciences, there remains the task of "comprehending". In spite of the fact that the views of philosophy sway from one system to another, we cannot dispense with it unless we are to convert knowledge into a meaningless chaos.

Even in our times, after quantum-relativistic theory has still been further developed, Weyl's assertion is still absolutely valid, especially his warning of the risk of diving deeper and deeper accompanied solely by the mere instrument of mathematics. However, Weyl, as well as Einstein, believed in a Creating Intelligence and their vision of the future of science was optimistic. A testimonial of this faith is given by the concluding sentence of Weyl's monograph cited above:

Whoever endeavours to get a complete survey of what could be represented must be overwhelmed by a feeling of freedom won: the mind has cast off the fetters which have held it captive. He must be transfused with the conviction that reason is not only a human, a too human, makeshift in the struggle for existence, but that, in spite of all disappointments and errors, it is yet able to follow the intelligence which has planned the world, and that the consciousness of each one of us is the centre at which One Light and Life of Truth comprehends itself in Phenomena. Our ears have caught a few of the fundamental chords from that harmony of the spheres of which Pythagoras and Kepler once dreamed.

But the position of man in front of the new vision of the universe was feeble at Weyl's times—and today is even feebler. The faculty of *comprehending* the innumerable implications of the mathematical argumentation cannot be founded on the mathematics itself, but neither can it be reached without mastering its more and more difficult language.

These arguments on the theory of the relativity had deep repercussions on the western culture. Its epistemological implications definitively razed to the ground the last bastions of the old positivism, but opened the road to the new philosophical relativism of the twentieth century. In fact, whoever was in a position to penetrate the deep meaning of the new ideas could not ignore their consequences on the philosophical ground. At the beginning of the 1920s, Paul Valéry [43] annotated in his Cahiers [17]:

Relativity: non-existence of simultaneity, that is to say necessity to define the place in order to fix the age—and the place is an observer: the existence and the presence of this observer have the effect to order, in connection with his nearest object—his body, for instance—the ensemble of the perceived objects. All the objects of this ensemble have the observer as a common property. I am, therefore, the common property of all present things. But in order that it be so, it is necessary that the various ages of the things be unrecognizable. I mean that the past of A, that of B and the present of C can coexist in some place and in some way for someone. It would be, therefore, necessary that the sight of body B, for example, be present as a sort of memory.

Still today there are physicists who are not convinced of the truth of general relativity and attempt to find alternative approaches, but up to now Einstein's relativity remains the only theory to elevate itself on the solid pedestal of new experimental data which never cease to confirm his theory. Moreover, his most important success was his ready inclusion in quantum mechanics, which was substantially corroborated by the theory of relativity.

---

[43] Paul Valéry (1871–1945), one of the most eclectic and brilliant personalities of the European culture of the last pre-war age, had various encounters with Einstein and other eminent physicists of his time, exciting, with his observations and criticisms, their interest and admiration. His intuitions in the field of mathematics and physics do often show a substantial originality. Unfortunately Valéry had not a professional scientific formation, so that he is often spoken about as a *génie gaspillé*, a wasted genius.

Yet the theory of general relativity only concerns motion in a gravitational field. In which way electromagnetic and nuclear fields interact with a gravitational one represents an open problem, whose solution appears to be extremely difficult and pressing, even today. In the beginning, the fascination of relativity theory resided in the conceptual simplicity of its foundations and in the mathematical rigour of its description, but this fascination is being constantly eroded by the necessary development of highly complicated and hardly accessible *unified models* that are intended to describe these interactions.

Chapter 3
Scientific Research as a Global Enterprise

The years immediately preceding World War Two saw a massive emigration of eminent scientists from Europe to the United States, where academic centres had until then played a relatively marginal role in the development of science. However, after the great depression, began in 1929, American researchers were galvanised to action by the federal government, which granted substantial funds to laboratories of major universities, not to mention subsequent and considerable financial investments in military research centres, such as Los Alamos, Oak Ridge and Hanford, where the best physicists and chemists of that time were employed. During the war, research was almost completely subordinated to military plans and was conducted under unprecedented conditions. As had never happened before, a number of brilliant scientists were called to live and work together in an atmosphere of strictly maintained seclusion, dictated by practically monastic conditions. The main purpose of this immense effort was the construction of the atomic bomb. The project involved an expansive research front, including, in particular, quantum physics and radiochemistry.

Their success was so rapid and prodigious that there was no time to meditate on the terrible consequences that the released nuclear energy could have provoked. However, after an initial enthusiasm for the victory which resulted, the tragedy of Hiroshima and Nagasaki shook the entire world which, for the first time in the history, became aware that the total destruction of mankind had now been made possible by weapons derived from recent discoveries in physics.

In the years to follow, although increasingly powerful atomic weapons continued to be constructed and developed as well as extending to other countries, a newly invigorated scientific research was now principally addressed towards a pacific use of nuclear energy.[1] Thus, quantum physics found in the area of a newly created nuclear technology fertile grounds from which one could finally obtain an experimental

---

[1] After the decisive test at Alamogordo, the construction of atomic bombs mainly became an engineering problem. In the years immediately following the war, the American government was unable to manage the research in nuclear physics which had greatly expanded during the war. Many of the theoretical physicists who had been active in military research centres were basically unemployed

C. Ronchi, *The Tree of Knowledge*, DOI: 10.1007/978-3-319-01484-5_3,
© Springer International Publishing Switzerland 2014

counterpart to new theories and attempt to validate and further develop quantistic models on a vast scale. In fact, in the post-war period, atomic and nuclear physics became, quite simply, the field where the great majority of public funds for scientific research were being invested. In addition to research-reactors, complex accelerators were constructed delivering particle beams of increasingly high energies.[2]

The most imposing phenomenon of the post-war period concerns, however, the increase of the number of people possessing enough scientific education to comprehend and further elaborate new theories. The USA is a case in point: in 1901 a total of 2,595 students were enrolled in scientific disciplines at the various universities and less than 10 received a doctorate in physics in that year [21]. In 2005 the number of students pursuing an advanced scientific education had grown to 930,000, 1,400 of whom obtained a doctorate in physics in that year. Consequently, what in the first decades of the twentieth century was a small group of scientists who worked within a few dozen universities, exploded in the years after World War Two to become a vast international scientific community (It is sufficient to mention here that in recent years 10–20 % of physics students at American universities are foreign citizens and since 2001 the number of doctorates obtained by these latter exceeds that obtained by Americans citizens).[3]

Nowadays advanced research is concentrated in a few countries (mainly in the USA, Russia, France, Great Britain, Japan and Germany), but researchers of international extraction are to be seen everywhere and the flow of information and know-how is practically unlimited and continuous. Furthermore, although a scientific hegemony of the Old and New Continents is still incontestable, the scenario is rapidly changing in a world where communication and information as well as development and production of knowledge have become of almost equal importance. In this regard, the spread and final development of English as a lingua franca has made this possible,[4] after some initial difficulties, an *au-pair* interaction of scientists in culturally disparate countries like Japan, India and China has taken place, countries which have supplied scientists of the highest rank in the last few decades.

The heritage of the great discoveries and innovations of the first decades of the twentieth century were thus transmitted to new generations that began to work within an unprecedented social and cultural context.

(Footnote 1 continued)
for a certain amount of time. Richard Feynman, one of the best brains at Los Alamos, recounts how he had studied on his own initiative for two years locking mechanisms of safes and mathematical methods needed to find out their combinations. This anecdote is reported in his book of memoirs [22]. It was only when universities and civil enterprises were involved that nuclear research embarked on a new epochal stage.

[2] Today the Tevatron of the Fermilab in Illinois can accelerate protons to an energy of 2 TeV (2^{13} eV) while that of CERN in Geneva has achieved, in its new 2008 version, 7 TeV energy (the ring of the machine has a circumference of 27 km!).

[3] Data reported by the American Institute of Physics, Statistical Research Center, 2008.

[4] Only scientists who have personally experienced this cultural transition can fully appreciate its importance. Actually, linguae francae have always played an essential role in the establishment of great historical cultures, like Greek in the classical world, Arab in Islamic countries and Latin in modern Europe.

3.1 New Horizons and Old Questions

The dissemination of quantistic theories and models in this new global milieu initially encountered a number of difficulties. In fact, the new theories not only appeared to be much more complex than the classical ones, but, more alarming still, they were hardly intuitive and were expressed in an increasingly specialised language. In fact, while in the past a traditional propaedeutic instruction in mathematics was sufficient in order to comprehend the language and models of classical physics, the formulation of advanced quantum physics demanded knowledge of new mathematical methods and, except for the simplest cases, numerical applications were only possible using expensive, powerful computers together with adequate levels of programming. All these changes contributed to encapsulate a vanguard of physicists dealing with quantum theory in specialised research institutes and laboratories where they maintained a few tenuous contacts with other branches of physics.

On the other hand, a great number of articulated research objectives were launched which demanded fresh intellectual forces which were to be found only in new generations of young, well-trained scientists.

To solve this problem, courses in mathematics at the various scientific faculties were radically reformed with a view to preparing students to be able to master the new disciplines, which were rapidly expanding their application fields, covering— just to mention the most important topics—elementary particles, atomic and nuclear structure, solid-state theory, astrophysics, cosmology, thermodynamics and chemical physics.

New quantum effects were investigated which showed great relevance, especially in solid-state physics. Discoveries in this field have led to technological and industrial applications of exceptional importance that have in turn expanded research and development in new areas (e.g., electrical superconductivity, nuclear magnetism, structure of semiconductors, interaction of matter with radiation, and, finally, non-linear optics with the extraordinary development of modern solid-state lasers).

These successes gave a new impetus to advanced research and aroused great public interest. However, the other side of the coin presents some problematic aspects. The frontiers of quantum physics have in fact been increasingly extending to fields that are very distant from those of our natural surroundings and can be reproduced only by using complex and expensive laboratory equipment. Most of the phenomena investigated under these conditions may have great relevance from a theoretical standpoint, but, in most cases, it is unlikely, at least in the near future, that they will be of any practical or technological interest except perhaps for spin-off applications resulting from the development of advanced equipment and instruments.

We are, therefore, faced with an equivocal expectation from research activities carried out at the frontier of science. On the one hand, there is our society, which is setting its hopes on a constant flow of beneficial scientific discoveries; on the other hand, there are specialists whose research work is projected toward distant horizons with ephemeral connections to our world. This might give rise to a misunderstanding on the purpose and obligations of modern research, an enterprise where a considerable part of public revenues are invested and where rising generations of young scientists have finally been persuaded that knowledge may be useful only when framed into a system of existing research objectives and activities.

We shall repeatedly come back to this crucial aspect of modern research in the next chapters. But in order to understand the cultural importance of this problem, we should start by casting a glance at the frontier of physics.

3.2 The Matter of Matter

In the first decades of the past century, new quantum theories seemed to give a final order to the corpuscular model of matter: atoms should be formed by a nucleus composed of protons and neutrons, surrounded by a spatial distribution of electrons. A new concept of nuclear forces had to be introduced to explain the cohesion in the atomic nucleus of protons of equal charge and neutrons, but quantistic models of atoms were mainly based on electromagnetic interactions and resulted in a simple and elegant model. The great difficulty encountered in calculating the electronic states of heavy atoms, where complex relativity effects must be accounted for, seemed to be of a practical nature only. Thus one presumed that, once adequate instruments for numerical processing were made available, one could calculate the states of any atom or compound present in nature. Furthermore, production of intense neutron beams in nuclear reactor facilities made it possible to create new artificial heavy elements (so-called Transuranium Elements) from subsequent neutron captures starting from uranium and thorium.[5] The new discipline of radiochemistry has made it possible to extend chemical manipulation from the 92 elements naturally present on Earth to thousands of nuclides representing the set of their stable and radioactive isotopes.

Yet the dream of having attained a final goal didn't last very long. Around 1930, experimental confirmations of Paul Dirac's predictions were obtained which proved the existence, for every elementary particle, of an *anti-particle* of equal mass and opposite sign. First the pair electron/positron was discovered, to which followed proton/anti-proton. Then the concept of antiparticle was extended to neutral particles, attributing to the preposition *anti-* a change of sign of the baryon quantum number; thus neutron/anti-neutron pairs were found, and, finally, anti-states were defined, corresponding to a general concept of "anti-matter". The collision of a particle with its antiparticle provokes a disappearance of their mass and its conversion into (electromagnetic) energy, in accord with the theory of relativity. However, experimental observations of processes of mass annihilation substantiated the expectation that energy could materialise in other particles smaller than protons,

---

[5] About 30 transuranic elements (TUs) have been created so far. All isotopes of these elements are radioactive, but lighter TU nuclides have a sufficiently long half-life so as to be found in nature—albeit in minimal traces—as products of natural nuclear reactions. The heaviest ones have very short half-lives—e.g., the most stable isotope of Rutherfordium ($Z = 104$) has a half-life of a few seconds. However, theoreticians predict the existence of "islands" of super-heavy Elements of higher stability.

neutrons and electrons. The confirmation came some years later when, by studying the phenomenon of beta radioactivity, it was observed that a neutron could transform into a proton/electron pair. Therefore, under certain conditions, the resulting neutron was unstable, which was already alarming enough, but, what's more, a remarkable deficit in the energy balance was observed in a neutron decay reaction. Niels Bohr, at that time the highest authority on the subject of atomic models, postulated that for this type of sub-nuclear reaction the law of conservation of energy could be violated. However, a young Austrian physicist, Wolfgang Pauli, rejected this hypothesis and proposed the existence of a particle of infinitesimal mass and zero charge, called neutrino, which would participate in a beta decay reaction. A clear experimental confirmation of Pauli's hypothesis was not late to come.[6]

At this point a vast, unexplored field was now created. New frontiers of physics were established where great experimental and theoretical efforts were concentrated. The history of the ensuing discoveries is too rich and complex to be summarised here. It is, however, instructive to briefly examine these developments to understand the type of difficulties modern physics is encountering and the limits that might be imposed in the future.

A race to discover sub-nuclear particles soon began: first in astrophysics with the observation of cosmic radiation, and then in experiments of accelerated-particle collisions at high energies. The main target was the proton: the most stable elementary particle. In the great accelerators constructed in the post-war period, one produced beams of protons colliding with electrons, neutrons and deuterons at higher and higher energies. What happened is well known: the effect was such as though one had broken through the wall of a room, thus opening a window on an external world. A true firmament of sub-nuclear particles was revealed which was of such variety that their classification soon demanded a unifying model. This was developed in 1964 by two American physicists, Murray Gell-Mann and George Zweig, who assumed the existence of particles, called *quarks*, which would represent the constituents of the great family of stable and unstable particles called *hadrons*. This family is divided into two groups: the first, *baryons* which include protons and dozens of similar particles, and the second group consisting of *mesons*. A baryon is formed by the confinement of *three* quarks whilst a meson consists of *one* quark and *one* anti-quark.

Quarks have never been observed individually, but hypothesising their existence has made it possible to coherently interpret phenomena of fragmentation of sub-nuclear particles. However, their properties are in striking contrast with those of other particles.

First of all, their mass under conditions of confinement is much lower than that in a free state. This difference of mass would explain the fact that they cannot beproduced

---

[6] Initially the neutrino was assigned a mass of zero; however, subsequent experiments showed that this particle indeed possessed the smallest of masses: five million times smaller than that of an electron. Experimental confirmation of the existence of neutrinos soon came from various directions. In particular, in connexion with cosmological models, this particle assumed a fundamental importance in the formation of matter in the universe (today we know that cosmic space is crossed by an intense and constant flow of neutrinos.)

in laboratory collisions since the collision energies needed to measure this effect are presently unattainable. Furthermore, quarks have never been observed in cosmic radiation as residuals of the Big Bang because their mean lifetime is too short. This explanation is reasonable, but some criticisms were raised since, under these conditions, it is unclear what it is meant by saying that quarks are hadron's "components". In 1976 Werner Heisenberg made a telling remark [18] on this subject:

> *"The development of theoretical particle starts with a mistaken question. They say us that a proton is formed by three quarks, or something like that, none of which has been observed. The initial question was: Of what are protons consisting? But who has put the question seems to have forgotten that the word "consists of" has a sense clearly enough only if the particle can be divided in parts by using an energy amount that is much smaller than the energy of the rest mass of the same particle."*

To this objection Stanley Drell, at that time director of the Stanford Linear Accelerator Center in the USA, answered, with a delicate subtlety, by citing a Biblical passage from the Epistle to the Hebrew (11, 3) [19]:

> *"We know by faith that the (elapsing) centuries were adapted by God's word, so that invisible beings became visible."*

This is not merely a witty remark, but has profound implications. Actually, in quantum physics it is generally accepted that there are states of matter that, intrinsically, we are not allowed to observe, but that can be described starting from what we know about a limited number of states that are accessible through physical measurements. Thus, even though the existence of quarks cannot be proved directly, their effective properties explain and even predict observable effects. This is certainly a solid argument in favour of their reality. Moreover, quarks might be present in the universe under other aggregate forms that are presently unknown. From this perspective, attempts have been made to construct a thermodynamic equation of the state of a quark fluid inside which a process of aggregation would take place at sufficiently low temperatures, similarly to what happens during the formation of solid phases from a liquid (for example, in neutron stars the hadrons can be described as a phase formed by three quarks aggregated by strong nuclear forces).

The fundamental problem resides rather in the meaning of the four force fields through which the interactions of particles are unfolded, i.e.: (1) gravitational interactions, (2) electromagnetic interactions, (3) strong nuclear interactions (for instance, the attraction between quarks) and (4) weak nuclear interactions (for instance, neutron decay and radioactivity).

Eminent theoretical physicists, from Einstein on, have in vain tried to unify at least the first two forces in a theory that also explains the laws governing macroscopic phenomena. On the other hand, in the microscopic field of quantum physics it is possible to neglect the weak effects of gravitation and to concentrate on the other forces.

In the last decades important successes have been obtained. First, a strict analogy has been demonstrated between electromagnetic and weak nuclear forces, whereby these latter are permitted to violate the laws of conservation of symmetry (*parity*), which gravitational and electromagnetic interactions do strictly obey. They would be, therefore, able to change "*colour*", i.e., the quark's quantum number, causing a transformation of hadrons (for example, of a neutron into a proton plus an electron) accompanied by the emission of quanta, whose mass, for collisions at high energy, can assume values greater than that of a proton.

For strong nuclear forces the difficulties are more challenging. Attempts have been made to explain these forces by assuming symmetry properties of an electromagnetic field (as for weak forces) as well as a symmetry violation that involves the creation of quanta, *gluons*, having a large mass. The fundamental aspect of strong interactions is the preservation of the neutrality of the particle's "colour", contrary to the property of the weak forces. If quarks, taken singularly, may have different "colours", they exist in nature in combinations in which the sum of the colours is neutral (i.e., correspond to "*white colour*", as defined by possible combinations of different colours).

The development of models for strong interactions is a frontier for modern theoretical physics where formidable difficulties of every kind (from the formulation of the equations to their solution) have been hindering any rapid advancement.

3.3 The Sub-Nuclear Universe

There is no doubt that during recent years, the investigation of elementary particles has occupied a privileged position. The particles' properties, investigated through experiments of collision at increasingly high energies, have been continuously revealing that those that were first considered indivisible can be disintegrated into smaller components. The more collision energy increases, the more types of particles appear, whose arrangement in a sub-nuclear universe requires theoretical developments of an ever profounder nature. We are dealing here with a complex picture in which a variety of particle types and sub-types are interacting through different forces. More than 400 different particles have been discovered in the last decades. They have sometimes been called "elementary", but for a long time nobody has believed that matter is made of such a great number of fundamental constituents.[7]

The two main classes of *fermions* and *bosons*, which in the past distinguished particles of fractional spin from those of integer spin, and represented, respectively, the constituents of matter (electrons, protons and neutrons) and the carriers of energy (photons), have been expanded in order to account for new types of particles. Among those particles known as fermions, some, called neutrinos or mesons μ and τ, are subject to conventional electromagnetic interactions and are classified as *leptons*. In addition, the class of bosons has been extended to include a sub-class of bosons W and bosons Z (particles of great mass and weak interaction) and, in another, eight

---

[7] One should remember that an aggregation of protons, neutrons and electrons alone results in a variety of forms in the universe as we are able to perceive it: 118 elements, of which 94 naturally occur on our planet, with approximately 3,000 isotopes. The number of compounds is immense: the molecules produced by our industrial activities alone are more than 8 million in number, without accounting for their various thermodynamics phases.

particles of different type: *gluons*, which regulate the intense energetic exchanges between quarks of zero mass and charge.

Some physicists think that fundamental particles are indeed quarks which form a new class: they are of six types, have relatively large masses and interact with nuclear forces of high intensity (much stronger than electromagnetic ones); they are the only ones to have fractional charges ($\pm 1/3$ or $\pm 2/3$) and appear in groups of three or two.

In the field of sub-nuclear particles, research has mainly been based on collision experiments whose results are used to develop models describing interactions which transform wave functions of incident particles into those of the collision products. The formalism of the analysis is very complex and special mathematical methods had to be devised and constructed *ad hoc* by describing particle interactions using a series of perturbative terms whose convergence sometimes creates serious difficulties.

In order to hint at the complexity of the problem, we outline briefly here the general terms for a simple collision.

Let the experimental state be that of two initially distant, non-interacting "incident" particles whose momentum and energy are known. During the experiment the particles come closer and begin, in fact, to interact: their dynamic properties and, more generally, their number and type may change during the interaction. Then, after a sufficiently long time, the wave function evolves towards a state of a number of "emergent" particles, possibly different from the initial ones.

If these particles are of the same type and number as the incident ones, we have an *elastic* scattering, in other cases, an *inelastic* scattering. The mathematical problem is, in its simplest terms, as follows: given an initial ($t = -\infty$) state of incident-free particles one must establish the probability that a final ($t = \infty$) state of emerging free particles is observed. Since energy and momentum of the initial state are determined and conserved though the whole process, in the final state, there is a *complete indetermination* of the time, t_0, at which the effective interactions starts. During this time the evolution of the Hamiltonian of interaction from time t_0 to time t transforms the non-interacting states into interacting states.

There are several problems inherent in this procedure. First of all the fact that "naked", non-interacting particles are mere mathematical objects, since, in reality, every particle is subject to self-interactions (for instance, an electron produces an electromagnetic field and in quantum mechanics this should be described as a particle that continuously emits and re-absorbs photons). Furthermore, the time-dependent short-distance interactions during the collision are difficult to formulate and the resulting wave functions may show anomalous features, sometimes incompatible with the requirements of the space to which they should belong.[8] Finally, the results often

---

[8] We have mentioned in the preceding chapter that the wave functions of quantum physics should constitute a rigorously defined metric space (Hilbert space) and within this space, differential operators of quantum physics are applied. These operators, which represent processes of physical measurements, transform a wave function into another belonging to the same space. In fact, if one considers their significance, the wave functions must be square-summable, i.e., the integral of their square (which represents a probability distribution), extended to the entire space, must be finite and normalised to one. However, in certain cases, especially those dealing with scattering calculations, wave functions are obtained that sometimes do not satisfy this fundamental property of metric space

show that current models are unable to reproduce experimental results and alternative theories must be simultaneously developed and compared.

One can see that this and other problems are very insidious, especially in analysing and interpreting pioneering experiments.

For instance, one recent theory is based on the hypothetical existence of punctual particles, called *preons*, of which quarks and leptons should be composed.

Another, more revolutionary one, is that of the so-called *strings*. This theory was incredibly successful[9] inasmuch as it enables the description of elementary particles to be reinserted in a comprehensive physical context of general relativity. Strings are not defined as moving points whose encounter in space establishes the property of the collision processes, but are treated as extended linear entities with their own topological properties and vibration modes. The strings, as they approach one another, do gradually develop their own interactions in a complex topological space. Their description involves a complex treatment of space that can require up to 26 dimensions (compared to the four of general relativity). To be sure, it is not an intuitive theory, and in fact its development and some of its fundamental laws derive from applying abstract mathematical objects, which were already known in the past.[10]

There remains, therefore, a legitimate doubt, already expressed by Einstein with regard to quantum mechanics, that the theory is nothing but a mathematical artifice which in some way reproduces some aspects of a still unknown reality. Supporters of string theory claim, however, to have achieved a novel approach to a global description of nature, from which high energy physics had been increasingly estranged. Its superiority with respect to previous models resides in its ability to unify the treatment of gravitational, electromagnetic and nuclear forces. Yet, although string theory refers to cosmological topics, it is a fact that this theory is not in a position to give any useful information on the laws of nature as we can perceive them. This is made worse by the present impossibility of any experimental validation since this would require analysing phenomena produced by particles at exceedingly high energies, much higher than those attainable in modern accelerators. Moreover, since there are several possible mathematical models for strings, and no criteria for choosing one are available, the theory also seems to irrefutable—but in a negative sense that physics gives to this attribute.

---

and are, therefore, submitted to correction procedures, called *re-normalisation*, where one has to manipulate objects which have no clear mathematical significance.

[9] From 1980 until now, more than 30,000 articles have been published on this subject.

[10] The concept of string was born in the 1920s when some physicists examined the possibility of extending the theory of relativity from four to five dimensions in order to be able to explain the extreme weakness of the gravitational field compared to the electromagnetic field by assuming the disappearance in a fifth dimension of hypothetical particles called *gravitons*. These concepts were picked up again in modern string theory.

3.4 Cosmology Today

The origin of the Universe represented the principal interest of naturalist philosophers in classical antiquity, whose speculations were always in search of the "seeds"
of all beings. Their visions, however, encompassed both the microcosm and the
macrocosm, as they were convinced that the same general laws ruled their genesis.
Strangely enough, with the birth of modern science, physicists no longer had the
same interest in cosmogony that their ancient predecessors had.

Since the Middle Ages, astronomers only focused on the motions of celestial
bodies that were believed to be perfect and eternal, and hence immutable. Until the
beginning of the past century it was, in fact, commonly believed that the universe was
infinite in dimension and stationary in time. Even Newton was convinced that the
cosmic effect of gravitational attraction was null because of the uniform distribution
of cosmic matter. Some physicists, among them Cartesius, tried to explain gravity
by assuming the existence of whirling motions of aether, but the total lack of data
on the structure of matter did not allow these models to be developed in any real
depth beyond those of classical antiquity. Actually, a salient feature of modern science is rather a pronounced tendency to follow analytical methods, where the object
investigated is repeatedly dissociated into smaller and smaller components. This analytical polarisation of scientific research has consequently directed all efforts toward
investigating a sub-nuclear reality, whose physical models are exclusively applied to
microscopic dimensions.

A modern approach to cosmology can hardly pursue the same investigation methods. Analogical models are instead considered in which phenomena are described
according to the most recent views on physical reality.

In fact, modern cosmology began with the consolidation of the theory of general
relativity and quantum physics. Indeed for the first time, physical models yielded
predictions not only within a scenario of ordinary observations, but in times and
dimensions of cosmic magnitude.

It was, once again, Albert Einstein who opened the way. In 1917 he published a
short article in which he introduced an evolution model of the state of the universe
based on a hypothesis of a uniform distribution of matter. He demonstrated that the
presence of a space curvature, as entailed by general relativity, was equivalent to the
presence of repulsive forces that counterbalanced gravitational attraction. The model
was simple and predicted that the universe was finite.

It was Alexander Friedmann, a Russian astrophysicist, who, in 1922, by assuming
the universe to be a fluid sphere, obtained the exact solution of the relativistic equations of its evolution and proved that, depending on the values of the space curvature,
the cosmic sphere is expanding indefinitely or, after a period of expansion, enters
into a contraction stage.

A few years later, in 1927, Georges Lemâitre proposed that the universe was
initially in a state of highest density (the primordial atom), in which the space curvature was sufficient to overcome gravitational attraction to yield an expansion in two
stages. It was the first time the Big Bang Theory was formulated. There remained,

however, the conceptual difficulty, already discussed by Friedmann, that, where in a point of the Universe matter density diverges, a singularity is created in space-time, in contrast with the basic axiom of general relativity which asserts that space-time does *not* contain singularities. Several hypotheses introducing a possible anisotropy of an initial state have been produced to circumvent this difficulty, but none can be proved to be definitively valid.

An important step forward was made in 1965 with the discovery by Arno Penzias and Robert Wilson of background radiation of the universe. This electromagnetic radiation, of a wavelength of around 9 cm, corresponds to the spectral thermal emission expected by a thermodynamic system at a temperature of 3 K. This temperature is in fact close to that calculated for the present state of our universe after the expanding initial singularity has cooled down. This discovery, corroborated by the following experimental observation of an almost perfect isotropy of the spatial distribution of this radiation, supplies a strong argument in favour of models of homogenous expansion of the universe, even though the question of the indefinable initial singularity remains unsolved.

In the last few decades the study of singularities in space-time received a decisive push forward by the theoretical work of Roger Penrose (1965) and Stephen Hawking (1974). The aspect explored by these astrophysicists was the gravitational collapse of giant stars that had exhausted their nuclear fuel and were rapidly cooling down. If their mass was large enough, their collapse became inescapable. States of highest mass density are consequently produced, which are called "black holes". Only giant stars, whose mass is at least three times larger than that of the sun, can collapse to become black holes.

Black holes are subject to quantum interactions, but behave like thermodynamic systems: they possess a defined temperature, emit a corresponding thermal radiation, like classical "black bodies", and are subject to evaporation phenomena. Their temperature diminishes with the increase of their mass, meaning that the heavier they are, the stabler they are, while the lighter ones tend to rapidly evaporate. Therefore, all black holes are becoming progressively smaller and, when they decrease below a certain magnitude, their evaporation assumes the character of an explosion, accompanied by the emission of matter and γ-rays. It is thanks to this feature that the properties of black holes can constitute the point of departure for a founded theory of the initial stage of the Big Bang. Up to now it has not been possible to observe explosions of black holes in the universe because their initial mass must be so large that their mean lifetime is something like 54 orders of magnitude longer than the age of the universe (which is estimated to be approximately 13 billion years).

Today the main problem in studying the evolution of the universe concerns its initial state and the mechanisms of expansion within fractions of a second following the Big Bang. The only way to explore these properties is to reproduce certain events, like black hole explosions, in laboratory, using mass particles accelerated to energies which might be attainable in accelerators of the next generation.

Perhaps the answer must indeed lie in quantum field theory currently developed in the study of elementary particles, as well as in the study of thermodynamics properties of systems at high energy-density. It is, however, difficult to anticipate whether these instruments will be enough to remove the obstacle of describing the initial singularity of the Big Bang or new laws, presently unknown, must instead be introduced in the corpus of theoretical physics. It is, however, fascinating that the interpretative key to the expansion of the universe, from the formation of galaxies to the ordering of matter in all its manifold aspects, is probably to be found in a microcosm, the properties of which we known only in to a minimal degree and which continuously reveal an unexpected variety of forms.

3.5 The Enigma of Physical Laws

Physics can be compared to a building, whose bricks are the quantities and the laws the architecture. How many types of physical quantities exist? As many as we want to define; however, historically, three[11] fundamental quantities were chosen in classical physics: length, mass and time, which establish together a measurement system whence all other quantities are derived, and for every quantity an experimental procedure is given that allows us to measure it, directly or indirectly, following well-established standards. But what are fundamental quantities? This is a deeply-rooted question which concerns the definition of physical objects.

The choice of length, mass and time was due to the "feeling" that these quantities were real, simple "things" so that we can materially show a unit of length and mass, as well as of time. All other quantities are, in one sense, "constructed" with length, mass and time, which represent their metrological *dimensions*, e.g.: [velocity] $=$ [length][time]$^{-1}$, [flow] $=$ [mass][length]$^{-2}$ [time]$^{-1}$, etc.).

The choice of the three fundamental quantities is, therefore, only based on a specific abstraction of our senses. We can, in fact, imagine a blind observer, provided only with a refined sense of touch, who might define temperature and pressure as fundamental quantities and arrive at a concept of length and mass only through complicated physical relations. The fundamental quantities thus represent the dimensions of reality as we believe to perceive them "directly".

Yet, from a modern point of view this might seem too archaic a concept. For instance, one could object that mass may be expressed as energy since according to relativity mass is not an invariant character of an object, or, to give an another example, absolute temperature should be considered as a fundamental quantity—as is actually the case in thermodynamics—otherwise we have to adopt Boltzmann's model of statistical mechanics to express temperature as a function of the molecule's mean velocity and mass, and so on.

---

[11] The oldest system called (MKQS) was based on *four* fundamental quantities since electrical charge was also added. However, in historical terms, systems have been defined with only three fundamental quantities: length, mass, time, whereby certain units are chosen *ad hoc* by fixing the value of given physical constants in order to operate in selected fields (electrostatics, electromagnetism, thermodynamics, optics, nuclear physics, etc.) of convenient magnitudes.

Finally, we must recognise that in modern physics we cannot clearly distinguish "things" from "laws", quantities from the phenomena through which they are perceived and measured.

Furthermore, the definitions of the units and of the method of measurement are objects of metrology, a somewhat complex discipline. In fact, one can easily understand the difficulty in conceiving a measurement method that is to operate not only in the various fields of physics but also on a scale that may cover tens of orders of magnitude. This would hardly be possible without the aid of physical laws, in which relations between different quantities appear as well as (measurable) universal physical constants (e.g., gravity constant, Avogadro's number, Planck's constant, etc.). At present, the universal constants are altogether only a dozen in number, but if we include those of atomic and nuclear physics we arrive at roughly 350 constants.[12]

As for the physical laws, they express universal relations between physical quantities of objects. These relations are mostly described mathematically, but they are the fruit of experimental observations and their number might well be as great as that of the phenomena in nature. However, most laws are consequences of others, in the sense that their logical-mathematical formulations are concatenated. Until the first half of the 1800s, one was persuaded that the whole system of physical laws could be extracted from a small number of fundamental laws: the law of inertia (Galileo), the laws of acceleration (Newton), the law of gravitation (Newton), the law of electrostatic interactions (Coulomb) and the law of electromagnetic induction (Maxwell). In fact, although it is true that many of the phenomena investigated in classical physics can be analysed with models exclusively based on these fundamental laws, when the majority of phenomena are analysed, insuperable difficulties are encountered requiring further assumptions.

For example, atomic interactions in many solids and liquids, and in high-density plasmas, albeit of pure electrostatic nature, cannot be described by a combination of simple Coulombic interactions. The close inter-distances of charges of finite dimensions require defining electrostatic potentials which have approximate empirical formulations, whose validity is limited to specific types of application. In other words, the Coulombic formula has no general validity, but is rather bound to certain ide-

---

[12] Currently most fundamental physical constants were experimentally determined with an accuracy of the order of a few parts per million, but their accuracy is not a problem of merely adding some decimal points: the central issue is rather to check how consistent the links are which they provide among different branches of physics through experiments which are meant to confirm the consistency of basic theories.

For instance let us take the Josephson effect:

It has been theoretically predicted that two weak interacting solid-state superconductors, if maintained at a dc-potential difference, V, produce an alternating current of frequency $\nu = 2eV/h$. This effect provides a direct method of measuring with high precision the ratio between the electron charge e and Planck's constant h. The same ratio can obviously be measured from the minimum wavelength produced by an X-ray tube working at a given voltage. Yet one can expect perfect agreement in the two measurements only if (i) Josephson's theory is true and (ii) an absolute unit is used in both cases for the measured voltage and (iii) for the X-ray tube, the voltage applied is correctly converted to the effective energy of the electrons which eventually excite the X-ray photons. The three issues imply knowledge in completely different areas.

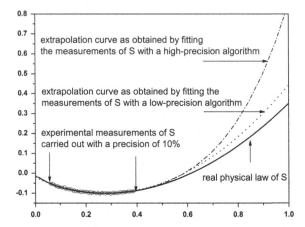

Fig. 3.1 The graph illustrates how, in phenomenological models based on mathematical fitting of experimental data, the reliability of extrapolations is often better in less precise models. This problem, known as *"overfitting"*, occurs in a variety of cases where the mathematical description of a physical process can only be calibrated with a limited set of experimental data

alised conditions. Therefore, whenever these conditions are not met, phenomenological relations must be obtained from specific experimental observations and from estimations based on *ad hoc* hypotheses on the nature of the systems investigated.

In all fields of physics phenomenological models are used which are valid and reliable only under restricted conditions. Actually, their intrinsic weakness is in representing the properties of ideal systems consisting of simplified *representations* of reality in a mathematical form. One of the dangers inherent in the use of phenomenological laws is that their accuracy does not guarantee the quality of extrapolations, even slightly beyond their experimentally determined limits. This also holds true in a strictly comparative context: a phenomenological model of low precision may supply better extrapolations than another of higher precision. Expressed in mathematical terms, this property is illustrated in the simple example of Fig. 3.1. The full curve of this graph represents the theoretical trend of a hypothetical physical phenomenon: in correspondence of a limited segment one supposes that experimental measurements exist, but subject to a typical random error. Since the "real" curve is unknown, the usual procedure consists in constructing a mathematical function whose parameters are fitted to optimally reproduce the experimental data. In our case a polynomial of second degree was chosen (with three fitted parameters) and one of fourth degree (with five parameters). This latter reproduces experimental data with better precision, but its extrapolation turns out to be much worse.

This, in a certain sense, trivial example is only meant to show that the deduction of a physical law is not only a problem of perfectly reproducing observations, but of "guessing" the "true" relation between the quantities involved, sometimes even by accepting that one is not in a position to reproduce the data perfectly. On this subject we cannot but recall once more the admonishment of the Aristotelian school in the face of the dangers of numerical precision in describing physical phenomena and its repeated insistence on the importance of an *analogical* agreement between observations and predictions of nature's laws.

Is it possible, however, to discover one particular physical law without revealing them all? In practice, it is as if we were asked to arrange the pieces of a puzzle without knowing the picture it should eventually represent. But before accepting the challenge, we feel it crucial to ascertain that this picture exists and that, to reproduce it, all pieces can be arranged in only one way.[13] Unfortunately, we would be able to obtain merely the first half of the proof only when the last piece perfectly fits in the last empty space of the puzzle. That, at least as far as modern physics is concerned, we are still immensely far from this endpoint is certain. Some even doubt that we will ever reach it, inasmuch as the missing pieces of the puzzle—that is to say, new experimental observations—are not simply at hand and become increasingly expensive and difficult to obtain.

In the face of these facts, the dream of Laplace and of certain philosophers and naturalists completely vanishes: namely, to be able, at least potentially, to write, if not solve, the equations of the evolution of the universe. Moreover, our criticism proceeds by asking to what extent do fundamental laws supply an answer to the question: "Why are they as they are?".

For example, why does the force of attraction between two masses m and M at a distance r have the mathematical form:

$$F = g\frac{mM}{r^2}?$$

What has fixed the value of the constant g? Why is the force inversely proportional to the square of r? We actually feel a certain unease when faced with the concept of "action at a distance" between two objects, inasmuch as it is unknown who and how is in a position to measure this distance and thence establish the resulting interaction force. For instance, this dependency could remind us of the emission or the absorption by every mass of "something" pre-established that is distributed on the surface of a sphere of radius r, whose surface density would, consequently, diminish with r^{-2}; but what is this hypothetical, mysterious agent? In this way, someone has proposed that the interaction force is due to an exchange of sub-nuclear particles, as illustrated by an analogy due to Denys Wilkinson [34] schematised in Fig. 3.2. In the case of nuclear forces the particles exchanged could be quarks or mesons. But for gravitational and electrostatic forces explanations do not exist. One supposed the existence in space of a continuous isotropic particle flow and that gravitational attraction results from the mutual shielding of two masses facing each other and, therefore, are hit only by particles that arrive on the opposite face. This, however, contradicts

---

[13] The uniqueness of the arrangement of the pieces of a puzzle provides the guarantee of an essential biunivocal correspondence of their mutual positions (given by their respective shapes) with their content (the fraction of the whole image printed on their surface). This correspondence is the *"conditio sine qua non"* which makes science possible.

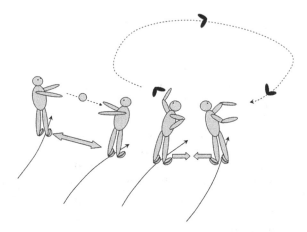

Fig. 3.2 Some people think that the interaction forces between two distant physical bodies are caused by exchanges of (unknown) particles, which take place continuously

the law of inertia since these particles would slow down every mass in inertial motion.

There is a great number of more or less fantastical "explanations" of this kind, but none up to now has proved to be valid.

Actually, the attempts to intuitively and visually[14] explain the cause of the law of gravity date back to the time of its discovery. Someone directly asked Newton what this meant. "*It means nothing and tells us nothing.*"—he answered—"*It tells us how is the motion and this should be enough. I have told you how a mass moves, not why.*" Furthermore, it is worthwhile reading what he wrote on this subject in his "*Principia*":

> "*When I make use of the term attraction…I do not have in mind a force in the physical sense, but only in a mathematical one; therefore, the reader should forbear from imagining that with this word I mean to designate a cause or a physical reason, or that I want to attribute to the attraction centres real physical forces, because in this treatise I take in consideration only mathematical quantities and proportions, without having to do with the nature of forces and physical qualities [73].*"

It is amazing that Roger Cotes (1682–1716), the mathematician who edited the same "*Principia*" expressed in the preface (p. 33) the opposite opinion, asserting that gravitation was "the ultimate cause", beyond which one could not proceed further. Even more curious is that, when one drew Newton's attention to this contradiction, he merely replied that he had not read (*sic!*) the preface by Cotes.

We must, at any rate, consider how arduous a task it was for a physicist of the seventeenth century to speak of attractions at a distance, a concept which appeared

---

[14] It is curious enough that the notion of mechanical *impact* is often more easily accepted as ultimate cause than that of force field. In reality, from the modern point of view, the impact between two particles (scattering), with all its possible outcomes, is one of the phenomena more difficult to explain, especially in quantum mechanics, and demands precise knowledge on the most complicated short-range force fields.

much less realistic than Aristotle's ethereal mobile spheres. After the discovery of analogous interactions of electrical forces, we know something more: these interactions are not instantaneous, but propagate in empty space with the speed of light. Later on, one obtained the confirmation that gravitational forces also propagate as waves at finite speed. Today we are accustomed to the idea of force fields, but their nature remains a mystery and while formulating several "whys" becomes more and more intriguing as physics progresses, the answers always remain suspended. In a scholium to the abovementioned edition of his "*Principia*", Newton wrote the famous phrase: "*Hypotheses non fingo*", but even in the absence of hypotheses, his law of gravitation explained both the experimental data on the fall of bodies, and Kepler's laws of planetary motion, in which the formula of gravitational force is implicitly contained. This has been proved valid for the fall of one rain drop as well as for the formation of galaxies. It was one of the greatest intuitions in the history of science, but no reasoning would have ever led to it if the equations of motions had not existed in a mathematical formulation of elliptical orbits. Similar considerations can be made for all laws of physics that are revealed in natural phenomena following a logical-mathematical construct. Because of the now commonplace textbook application of these laws, we are often not aware of the deep mystery of this correspondence. Richard Feynman observed his usual acute discernment [23]:

> "*I find amazing that it is possible to predict what happens by using mathematics, which simply consists of following rules that have nothing to do with what is happening in the original objects.*"

Is it reason that determines truth or is the perceived truth that obeys reason? We are always being confronted with the same initial question, whose possible and contrasting answers have been faced for more than two millennia. However, whatever the true answer may be, it is necessary to keep in mind that even mathematics and logic have limits inherent in abstractness or, if we prefer, to the emptiness of their content. Gilbert Chesterton wrote in one of his brilliant Elsevier's

> "*You can only find the truth with logic if you have already found truth without it from the daily news.*"

This aphorism evokes an image where logic appears as an ordered armed force, an *acies ordinata*, which proceeds to conquer an adventurous, unknown land, following mysterious scouts who precede it. Therefore, scientific knowledge will always appear as the fruit of an adventure, a product which can be viewed from many aspects as risky, a goal and at the same time a line of departure for an endless quest. Perhaps the most dangerous perversion of man consists in believing that he entirely possesses scientific knowledge and that he uses it as a universal key to open all doors he happens to encounter.

Chapter 4
The Crisis of Growth

In our society serious questions regarding the future of science reflect a widespread fear that in certain important areas scientific progress is going down a blind alley.

The first fear concerns research in general, which, in the last decades, has been converted into an industrial enterprise, whose objectives are established according to a sort of market dynamics where increasing public investments tacitly imply that the results must satisfy the demand, a situation which may eventually become a matter of life and death for scientific research. The second, perhaps more problematic concern, is that the current scientific language is losing its universal character and is being split into a gamut of "dialects" practised within narrow circles of specialists while being elsewhere almost incomprehensible. This is obviously in contrast with the above-mentioned tendency towards public control and management of research objectives.

For instance, if we consider classical physics, we realise that the adopted mathematical formalism has produced substantial simplification in formulating fundamental laws and has made describing the behaviour of even complex systems possible, by clearly analysing the phenomena involved.

For quantum mechanics, however, the process has taken a reverse direction: initially the existence of *quanta* was not considered to be a basic tenet of a diverse approach to reality; indeed, certain aspects possibly became simpler than those described by assuming a continuum of space and time. On the contrary, attributes of space become increasingly removed from any geometric representation. Let us consider, for instance, the wave function of a free electron in an absolute vacuum: this function extends over a continuum, infinite space and, only in the case of strong interactions, like, e.g., those binding an electron to an immobile proton, does the wave function assume localised shapes, corresponding to discrete values of the energy of the electron. We know that the sequence of the *eigenvalues* of energy is rigorously defined mathematically, while in every possible energy state the position of the particle remains indeterminate. Thus energy, a fundamental, measurable physical quantity, is reduced to a mathematical parameter resulting from a complex integration procedure of a wave equation and has, at least apparently, no direct connection to the classical definition based on the acceleration and motion of a particle in a potential field.

C. Ronchi, *The Tree of Knowledge*, DOI: 10.1007/978-3-319-01484-5_4,   125
© Springer International Publishing Switzerland 2014

Understandably, more than a few physicists have wondered whether the complex intermediary mathematical objects and procedures upon which quantum mechanics is based could be, in one sense, redundant. The recurrent questions have revolved around doubts that are as old as science: are quanta perhaps an indication of the discrete nature of all constituents and processes in the Universe? Could it be that the concept of *continuum* constitutes a mental folder suited merely to contain and organise only part of our experience, but ill-suited to describe a possibly simpler atomic and sub-atomic reality, the essential properties of which are, however, still unknown?

For this reason, some contemporary theoreticians are indeed of the opinion that all physical entities and the evolution of their properties should be described in terms of *rational* numbers. In particular, the measurement unit of a physical quantity should *not* be arbitrary, but expressed by an integer number of elementary, *physically pre-established* units (see, e.g, the arguments of John Ehlers [33]). One could simply have replied that such objections do not have an effective foundation: actually, in quantum mechanics calculations, complex differential equations are solved analytically only for a few simple cases, while they are currently solved numerically using electronic processors that operate, after all, only with integer numbers. The problem is, however, more subtle because physics equations are necessarily bound to operations, like passage to the limit, differentiation, integration and convergence, etc., which are not applicable to a set of integer or rational numbers. For instance, if an analytical solution has a singular point, the numerical integration programme must contain *ad hoc* instructions to circumvent a singularity domain where regular operations would lead to, e.g., to an overflow. On the other hand, if equations of physical laws were rewritten using integer terms, we would face the problem, which in most cases would be insolvable, of searching for exact solutions. What would a lack of solutions in such cases mean in an equation containing physical variables? Would it have a predicative value for the existence of the corresponding state? Or would it simply indicate a problem of numerical precision[1] of the equation coefficients?

The first issue concerns our conception of space-time. Is this space only a formal entity or does it contain the principles of physical phenomena? The recurrent apories arising from applying alternative concepts of *continuum/discretum* and *global/local* to physical space are only solved if the space structure has first been established, and this condition depends on the answer to the preceding question.

One should bear in mind that quantum mechanics was initially conceived in Euclidean space and only after merging with the theory of general relativity was it extended to a curved space, described by the differential geometry developed by Riemann. Yet, the metrics of this space, whose curvature is related to relativistic dilations/contractions of physical variables, presupposes a *continuum* hypothesis in a stronger type than does the metrics of a Euclidean space.

---

[1] For instance, we should remember that in computer programming of numerical calculations with two variables x and y, representing "real numbers", it is *not* allowed to use decisional statements of type "**if** $(x.$**equal**$.y)$...**then** ", because the electronic processor is *not* in a position to decide on the equality of two real numbers. A serious limitation, for this indecision might be significant.

Moreover, in recent theories of sub-nuclear particles, and, in particular, in String Theory, space is shaped by even more complex topological properties than those of the relativistic *chronotope*. According to some people, the topology and the differential geometry of these spaces represent a problem in itself, which, once solved—certainly by means of a mathematical formalism of extreme complexity—will provide a way of reformulating the physical laws in a suitable space with definitive dimensions and properties. Considerable work has been carried out in this field, but, owing to the extreme flexibility of this approach, an uncontestedly unique choice of the type of space on the basis of the available experimental data seems, at least at the moment, hardly possible.

Furthermore, if progress proceeds in this direction, modern physics appears definitively bound to sophisticated instruments of infinitesimal analysis, which are irreconcilable to common sense understanding, contrary to the opinion of many epistemologists who consider agreement with common sense to be absolutely essential for our knowledge of reality. Actually, in a scenario where phenomena would be nothing but occasional signals emerging from a universe inscrutable in its essence and impenetrable in its dynamics, it is justifiable to ask ourselves whether the progress made by science should represent, as in the past, a quest for *ultimate* causes.[2] If this is *not* the case, we should expect a radical change in the scope and definition of science.

4.1 The Hypothesis of *Continuum*: Instrument or Impediment?

The dawn of the past century found natural sciences in a state of full maturity. Their contents were perfectly organised in models based on plain and simple laws formulated on an easily teachable and approachable mathematical platform, which was expected to make further progress possible in a vast area of exploration of still unknown phenomena.

Quantum mechanics had provoked much criticism and dispute, but not a great deal of uneasiness since in the opinion of experts the foundations of classical physics were not being undermined in terms of all practical aspects (an opinion which was not completely true) and general relativity appeared, in its definitive consolidation with quantum physics, to be an extension of classical mechanics, which, while invalidating the generality of Newton's laws, confirmed their effective expedience within space and time limits relevant for man.

A century later, the situation had radically changed: science had grown continuously in content and complexity, branching out into numerous, special disciplines. The traditional language of physics dissolved in a variety of "jargons" where the

---

[2] The risk of losing a common sense of reality through scientific analysis was already clearly perceived in antiquity. Plato in his "*Laws*" recommended "*not pushing too far the search for causes because this was not allowed (οὐδ'ὅσιον) by divine law*".

terms of language assumed a different meaning depending on the specific context. In addition, modelling and research methods rapidly evolved and differentiated in various areas, in particular in those where pioneering methods of investigations are being pursued.

Nuclear physics became more and more complex with the discovery of an increasing number of sub-nuclear particles. The concepts of mass and electrical charge, taken from classical physics, still furnished the load-bearing pillars for a concept of the constitution of matter, but their nature remained obscure and the interaction of electromagnetism with gravity still represented an enigma—even by allowing for merely qualitative conjectures. The bipolarity of charge, the existence of antimatter and the equivalence between electromagnetic radiation and mass, allowed us to suppose a possible constitution of mass and charge of more elementary entities. But the only way to construct more general models is through the development of a unified theory of quantum physics and general relativity which explains the interactions of the force fields discovered up to now (gravity, electromagnetism, weak and strong nuclear forces).

During the past decades, the analysis of pioneering experiments is based on models formulated by very complex equations, whose solutions and predictions can be carried out thanks to powerful processors. However, the dimension of the problems has grown at the same pace. It has always been said that a better algorithm is to be preferred over a faster computer, but even from this perspective we have reached a critical point since programmes are available that develop algebraic manipulations that a man could only execute with enormous difficulties and risk of error.[3] Moreover, while mathematical methods and calculation apparatuses become increasingly sophisticated, the area of the phenomena to which they are applied continues to shrink with every advance in analysis. Yet, however, arduous the way from a macroscopic to a microscopic world may be, it is infinitely easier to trace and to cover this path than go the opposite way, which, having reached the sub-atomic world, would bring us back to reconstructing the world of our perceptions in a variety of its forms and interactions. Many people wonder if quantum physics in all its diversifications will be ever in a position to describe systems and phenomena of ordinary complexity.

The main limitation is due to the formal burden entailed by the necessity to operate using a number of mathematical entities (matrices, tensors, functions, operators etc.), which intuitively appear to exert a mediatory role, but, in terms of the final results, the data they carry are sometimes marginal and the information they provide almost worthless. Metaphorically, it is like trying to speak using a vast vocabulary, but fundamentally disconnected from the subject at hand. Today there is a vaguely held opinion that, if we do not succeed in finding a more adequate mathematical language, then quantum theory will remain applicable only within a limited sphere, and physics will have to renounce its aspirations for generality and self-consistency.

---

[3] We note here as examples for the reader who has some familiarity with these problems the cases of the calculation of Dirac's matrices in the most complex Feynman's diagrams or in verifying Ward-Takahashi correlation-functions identities in problems of quantum gravity.

We summarise here some aspects of one of the essential problems on which the most radical criticism has been focussed.

The fundamental structure of the physicist's language is bound to an intuition of a space-time characterised by a *biunivocal* correspondence between measures of physical quantities and numbers. An assumption of a field of real numbers, **R**, is not due to physical or metrological reasons, but represents the necessity of presupposing the development of infinitesimal analysis and differential calculus, which, if applied to physical laws, allows (at least in classical physics) their exhaustive and compact formulation. However, one can object that **R** contains infinitely more irrational than rational numbers, whereby only the latter are consistent with a concrete measurement and, conceptually, with a corpuscular image of matter. Since antiquity one has recognised that a real-number *continuum* is an abstraction and that it is uncertain if all its properties are applicable to concrete quantities, but the main argument in its defence is that the field, **R,** contains a sub-set of rational numbers, **Q**, and that, in any algorithm, a real number can be approximated by a rational number with an *arbitrarily small* error. Moreover, for rational numbers, as for real numbers, the infinite divisibility of a finite interval is still valid.[4] A substantial difference consists, however, in the fact that the set of rational numbers is enumerable (i.e., can be written as a sequence), while that of real numbers is not.[5] On the other hand, a possibility of *counting* objects, no matter how large their total number is, is a necessary condition for any corpuscular ensemble. Moreover, in dealing with integer numbers, properties exist as even/odd, successive number, divisor and prime number, as well as complex applications, e.g., combinatory relations and cyclical groups, which allow us to formulate mathematical structures and functions that have absolutely no correspondence to the field of real number.

Problems exist that can be solved in the field of integer numbers owing to their property of forming a "*well-ordered*" set. We show here, for example, an application of the "drawer's principle" or "Dirichlet's drawers" which is illustrated by the following problem:

Let the following theorem be thus demonstrated:

Thesis: In London there are at least two persons who have exactly the same number of hair.

Demonstration: It is known that the number of hair of a person is, within a generous overestimation, less than one million. In London there are more than one million inhabitants. We begin to ask if one has one single hair, then if one has two, and so on. When we have asked if someone has one million hairs, we have certainly exhausted all possibilities. From this point on, the remaining inhabitants will have a number of hairs that will have already been found in at least one other inhabitant. The result is

---

[4] The thesis of the theorem says that "Given two *different* rational numbers $p < q$, there always exists a third number, r, such that $|q - p| > |r - p|$".

[5] In set theory, a parameter, called cardinality, is defined and indicated with aleph ($\aleph$). The cardinality value $\aleph_0$ (aleph-zero) is that of sets of infinite, countable objects. Since a real number is defined by two successions (sub-sets) of rational numbers the cardinality of continuum is $\aleph_1 = 2^{\aleph_0}$ (aleph-one). This subject is taken up in the next section .

certain and *not* merely probabilistic as would be the case if we had tried to obtain the demostration by operating in a real-field.

The power of this type of procedure resides in the possibility of establishing a hierarchical criterion in compliance with which all objects are ordered, without exception. Moreover, the set of integer numbers has an important property which, in correct mathematical language, is expressed as follows: given an irrational number, a, it is always possible to find two integer numbers, n and m, for which $|na - m|$ is arbitrarily small. Synthetically, the theorem is so formulated: "the set $\{[na]$: with n integer$\}$ is dense in $[0,1]$".

Back to physics: within a simple corpuscular concept of matter, if we think of a physical line having a length L, we expect its measure to be equal to an integer number of units corresponding to the length of the elementary particle of which the line is composed. Even choosing as a measurement unit an arbitrary multiple k of this length, we would always have the length of the line being expressed by a rational number $L = n/k$.

However, the **Q** field does not possess complete metrics, in the sense that converging successions of rational numbers tend to irrational numbers: on the other hand, the completion of **Q** with these limit-numbers corresponds, indeed, to the field of real numbers **R**.

To reformulate the physical laws in a **Q**-field would be problematic since we would lack the objects that constitute the essential terms of the current language of mathematical physics (all transcendental functions, universal constants π, e, etc.). Moreover, the formulae should be written in terms of rational functions only and for algebraic equations rational roots would have to be found, considering that it has been mathematically proved that in an infinite number of cases these do not exist.

Furthermore, from a physical point of view, we can see the incongruity of the concept of uniform mass since mass is concentrated in atomic nuclei, whose dimension is a million times smaller than that of atoms. We know that the space inside real bodies is characterised by electrostatic interactions between atomic nuclei and electrons and is almost "empty" and accessible to neutral or energetic particles. Furthermore, the universe of sub-nuclear particles discloses the atomic nucleus as an open and complex system where simple atomic theories are inadequate to define the various aggregations of matter with their multifarious packages: from atoms to the largest cosmic bodies.

4.1.1 The Objections of Physicists

Let us suppose it were possible to establish a universal unit of minimal length and hence all measures be expressed by integer numbers of this unit. In concrete terms: let 10^{-15} m (a proton diameter) be the assumed smallest indivisible length and 10^{26} m (an estimate of the diameter of the Universe) the greatest conjecturable length[6]: any physical length could, therefore, vary over 41 orders of magnitude. In an extreme case, a length of the order of magnitude of 10^{26} m would be expressed by 41 digits, one of the order of 1 m by 15 digits and one of 10^{-15} m by one single digit. Using a modern computer, calculations with numbers of 41 digits are feasible, but the problem becomes crucial when we observe that a set of integer numbers, **Z**, does not constitute a *"field"* with respect to operations of multiplication and its inverse, division. For instance, in **Z** an odd measure cannot be divided by two; or we might find that it was often impossible to obtain the length of the hypotenuse of a right-angled triangle in terms of rational fractions—in fact, Diophant would teach us that this only happens for particular dimensions of the two catheti and the hypotenuse, corresponding to the so-called Pythagorean triplets: (3, 4, 5), (5, 12, 13), (7, 24, 25), (8, 15,17), etc..

It could be objected that these difficulties derive from the obvious fact that a particles' agglomerate of finite extension cannot perfectly fill the space, but necessarily produces packing/stacking features that, dependently on the local agglomeration symmetry (cubic, tetragonal, hexagonal, etc.) or on the local statistics of the particles' clusters, must be conveniently treated—e.g. as one does in crystallography. However, in the chosen example (of the right-angled triangle) one must recognise that this simple geometrical figure can be exactly "materialised" only for selected, discrete dimensions of the catheti, where the required conditions can be visualised by the procedure of *gnomon* addition illustrated in Fig. 1.6.[7] In this context, one can see, again, the complex limits of the analogy between ideal geometrical figures and a discrete structure of matter.

A consideration of epistemological nature is here in order. For every physical quantity, the existence of an indivisible unit is not the sole immediate consequence of a strictly corpuscular character of matter, but it implies that space and time also have a discrete structure, which might not be immediately intuitive in our customary model of reality. When, in practice, we execute a physical measurement of a quantity, the indivisibility of the unit requires adopting a *defined* standard and measurement method on which the significance of the measure itself depends—and not only for practical reasons.

For example, in the famous Zeno's paradox, Achilles will never catch the turtle because he falls into the pitfall of wanting to *count* all intermediate stages of distance he sees and has to recover, holding in suspense the measurement units of space and time, so that thinking alone freezes him definitively at the starting block. The obvious remark is that Achilles does not need to think, it would be enough for him to run in order to catch up with the turtle. But this was just what Zeno (a disciple of

---

[6] Considerations on relativistic quantum physics indicate that for lengths smaller than 10^{-35} m (called Planck length) and times shorter than 10^{-44} s (Planck time) use of *continuum* properties involves a certain theoretical incongruence. However, these figures are obtained from a dimensional analysis of universal constants and we don't known what they really mean.

[7] The equation $n^2 + m^2 = p^2$ can be formulated in this manner: given a square of side n, construct with all elements of a square of side m the *gnomons* which extend the square of side n to that of side p. One can easily see that with this method general formulae are found which make it possible to calculate the Pythagorean triplets.

Parmenides) maintained, i.e., that movement is a mere illusion of our senses. But if Achilles, during his race, closed his eyes at regular intervals and stopped to think for an instant, at a certain point he would find he had surpassed the turtle and would get a measurement of the time elapsed and of the distance covered from the number of intervals he was able to count, but with a limitation that he would not know, exactly, when and where he had caught up with the turtle. He could open and close his eyes at shorter and shorter time intervals and hence diminish the uncertainty concerning his position and time of overtaking, but he would never be able to eliminate it completely. In theory, he could do this if he were in a position to open his eyes exactly at the time of passing the turtle or at intervals corresponding to its submultiples. Yet, to do that he would have to synchronise the time unit with his particular case and the time measurement would consequently lack any general significance.

In conclusion, independently of the magnitude of a physical object, once a unit is given, our mind can conceive of its measurement only through a sequence of contiguous and not overlapping units that cover the object in defect or excess.

An indivisible unit is, therefore, an entity that guarantees the generality of a concept of measure, but, at the same time, implies accepting a possible indeterminism that demarcates a dark zone inside which nothing can be said.

4.1.2 The Objections of Mathematicians

The concepts of infinitesimal and infinite numbers have today been incorporated and well-organised in differential calculus, whose formalism allows us to deal with them as mathematical objects, and whose use requires a certain skill and caution wherever intrinsic dangers must be circumvented by reasoning and applying suitable procedures. Nevertheless, infinity has never obtained an undisputed right to citizenship in the domain of mathematics. Actually, since classic antiquity, objections have been raised on the rightness of introducing infinity in mathematical procedures and accepting the risk of running into consequent apories.

The dispute concerning infinity and continuum lasted for more than two millennia. From antiquity through the Middle Ages until the end of the 19th century it involved mathematicians, philosophers and theologians and was in fact only concluded with the work of Georg Cantor (1845-1918). In 1870 Cantor started a conclusive analysis of Aristotle's criticisms on infinity and in the following three decades developed a modern theory of transfinite numbers [97]. He considered Aristotelian definitions of *potential* and *actual* infinity, from a standpoint of manifold theory and asserted their existence both *in abstracto* and *in concreto*.[8]

He called *improper-infinite* a quantity which is finite, but can increase at will. Then, given a finite manifold, Cantor defined *cardinality* as generalising a notion of

---

[8] During his work Cantor often kept in touch with philosophers and theologians in order to link his theory, and, especially its axiomatic aspects, with metaphysical speculation. Among his correspondents were Aloys von Schmid, Joseph Hontheim and Constantin Gutberlet, who were involved in the discussion concerning thermodynamics and cosmology. Tillmann Pesch [98], one of the more

finite cardinal numbers. This corresponds to a consistent—not demonstrable—notion
of a given plurality. The finite cardinals are generalised by defining *proper-infinite*
(*transfinite*) cardinal numbers. These numbers possess the same reality and clear
definition as finite ones and operations of addition, multiplication and potency can
be defined for them. Infinite cardinality numbers are indicated by the letter $\aleph$ (aleph).[9]
The first one is $\aleph_0$, which corresponds to cardinality of a countable set (e.g., that of
integer and rational numbers). The next one, $\aleph_1$, is a cardinality of a continuum or
real numbers. The cardinality of infinite sets of real number sets is $\aleph_2$, and so on.

A crucial point in this definition is that, without any further axiom, it cannot be
demonstrated that there does *not* exist any intermediate cardinality between $\aleph_0$ and
the cardinality of a continuum.

Finally, there is a very important consequence of cardinality theory, which tells that
there does not exist a sufficiently small linear quantity ζ so that, given an arbitrarily
large integer n, the inequality holds: $n\,\zeta < 1$.
The reason is that multiplication of ζ by a transfinite number cannot be made finite
and hence ζ cannot be an element of a linear quantity. This demonstrates a fundamental aspect of Archimedes' axiom, which turns out to be an intrinsic property of
a continuum.

The work of Cantor was of fundamental importance for, in turn, the most remarkable definition of real numbers which was developed by Richard Dedekind (1831–
1916), using the notion of a "*Dedekind cut*".[10]

In spite of the consolidation of continuum theory brought about by defining transfinite numbers, criticism began to appear by reiterating that the fundamental object
of mathematics is a natural number and its only acceptable, intuitive specification is
that of an ordinal integer, which must refer to primal concepts of *equal* and *subsequent* numbers. Natural numbers are unlimited, but critics rigorously argued that it
was necessary to admit that every natural number is *properly* finite and has a well-
defined position in any sequence of numbers; thus an infinite integer number can

---

(Footnote 8 continued)
important representatives of neo-scholasticism, contributed greatly to the diffusion of Cantor's ideas
of transfinite numbers in philosophic circles.

We should, however, mention that these interests provoked repeated violent attacks on Cantor
from the aggressive positivistic milieu, attacks that hindered acceptance of his theory—now a pillar
of modern mathematics—and, finally, ruined his health. Pesch was even arrested several times in
Germany under the charge of being a Jesuit. A telling comment on the climate of European culture
at the end of the 19th century.

[9]   Note that this definition is different from that of infinite (∞) as used in infinitesimal analysis,
where it pertains to any kind of number. Cardinality is meant to define different "sizes" of different,
infinite sets, but in a distinct way. Thus, for instance, sets of natural numbers and sets of prime
numbers are both infinite and the latter type is a subset of the former; nonetheless the two sets have
the same cardinality.

[10]   A Dedekind cut is a partition of *rational* numbers into two non-empty sets A and B, such that
all elements of A *are less* than *all* elements of B, and A does not contain any greatest element. If
B has a smallest element, the cut corresponds to that rational number. Otherwise, the cut defines a
unique irrational number, which is in neither set, and "fills" the gap between A and B. The set of
irrational numbers is thus defined as the set of the set pairs of all rational numbers and hence has
the cardinality $\aleph_1$.

only be conceived as *potentially* reachable (*improperly infinite*), but never treated as a *given* entity, usable for any reasoning or procedure. Despite this fundamental limitation, critics assert, modern mathematics is the fruit of serious transgressions by improper use of infinity, beginning from a definition of *continuum* and irrational numbers which implies the use of passage to infinity as a procedural method. For example, an irrational number, according to critics, is ill-defined by two converging successions of rational numbers. In other words, the definition tells us how an irrational number can be *in principle* identified, but, *in practice*, nobody would ever be able to construct it.

Now, towards the first half of the 19th century, just when infinitesimal analysis was maturing and being consolidated, a heretical tradition became prominent by asserting that mathematics constructed on real numbers must be refused as a whole and a new formulation based only on a concept of integer numbers was predetermined. These ideas, grouped under the label of "*intuitionism*", were first proposed and made prominent by Leopold Kronecker (1823–1891) and taken up, until now, by a group of mathematicians, among whom the most influential was Luitzen Brouwer (1881–1966), a brilliant Dutch mathematician and philosopher, who delved deeply in discussions of gnosiological aspects of intuitionism.

Intuitionism is actually a form of "*constructivism*", a concept of mathematics according to which every object can be accepted only if a way is defined to construct it with a *finite* sequence of instructions. One of the more important consequences of this requirement is that logic cannot be applied "*in toto*" to mathematical reasoning. For instance, a method of demonstration "*ab absurdo*", which actually circumvents constructivist principles, is *rejected*. The same applies to the principle of "*tertium exclusum*", found in traditional logic, because propositions can exist that are neither true nor false.

Intuitionist theories have been opposed by the majority of mathematicians and have given rise to an implacable war between supporters of intuititionism and notable thinkers, such as Georg Cantor and David Hilbert (1862–1943).[11]

From an application point of view, the main problem of intuitionist mathematics resides in the necessarily recursive character of its calculation methods, even for simple operations like addition and multiplication (in fact, all calculations must

---

[11] The war of intuitionism was indeed a strange one since, curiously enough, viewd from a distance of some decades, the sides and underlying interests were reversed. The first violent clash took place between Kronecker and Cantor, the creator of modern set theory and of transfinite numbers. Cantor represented the standpoint of the majority of mathematicians against intuitionism, but he was also working on pioneering theories which were being opposed to by the academic world. On the other hand, Kronecker was an influential member of the Berlin Academy and was personally against Cantor's new ideas. The battle with intuititionists was very hard-fought and lasted more than ten years, finishing with the victory of Cantor, in spite of the incomprehension and the attacks on the part of some mathematicians of his time, a situation which ruined him mentally. The second battle broke out half a century later, between Brouwer and Hilbert. The latter, who represented the compact front of the academic establishment of his time, accused Brouwer of casting away all the precious fruits collected during three centuries in order to provide safer passage for, in his questionable opinion, the boat of mathematics. This time it was Brouwer, the great perfectionist of intuitionism, who lost his mental health in the battlefield.

essentially be based only on operations of comparison of two numbers and by passing from one number to the successive one). At first glance this appears to be prohibitive for any kind of application. However, in the last few decades, with the development of electronic calculation, intuitionist/constructivist mathematics has been gaining ground. New formalisms, such as those called PRA (Primitive Recursive Arithmetic) are being used more and more , although their use is limited to specific problems.

Today in the acrid war of intuitionism against traditional mathematics an armistice has been reached, but contentions remains, with both sides maintaining their positions. It is nonetheless to be hoped for that, without having to reject the fruits harvested in the past, new ones will mature on a tree which sprouted and grew from the seeds of intuitionism (one of them, that of *p*-adic numbers, is presented in Sect. 4.3).

4.2 Real and Complex Numbers

If, on the one hand, intuitionist criticisms of *continuum* insist on reconstructing mathematics using as a base integer numbers only, conversely, towards the middle of the 1800s, in mathematical physics one began to use a continuous numerical field of greater generality than that of real numbers, which allowed a more comprehensive and powerful *representation* of physical laws: the field of complex numbers, whose theory had first been developed by mathematicians of the late Renaissance and was completed in the past century. It is worth examining in some detail these numbers since they have proved themselves to be necessary instruments for completely deploying algebra and, at the same time, from a new point of view, reveal the problem of the correspondence between physical quantities and numbers which describe their space.

We know that solving algebraic equations in the field of real numbers is subject to serious limitations with regard to the existence and the number of solutions.

When, by the mid-XVI century, Gerolamo Cardano, while studying general solutions of algebraic equations of third degree, discovered that, for certain values of coefficients, the equation clearly admitted three real solutions, but that these could *not* be expressed with radicals. Furthermore, he realised that all procedures he tried to use always ended up with the expression of a number whose square was -1, which appeared to be absolutely insignificant. Cardano finally decided to proceed with an algebraic treatment, carrying forward in all successive steps the root of -1, without asking himself what its meaning could be. Thank to this stratagem he succeeded in finding the three real solutions to the equation.[12] Evidently, through this procedure,

---

[12] The most famous case is that of the equation $x^3 + px + q = 0$, whose algebraic form is said to be *"irreducible"*. When its coefficients are such that $27 q^2 + 4 p^3 = 0$, the resolutive formulae yield three real solutions, while if this expression is greater than zero one of the solutions cannot be expressed with radicals of real numbers, even though the equation has in fact three real solutions (for example, when $p = -2$ and $q = 1$), but require complex numbers to express them in radicals. All this appeared odd, to put it mildly, since the results were in contrast with the geometric image of the problem.

Fig. 4.1 Vectorial representation of complex numbers and of their operations

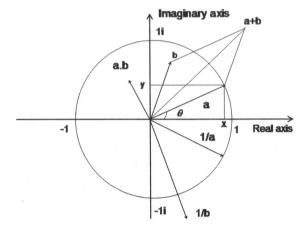

a mysterious entity, called an imaginary number, was playing a mediator's role thus making it possible to obtain the correct solutions. These results aroused the interest of several algebraists of the Renaissance (in particular, of Raffaele Bombelli), who took Cardano's artifice as a starting point for defining a new numerical field: that of *complex numbers*.

We are now faced by a wonderful generalisation process, by which meaning was given to an expression, the square root of -1, which previously didn't have one and while leaving behind an arithmetical difficulty in order to free generalised operations on algebraic objects from constraints inherent in their contents. Algebra as a science of abstract operations was born.

We cannot enter into arguments of the historical development of algebra, but, in order to appreciate the importance of complex numbers, let us consider briefly their main feature, which brought algebra to a state of perfection. Here we briefly introduce complex numbers using modern formalism.

A complex number consists of an ordered pair of two real numbers (x, y), which, in a system of Cartesian co-ordinates, can be represented as a vector in a plane (Fig. 4.1). Like all vectors, a complex number, a, can be represented as the sum of its components along the reference axes:

$$\bar{a} = \bar{x} + \bar{y}$$

The x-axis is called *real* and the y-axis *imaginary*. After fixing this convection, the vector signs can be omitted and the complex number is written in the form:

$$a = x + iy$$

where y is the value of the co-ordinate of an imaginary axis "i". In its general form a complex number has, therefore, a real and an imaginary component. If the imaginary component is zero, the complex number is a real number, while if the real component is zero it follows that the number is a pure imaginary one. The operations of addition and subtraction of complex numbers are defined from well-known properties of operations with vectors, and are illustrated in Fig. 4.1.

Two centuries after Cardano, Leonard Euler completed the formalism of complex numbers using formulae that express a relation between the trigonometric functions of sine and cosine and an exponential function. A complex number could, therefore, be expressed, in addition to the above- mentioned *vectorial* form, in a *trigonometric* form[13]:

$$a = x + iy = r(\cos(\theta + i\sin(\theta)))$$

and, using Euler's development of trigonometric functions, in an *exponential* form:

$$r(\cos(\theta) + i\sin(\theta)) = r\,e^{i\theta}$$

where $r = (x^2 + y^2)^{1/2}$ is a *real* positive number called "*module*" and θ the angle, called "*argument*", between the vector and the direction of the x-axis. The exponential form tells us that a complex number is represented by a segment of length r positioned by a rotation in the plane around the origin of the axes of an angle θ. When this angle is zero (or multiple of π) we have a real number (respectively positive or negative), when it is equal to an odd multiple of $\pi/2$ we have a pure imaginary one. For any other angle, the number has both a real and an imaginary part.

If addition and subtraction between two complex numbers correspond to those defined for two vectors (and can be easily executed if the two numbers are expressed in vectorial form), in order to fulfil the requirements of a *numerical field*, the operation of multiplication (and division) between two complex numbers must be defined in such a way that the properties (distributive, commutative, etc.) possessed by these operations in the field of real numbers remain valid. It can be demonstrated that this happens if the multiplication of two complex numbers is defined as formally resulting from their multiplication in exponential form. Therefore, the product of a complex number $a = r\,e^{i\theta}$ with an other $b = s\,e^{i\varphi}$ is simply given by $ab = rs\,e^{i(\theta+\varphi)}$; *i.e.*, the module of the product is the product of the module of a times the module of b, and the argument of the product corresponds to a rotation from θ to $\theta + \varphi$. In particular, for the nth powers of a, we have: $a^n = r^n\,e^{in\theta}$ and, in the case of a square root ($n = 1/2$), one obtains the mysterious Cardano number: $\sqrt{-1} = e^{i\pi/2}$. It can, therefore, be seen that various representations of a complex number allow us to comprehend their powerful role in solution methods of algebraic equations. Let us consider here a very simple example: The equation:

$$z^2 + 1 = 0$$

---

[13] We note that a trigonometric form is the simplest derivation of analytic geometry which, however, involves an important property: while in vectorial form a complex number is defined by two single real numbers, x and y, owing to the periodic character of functions *sinus* and *cosinus*, in a trigonometric form the same number is defined by its module, r, and infinity of values of the argument $\theta + 2n\pi$, with $n = 0, 1, 2 \ldots \infty$. We shall see that this property is of primary importance in certain operations.

does *not* have solutions in the field of real numbers. Yet, if $z = x + iy$ is a complex number, the equation is written:

$$z^2 + 1 = (x + iy)^2 + 1 = \left(x^2 - y^2 + 1\right) + 2ixy = 0$$

The complex roots of this equation require that both the real part and the imaginary part of the expression on the right-hand side be zero, i.e., we must have for the imaginary part:

$$xy = 0,$$

i.e.:

$$x = 0 \text{ or } y = 0,$$

from which we obtain equations for the real part:

$$-y^2 + 1 = 0 \text{ or } x^2 + 1 = 0,$$

where we see that only the first one is solvable in a real field and yields the solutions:

$$y^2 = 1, \text{ whose roots are } y = \pm 1$$

Finally, the two solutions, z_1 and z_2, of the initial equation are the two *pure imaginary* numbers:

$$z_1 = 0 + i1 \quad \text{and} \quad z_2 = 0 - i1$$

If the starting equation had been:

$$z^2 - 1 = 0$$

we would have found the two *real* roots, written in complex form:

$$z_1 = 1 + i0 \quad \text{and} \quad z_2 = -1 + i0$$

But full use of the properties of complex numbers in solving algebraic equations goes beyond this result. From its trigonometric form a complex number does not change if we add to its argument a multiple n of 2π. Then if we extract in an exponential form the pth root of z, we obtain:

$$\sqrt[p]{z} = \sqrt[p]{r}\, e^{i\frac{\theta}{p} + i\frac{2n\pi}{p}}$$

where one can see that there are p distinct pth-roots of z, having the same module and different arguments, corresponding to $n = 1, 2, \ldots p$.

The fact that a complex number, z, has p roots of order p is a consequence of the definition of the product of two complex numbers. This property is of fundamental importance because it removes asymmetries and exceptions from algebra, which appear in the extraction of roots in the field of real numbers, an importance that would have scarcely been imaginable *a priori*. This most remarkable result is condensed in the *Fundamental Theorem of Algebra* which says:

An algebraic equation of n-degree possesses n complex solutions, some of which can be real numbers, depending on the values of the equation coefficients.

If we think how the limits and the intricate casuistry come together in solving algebraic equations in a real field, where solutions do or do not exist depending on the degree of the equation and on the numerical values of its coefficients, we immediately understand that complex numbers offer algebra a basis from which it can deploy its intrinsic completeness.

The following step consists in considering complex functions of complex variables, which introduce even greater operative advantages to the theory of differential equations than those obtained for algebraic equations. One of the more important ones is due to the fact that their differentials have the properties of vectors and, therefore, are characterised by a direction and amplitude. Due to this feature, it is possible to choose integration or derivation paths and "circumnavigate" singular points, offering new, powerful integration methods for differential equations.

If we consider the long arduous path from the first, tenuous steps made by Cardano to completely defining the property of the field of complex numbers, we must necessarily conclude that this field, which opens the door to a perfect fulfilment of algebra, existed, in one sense, already for itself, as a mathematical object, and, therefore, could only be "*discovered*" and not arbitrarily "*created*" by man.

On the other hand, however, when one began to formulate physical laws using complex numbers one rightly asked what they really represented. The demonstration, obtained later on, that a complex field represents a completion (called "*algebraic closure*") of a real field and that this completion is unique, supplies the most important argument for justifying the use of complex numbers in describing physical phenomena. However, a rigorous and essential restriction is that, fundamentally, physical quantities must be represented only by complex numbers with an *imaginary* part *exactly equal to zero*, even though, in an operative context, a physical quantity can be decomposed into a real and an imaginary part.

Today in nearly all fields of physics complex numbers are used owing to the possibility of solving complicated differential equations using suitable variable transformations. There are mathematical procedures that are currently being used without questioning the precise physical significance of the variables in question. This is the case, for instance, of wave functions of quantum mechanics, where physical quantities are obtained with mathematical procedures defined in a field of complex numbers.[14]

---

[14] The physical significance of a wave function, z, is given by its square only if z is a *real* function. If function z is *complex*, and hence is written as: $z = x + iy$, its physical meaning is given by the product of z times its conjugate $z^* = x - iy$. The product $zz^* = x^2 + y^2$ is always a *real* number.

4.3 p-Adic Numbers: an Alternative to Continuum

Modern physics concerns itself with the fundamental question regarding the generality and the properties of the numerical field one needs to describe the world of phenomena. We have seen above that integer numbers are founded on a deep intuition of our mind and hence, we might say, are given to us by God. Furthermore, while the quantities corresponding to observable physical phenomena are expressed by rational numbers, which represent a direct generalisation of integers, the mutual relations between these quantities, as expressed by physical laws, involve a notion of space where these relations can completely unfold. If one examines in depth these notions one realises that they are based on implicitly assumed topological properties of Archimedean geometry, which, in turn, requires hypothesising *continuum* and hence metrics based on real or complex numbers. If, however, one searches for alternative foundations, we may need a different concept of numbers and the road must unavoidably traverse the complex and abstract discipline of number theory.

In modern mathematics several fields and classes of numbers exist that are defined for particular purposes. For instance, one of the most important is the field of algebraic numbers, which is composed of (real or complex) roots of polynomials with integer coefficients. Algebraic numbers occupy a central place in modern mathematics, in which important problems of geometry, topology and number theory converge. Algebraic numbers are a subset of real or complex numbers and possess intricate, interesting properties.

Criticism to a concept of *continuum* inherent in real, **R**, and complex, **C**, numeric fields, have found a resonance in the area of algebraic numbers and yielded some fruit which were slow to ripen, but which today are opening up new ways for mathematical physics.

These new ideas arose in circles of the Kronecker school, when one of his pupils, Kurt Hensel, at the beginnings of the 20th century developed a new mathematical object: the *p*-adic number field, $\mathbf{Q}_p$. This field represented a completion of rational numbers, **Q**, but was fundamentally different from **Q**-completions corresponding to **R** or **C** fields. Hensel developed the concept of *p*-adic numbers while he was studying algebraic numbers. From this point of view, the properties of these numbers are rather abstract and were first treated in the context of group theory. We briefly examine here the definition of *p*-adic numbers starting from a simpler standpoint than Hensel's.

(1) At the basis of *p*-adic numbers is a new definition of the *absolute value* or "*norm*" of a number and of the *distance* between two numbers. Let us first recall the traditional notion of *distance* of two *objects*: Generally, the distance between two elements x and y of a set E is defined as an application, $dist(x,y)$, which associates a *positive number* to any element pair (x, y) of E which obeys the following conditions:

$$dist\,(x, y) = dist(y, x)$$

Since the conjugate of a real function is its own function, in both cases their *norm* $\| z \|$ is generally defined as the square root of the integral of the product zz^* over space.

$$dist\,(x, x) = 0$$

$$dist\,(x, y) \leq dist\,(x, z) + dist\,(z, y)$$

(2) On the other hand, the *absolute value* of a *number*, x , is an application $|x|$, which associates a positive number to any element x of a numerical field K with the conditions:

$$|x| > 0 \text{ for } x \neq 0;\ |x| = 0 \text{ for } x = 0,$$

and, for any x, y

$$|x.y| = |x|.|y|$$

$$|x + y| \leq |x| + |y|;$$

under these conditions the *distance* between x and y is given by:

$$dist(x, y) = |x - y|$$

This definition of distance makes sense to us by telling us how *near* or how *remote* the two numbers x and y are. Definitions of distance and absolute value may look very trivial, but in fact their specification originates from a primordial intuition of space. Very important is the consequent metric condition stipulating that the distance between x and y cannot be larger than the sum of the distances between x and z, and z and y, where z is arbitrary:

$$dist(x, y) \leq dist(x, z) + dist(z, y)$$

For instance, let us suppose that E is a line and x, y and z are three of its points; one can easily understand that the above relation between the mutual distances of these points obeys the above relation by an equal sign. But if we have, for instance, three non-aligned points in a 3-dimensional Euclidean space, the sign "$<$" holds, implying that the three points define a plane and in this plane they correspond to vertices of a triangle of which the length of one side is always smaller than the sum of the other two.

These properties are consequences of the Euclidean conception of geometric space. In particular, they are bound to the principle of Archimedes, which states that, given two real numbers $a < b$, there always exists an integer number n for which, no matter how small a, is holds:

$$na \geq b,$$

a property we have already discussed in the preceding section. In particular, if a and b represent two physical quantities, we can make a sufficiently small so that b can be expressed as an integer number of units of length a, with an arbitrarily small error. We have seen force above, but also the limits of this principle in physics.

Now, the p-adic numerical field proposed by Hensel is, like $\mathbf{R}$, a completion of rational numbers, $\mathbf{Q}$, but is based on a different definition of the absolute value: for an *integer* number, m, the p-adic *absolute value* is defined as:

$$|m|_p = 1/p^r$$

where p is an arbitrarily predetermined *prime* integer that characterises the field $\mathbf{Q}_p$, and r is the highest exponent by which m is divisible by p^r. It should be noted that the p-adic absolute value of an integer is always a positive rational number ≤ 1, and large integer numbers may have very small absolute values.[15] Consequently, the absolute value of a rational number m/n is given by:

$$|m/n|_p = |m|_p/|n|_p$$

It is easy to see that rational numbers which are very close to each others may have a very large p-adic distance and vice versa.

At this point one might ask: how are these abstruse p-adic numbers useful, which, actually, were conceived for solving very particular algebraic problems? In fact, even for mathematicians the answer to this question was not found immediately. But some properties of p-adic numbers have no doubt shown that they were not just extravagant constructions, but in the field of rational numbers the definition of a p-adic absolute value represents the only alternative to an ordinary definition that can entail important developments. Today, many physicists do believe that p-adic numbers have an enormous, unexplored application potential, especially in quantum physics. Let us proceed with an examination of their properties.

Hensel demonstrated that there is a biunivocal correspondence between series with integer terms having a decreasing p-adic absolute value and p-adic numbers, so that a p-adic number can be represented by a series (called Hensel series) of type:

$$\sum_{i=-n}^{\infty} a_i p^i$$

where n is an integer, p, the prime number characterising $\mathbf{Q}_p$ and a_i are integer coefficients that can assume values between 0 and $p - 1$. This is an analogous formula to the usual one, by which a real number can be expressed in base 10 as the sum, for example, of thousands, hundreds, tens units, to infinitely small decimal fractions. But, while for a real number a sum starts from the highest power of 10 (the order of magnitude of the number) extending down to smaller decimal fractions,

---

[15] In this definition there is an analogy to modular arithmetic, often employed in periodic structures, where only positive numbers are used that are less than a chosen value, called modulus. In modular arithmetic, addition, subtraction and multiplication are like in regular mathematics, but there is no division. The notation used for equality expressions involving modular arithmetic is: $x = y$ (mod m). This means that x and y leave the same remainder when divided by m. For example, in time measurements a correspondence of 8 p.m. to 20 h can be expressed as $8 = 20$ (mod 12).

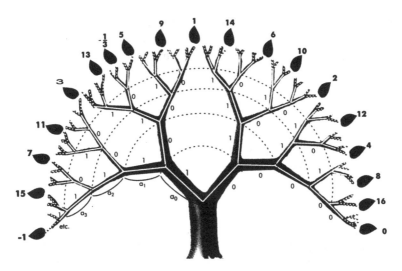

Fig. 4.2 The sketch follows Daniel Barski and Gilles Christol and was reproduced from Ref. [89]. It provides a graphic representation of some 2-adic numbers. If represented as a Hensel series $\sum_{i=0}^{\infty} a_i p^i$, where $p = 2$ and $a = 0$ or 1, a 2-adic number can be seen as a tree branch consisting of sub-branches starting from the trunk and passing through an infinite number of bifurcations. In our scheme we have assumed that in each bifurcation the right branch corresponds to $a = 1$ and the left one to $a = 0$. The length of the ith sub-branches corresponds to a 2-adic absolute value of 2^i (that is 0.5^i), which decreases with an increase of i. The graph shows that two p-adic numbers are near if, starting from the trunk, they have a long common path. Note that most of the adjacent integer numbers reported in the graph differ by powers of 2 (4, 8 or 16), corresponding to relatively low absolute 2-adic values. Thus, for instance, 15 and 16 are much more distant than 0 and 16. Closer integer number pairs would be those of type $(2^{n+1}, 2^n)$ with larger values of n. Note that the most distant integers are those whose difference is an odd number and have a 2-adic difference of 1, the largest absolute value an integer can have, whilst non-reducible fractions can have p-adic absolute values extending from 0 to infinity. Rational p-adic numbers can also fit in this kind of representation

for p-adic numbers the series starts from the *tail*, with the highest negative power of p ($i = -n$) proceeding limitlessly towards increasing powers of p (we must, in fact, remember that in a p-adic field these powers have decreasing absolute values) (Fig. 4.2).

Let us provide some examples of p-adic series assuming as characteristic prime $p = 2$.

– First of all we note that for all integers the Hensel series starts with $n = 0$.
– For the number 0 the coefficients a_i of the 2-adic Hensel series are, obviously, all zero.
– For the number 1 the coefficients of the series start from $a_0 = 1$ and are $a_i = 0$ for all other values of i.

– A little more complicated is the calculation of 2-adic series of -1. We write this number as $-1 = 1/(1-2)$. Since the 2-adic absolute value of 2 is $1/2 < 1$, the previous fraction can be developed in a geometric series:

$$1/(1-2) = 1 + 2 + 2^2 + 2^3 + \dots$$

hence the coefficients of the Hensel series of -1 are all equal to 1.
– From this result we can calculate the coefficients of $-2(= -1 - 1)$, that turn out to be: $a_0 = 0$ and $a_i = 1$ for all other values of i.
– Analogously to -1, we can calculate the series of 1/3, a rational number:

$$1/(1-(-2)) = 1 - 2 + 2^2 - 2^3 + \dots$$

where the coefficients are alternately equal to 1 and -1.

For all p-adic *integers* the coefficients a_i start from $i = 0$ and above a certain value of i are all equal, while for *rational* p-adics the coefficients start with a finite negative i and above a certain value of i are repeated periodically. Hensel series with infinite, non-periodic coefficients correspond to particular real numbers. Generally speaking, the calculation of Hensel series coefficients for a given rational number may turn out to be very laborious, but is always possible.

There is a straightforward relation between the traditional absolute value of a rational number $|x|$ and the p-adic absolute value, namely:

$$\prod_p |x|_p = \frac{1}{|x|}$$

where the product is extended to p-adic absolute values referred to *all* prime numbers. From this equation one understands that p-adic metrics represents, in a certain sense, the *inverse* of usual metrics. The first consequence of this definition is that the relation:

$$|x + y| \quad \leq \quad |x| + |y|$$

holds true in a strengthened form:

$$|x + y|_p \quad \leq \text{Max}(|x|_p, |y|_p)$$

which is called *ultrametric* condition. This condition which represents an essential property of a p-adic field, stipulates that the absolute value of the sum of two numbers cannot exceed the larger of the two, contrary to our usual intuition.

Actually, if one considers a p-adic number we must mentally eliminate the primordial idea of a numerical sequence from the usual concept of distance. On the other hand, expressing p-adic numbers by Hensel series suggests their representation in the form of a tree diagrams where each sub-branch corresponds to a term of the series and, in correspondence of each node, there are p sub-branches of length equal to the absolute value of the term. A diagrams of this kind is represented in Fig. 4.2 for $p = 2$.

The practical advantage of this tree-like structure is that the distance between two p-adic numbers x and y can be calculated by starting from the position on the tree of these numbers and adding the segments of the ramification one has to pass through until one reaches a common node.

The relation between the distances of three numbers has the ultrametric form:

$$dist_p(x, y) \leq \text{Max}(dist_p(x, z), dist_p(z, y))$$

In geometric terms this means that every one of the three sides of a given triangle is always smaller than (or at the most equal to) the greater of the other two. An *ultrametric* relation violates the Archimedean principle and invalidates fundamental properties of Euclidean space entailing other properties that challenge our intuition. For example, in an ultrametric space it can be seen that all triangles are isosceles, that every point of a disc can be its centre, that a disc and a circle are simultaneously open and close sets. We are, therefore, facing a space, which is essentially different both from ordinary Euclidean space and from the differentiable Riemannian manifolds that represent its generalisation.

The importance of these concepts for mathematical physics was only realised after a long period of incubation. Nevertheless, Bernhard Riemann, already in 1854, in his first lecture at the University of Göttingen raised the problem of the generalisation of metrics of physical space. He stressed that traditional metrics was founded on empirical knowledge, which, however, suggested that this could not be applied to infinitely small objects where the metric relations could, among other things, be inconsistent with the Archimedean principle (incidentally, on that occasion Riemann also examined the hypothesis of a physical space of more than three dimensions). We know that Albert Einstein came independently to the same conclusions when, in his theory of relativity, he defined space-time not as a Riemannian manifold, but as a pseudo-Riemannian manifold of signature (1, 3), a space in which metrics is dynamically determined by the physical conditions of the objects investigated.

But there is more. If, on the one hand, quantum physics was solidly constructed using the concept of continuous space, on the other hand, in the first half of the 20th century, it appeared increasingly obvious that the reality that underlies the concept of space forms a *discrete manifold*. The reasons are of a objective and not only of speculative or formal nature. For example, in quantistic gravitational theory one realised that quantum fluctuations implied that every physical measurement in the smallest regions of space-time was necessarily accompanied by infusions of great amounts of energy in infinitesimal volumes and that there was a length, called a Planck length, given by:

$$l_{\text{Planck}} = (\hbar G / c^3)^{1/2} \approx 10^{-33} \, \text{cm},$$

where G is the gravitational constant, and c the speed of light, below which any physical measurement should be impossible. Since physical measurements based on the Archimedean postulate correspond to the measurement and sum of customarily defined distances, the Planck length limit means that for very small distances this postulate must be abandoned. What this renunciation meant for mathematical physics remained, however, obscure.

It was only at the beginning of the 1980s that this argument received a new impetus, thanks to the work of a number of eminent mathematicians, especially those of the Russian Steklov Institute of Mathematics. Mathematicians there concluded that there were only two possibilities for constructing alternative mathematical fundaments of quantum physics: one was to replace the field of real numbers with that of p-adic numbers, the other to develop a new formalism based on *finite* numerical groups.

In the past decades, considerable efforts were made to develop instruments of calculation in the field of p-adic numbers that correspond and replace those of traditional calculus in a real and complex field. The results obtained are remarkable, but we are still far from being able to completely inscribe quantum mechanics in a p-adic field.

Furthermore, a hypothesis has recently been proposed that the space of physical phenomena is probably neither a continuous nor a p-adic one, but is fundamentally *adelic*, that is, composed of all possible completions of the field of rational numbers.[16] From this perspective, a description of reality would have two complementary aspects, a *transcendental* one, based on our intellectual perception of *continuum* and an *arithmetic* one, in which we can calculate certain important effects, each considered individually. These two aspects would be complementary as, for instance, conjugated quantities are in quantum mechanics.

With this hypothesis the difficulty of having to choose, more or less arbitrarily, a characteristic prime number p in order to construct a specific p-adic field is avoided. An adelic model, comprising, by definition, all possible completions of a rational field, would place all the prime numbers on the same level.

Recent developments in this direction[17] are supported by group theory and have led to important results in applications of quantum mechanics, in particular in string theory and in cosmology (mini-superspace). Other applications are considered in the field of hierarchical disordered systems (spin glasses), of diffusion processes in tree-like structures, of dynamics of macromolecules, of thermodynamics (p-adic entropy) and even of models of cognitive acquisitions through neuronal trajectories.

---

[16] An *adele* x consists of an infinite sequence:

$$x = (x_\infty, x_2, \ldots . x_p, \ldots)$$

where x_∞ is a *real* number, and x_p are p-adic integers except for a *finite* set of prime numbers, S. The adele ring **A** is hence given by:

$$A = \bigcup_S A(S) A(S) = R \times \prod_{p \in S} Q_p \times \prod_{\notin S} Z_p$$

This ring possesses an ample generality, including real and p-adic numbers encompassing all values of p.

[17] An excellent review of the current state of p-adic mathematical physics is given in Ref. [87]

It is, however, uncertain whether these ideas will converge into a dominant mainstream able to produce a revolution in physics leading to a new coherent and effective mathematical language. However, if this does happen, we know what the price will be: beyond a complete estrangement of a mathematical-physical space from primordial intuitions fixed in Euclidean space, even the concept of numbers will become more abstract, with new metric aspects that can be reconciled only with difficulty with initial intuitions that have rendered possible the parallel development of mathematics and physics.

We conclude here with a final remark of purely speculative nature. It has been mentioned above that models are being presently considered that try to explain the psychological functions in cognitive processes as a spread of mental states in a p-adic space, representing a neuronal *reticulum*.

From this perspective, some people think that the extent of p-adic space in mental states depends on the individual. Furthermore, two subjects A and B may in theory develop mental spaces with different values of p. We would then find ourselves facing a possible sort of neuronal Darwinism that would explain, among other things, the existence among humans of rare geniuses in mathematics (and in other specialities). Would it then be possible for man in the future to be able to collectively develop more complex mental spaces so that, even for the average individual, interpreting phenomena will be consciously perceived and organised in schemes more suitable for describing reality than the present ones?

Seeing that answering such questions is impossible, one may generally state that in physics every phenomenon must be studied and described within a predefined dimensional environment, if only for the simple reason that there is an intrinsic limit to the precision of every measurable physical quantity. Accounting for this argument would apparently exempt us from taking up a definite position regarding the actual rightness of the *continuum* hypothesis in nature and mathematics. This observation may be correct, but it does not remove the difficulties inherent in problems concerning what one calls "frontier zones".

Let us consider a simple example: the repulsive energy between two electrical charges of the same sign respectively situated in positions A and B is proportional to the inverse of their distance AB; when this is zero, energy is infinite. But, what does it a zero distance between two positions affected by an unavoidable indetermination mean? The question is not merely rhetorical, since it concerns, to cite an important case, interaction potentials between atoms in solids. Even for pure ionic bonding, short-distance interactions cannot be described by simple Coulombic law, but require substantial empirical adaptations. In order to explain all this, one is referred to quantistic effects, but this reference confirms the effectively discrete nature of space implied by a confinement of charges in atomic orbitals and hence a discretisation of atomic inter-distances. The concept of space becomes then difficult to define since it involves partial occupation of an ideal continuous space, as it happens, for example, in a crystalline lattice, where the atoms occupy discrete positions, depending on the symmetry of their bonding, with intrinsic lattice voids, etc.. Therefore, if all fundamental physical quantities (e.g., length, mass, time, electrical charge) consist of an integer number of indivisible units and we try to develop a consistent mathematical model of physical laws, we would face formidable problems that had already hindered the development of mathematical physics in antiquity.

It is incontestable that the *continuum* hypothesis, apart from philosophical considerations, has provided great flexibility in constructing physical models that would have scarcely been possible without making *tabula rasa* of the limits imposed by rational numbers. We have seen that today, new mathematical knowledge (for instance, in the field of group theory) makes it possible to formulate certain physical laws using different formal approaches, but it is difficult to foretell whether future progress will enable a novel general language to be developed for physics.

4.4 The Space of *Quanta*

As mentioned in the preceding chapter, the development of quantum mechanics has required accepting a new, more general concept of space, whose objects, wave functions, explain and represent the undulatory nature of matter.

While in Euclidean space $\mathbf{E}^3$, the trajectory of a material point is described by a three-dimensional vector varying as a function of time, in quantum mechanics a particle is described by a wave function, whose square (or, more generally, the product with its conjugate) represents the probability of finding the particle in a certain state within an infinitesimal space-and-time element

In Chap. 3 we have shown how a wave function can be mathematically represented by the sum of sinusoidal vibrations of increasing frequency. This property was generalised by demonstrating that a wave function can be defined as a linear combination of a group of pre-established functions, which defines a function-space with own operators and metrics. Analogous to a Cartesian point P = [x, y, z], represented by a vector equal to the sum of its components along the reference axes, any wave function is represented by a linear combination, $\Psi(x, y, z) = \sum a_k \varphi_k(x, y, z)$, of a group of ortho-normal functions, where a_k are constant coefficients.[18]

The decomposition of a wave function into a linear combination of basis-function simplifies solving the differential equations which the wave function must satisfy and

---

[18] The integrals of the *square* of these functions extended to the whole space (their *norm*) must be equal to one, and the integral of the *product* of two *different* functions, extended to the whole space, is zero, These functions, suitably chosen, constitute a *basis* for space and correspond, in a certain manner, to the *versors* of the axes of reference in Euclidean space. Actually,we are dealing with a generalised concept of vector with an infinite number of components.

A *basis* can, for instance, be a class of simple orthonormal functions of type:

$$\{\sin(n\pi x), \in Z, n \geq 1\} \quad \text{and} \quad \{\cos(n\pi x), \in Z, n \geq 0\}$$

where Z represents the set of integer numbers. In fact, every wave function can be represented as a linear combination of these basis-functions, analogous to a vector, which can be represented by the sum of its projections on the reference axes. The space of wave functions (called Hilbertian space) is therefore defined by all possible linear combinations of the basis-functions.

allows for the establishment of an important analogy with the property of equations of classical mechanics in vectorial space.

However, the function-space is *not* the space of physical phenomena, but rather a kind of meta-space.

Let us consider again the simplest case of a quantum particle. Since the square of a wave function corresponds to the probability density of finding the particle in a certain point and in a certain state, in order to be able to reproduce the position of a material point in the precise position $[a, b, c]$ (i.e., to obtain $\Psi(x, y, z) = 0$ for $[x, y, z] \neq [a, b, c]$ and $\Psi(a, b, c) = \infty$ it is necessary to add an infinite number of (periodic) ortho-normal functions that can be interpreted as interfering waves. One may understandably find this procedure cumbersome, far from intuitive and, perhaps, arbitrary. Consequently, the question arises whether the space of wave functions corresponds to a choice of a mathematical object among others or corresponds, somehow, to a necessity for physical laws. The answer is still pending, but one thing is evident: the function-space appears in the context of differential equations as an analogy to complex vector-space in the context of algebraic equations.

We know that wave equation formalism in quantum mechanics was formulated in 1925 by Erwin Schrödinger on the impetus of the discovery of photons (Einstein, 1905) [104] and of the principle of equivalence particle/wave (De Broglie, 1924) [105], which states that the energy of a particle is proportional to an inherent vibration frequency, ν, which defines the undulatory behaviour of the particle.
We show here, in very simplified form, how one comes to the Schrödinger wave-equation:

Let an isolated particle of mass m and energy $E = h\nu = \hbar\omega$ be described by a plane wave propagating in direction x as a function of time, t, with a momentum $p = \hbar k$. This wave can be expressed as a complex exponential of wavelength, $1/k$, proportional to the particle's momentum, and of the periodic oscillation frequency ω, where $h\nu = h\omega/2\pi = \hbar\omega$:

$$\Psi(x, t) = e^{i(kx - \omega t)} = e^{\frac{i}{\hbar}(px - Et)}$$

One can see that the derivative of Ψ with respect to x yields the equation of particle momentum $p = \hbar k$:

$$p\Psi = -i\hbar\frac{\partial}{\partial x}\Psi(x, t)$$

Similarly, the second derivative of Ψ provides an analogous equation for the *square* of p.
The partial derivative of Ψ with respect to time, t, yields the equation for energy, E :

$$E\Psi = i\hbar\frac{\partial}{\partial t}\Psi(x, t)$$

We have thus the wave equation for a free particle, which corresponds to the classical equation: $E = p^2/m$:

$$i\hbar\frac{\partial}{\partial t}\Psi(x,t) = \left(-\frac{\hbar^2}{2m}\frac{\partial^2}{\partial x^2}\right)\Psi(x,t)$$

A further step is to consider the *total* energy of a particle in the presence of an external potential V; in this case one must add the potential energy, $-V$, to the kinetic energy, expressed by the second derivative term. We finally obtain the Schrödinger wave equation in its simplest form:

$$i\hbar\frac{\partial}{\partial t}\Psi(x,t) = \left(-\frac{\hbar^2}{2m}\frac{\partial^2}{\partial x^2} + V(x)\right)\Psi(x,t)$$

These equations are well known in the theory of wave propagation in materials, but in our case we have no substance underlying these waves. Furthermore, when a potential field $V(x)$ assumes a complex form, the solution, Ψ, of the equation may be much more complex than that of a plane wave and its features can hardly be interpreted geometrically. However, every solution of a wave equation must be a member of Hilbert space. This function can contain singularities, but only to a small degree, so that its *norm* remains finite.

Wave functions were thus inspired by undulatory mechanics and were consolidated by properties of function-space, which makes it possible to give a statistical interpretation to any wave function. Although there exists neither a medium that supports these "waves" nor a concrete physical meaning of their periodical "vibration", we can see, at least in an analogical way, that solving wave equations represents a method of establishing a relation between the energy of a particle and its "distribution" (or its density of probability) in space, as is the case in optics for diffraction phenomena of light, produced by phase differences of spectral components.

As has been previously stated, an unsettling aspect of the concept of wave function is rather the correlation that this function involves between its values in all points of space.[19] A correlation through which an event in point A can imply, at least in theory, an immediate corresponding event in point B, even if it is infinitely distant from A. It is worth examining this effect in greater detail because it fundamentally contradicts our perception of causality as ordered in a time sequence. Numerous examples exist which demonstrate this paradoxical side to quantum mechanics.

Let us suppose that two electrons are created in a point O of empty space in a state of correlation (*quantum entanglement*)[20] with parallel spin axes, and let suppose that the two electrons move with opposite velocities, one towards point A, and the other toward point B. The initial wave function describing the two correlated particles is given by a linear combination of the two states, one with both spins in one rotational direction, indicated by $|00\rangle$ (spin down/spin down) and the other, with contrary rotational direction, indicated by $|11\rangle$ (spin up/spin up). The wave function of the

---

[19] A quantistic correlation concerns two or more interacting objects, in whose inclusive wave function their states are, so to speak, "overlapped" and the objects have lost their individual identities.

[20] This is described by a wave function completely defined by the interaction potential of the two electrons.

electron pair with parallel spins is written as: $|\Psi\rangle = 2^{-1/2} (|00\rangle + |11\rangle)$, that is to say that the state in which both electrons are in spin down has the same probability as that in which they are both in spin up; the states of anti-parallel spins are excluded from the chosen initial conditions. One should note that this state is not merely hypothetical, but can be obtained experimentally.

In order to explain the entanglement effect in a simple way, we describe here an example, repeatedly reported in the literature.

We imagine two observers, one at A and one at B, called, respectively, Alice and Bob. We suppose that Alice measures the spin of the electron arriving at A and finds it in spin-down. Bob, who is at B (at a distance from Alice—even, say, of several light-years), will also have to find his electron in spin down, because, after the measurement done by Alice in point A, the only possible state of the electron pair is $|00\rangle$, in spite of the fact that Bob had, a priori, an equal probability of finding his electron in spin up. Bob would, therefore, instantaneously obtain information on the result of Alice's measurement, no matter how great their distance is. Yet, this contradicts the theory of relativity, according to which no information can be transmitted at a speed higher than that of light.

This simplest of possible experiments does not directly supply a method for instantaneously transmitting information at a distance, but, with some expediency, this can be realised (see the case discussed in Fig. 4.3). Obviously, the principle of entanglement can be imagined as applied to a wave function, $|\Psi\rangle$, as being much more complex than that containing information on the spins of two electrons. For example, it might be a codified message consisting of a packet of binary states. Moreover, since in quantum physics there are states corresponding to creation and annihilation of matter (e.g., one photon that is transformed into an electron/positron pair) one could imagine experiments with instantaneous disappearance and reappearance of objects at any distance. These are obviously day-dreams, but possible applications of entanglement effects are currently being studied, for instance, in computers operating with quantum-bits.

These techniques, called *teleportation*, are today objects of pioneering experimentation. Apparatus to carry out complex processes of entanglement are presently not possible because particles in correlated states can only subsist within a space completely protected from external perturbations. For most particles, the first necessary condition is their confinement in a shielded absolute vacuum. The one exception is represented by photons which, for appropriated wavelengths, can be easily produced in correlated states and transmitted at great distances via optic fibres.

The concept of quantum computers is indeed based on the possibility to generate quantum-bits through two distinct states of polarisation of photons. The advantage of quantum elaboration of binary data of this type is that a limitless number of input data (identified by a sequence of q-bits, e.g., large databases) can be overlapped creating a quantum-state *absolutely* protected from any sort of non-authorised "decoding", but containing all information of the input. This state can be *immediately* transmitted to distant correlated machines, where the state can be "filtered" with a key, provided separately, in order to obtain precise information, for example to find a particular datum or to execute certain operations. There are problems of enormous mathematical

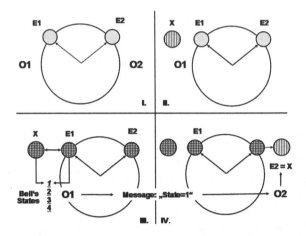

Fig. 4.3 Simple outline of instantaneous transport due to quantum entanglement. Let us consider the following experiment:

(1) Let two correlated photons, called E1 and E2, be present with undefined polarisation at distant positions and be possibly manipulated by two respective local observers, O1 and O2. The polarisation of photons can assume one of the two values, 0 or 1, corresponding to two perpendicular vibration directions.

(2) Let us suppose that observer O1 has on hand a third photon X of unknown polarisation and wants to transmit it to observer O2.

(3) To this end, observer O1 "overlaps" his photon E1 to photon X, correlating their states. Since these states are undetermined, there are four equally probable states (called *Bell states*):

$$\text{state1} \equiv 1/\sqrt{2}\,(|0\rangle_{E1} \otimes |0\rangle_{E2} + |1\rangle_{E1} \otimes |1\rangle_{E2})$$
$$\text{state2} \equiv 1/\sqrt{2}\,(|0\rangle_{E1} \otimes |0\rangle_{E2} + |1\rangle_{E1} \otimes |1\rangle_{E2})$$
$$\text{state3} \equiv 1/\sqrt{2}\,(|1\rangle_{E1} \otimes |0\rangle_{E2} + |1\rangle_{E1} \otimes |0\rangle_{E2})$$
$$\text{state4} \equiv 1/\sqrt{2}\,(|1\rangle_{E1} \otimes |0\rangle_{E2} + |1\rangle_{E1} \otimes |0\rangle_{E2})$$

Of these states, only the first corresponds to the case in which photons E1 and X have the same polarisation, (0,0) or (1,1). In the other three cases the directions of polarisation of the two photons are rotated one with respect to the other.

(4) At a certain moment, observer O1 determines experimentally in which of the four states his two photons are found: at the same moment, photon E2 will be found in the same state as E1.

(5) Observer O1 communicates in an ordinary fashion to observer O2 in which state he has determined his photon pair X-E1 to be (in the case illustrated one supposes that it is state. With this information, observer O2 knows that his photon E2 is an exact copy of X. If the state found by observer O1 were 2, 3 or 4, observer O2 from the result of the test done by observer O1, would know how he would have to rotate his photon E2 in order for it to coincide with X. One should point out that observer O2 *immediately* receives the information on the state of X at the moment in which observer O1 executes the test, and, for instance, can also elaborate it or transmit it to another observer. However, observer O2 can decode this information only after having received from observer O1 a message containing the result of his test; a message that can be transmitted only with finite speed. Note that the state of polarisation of X remains *unknown* until the entire procedure has been terminated but is absolutely *determined* at the moment when observer O1 has carried out his measurement

importance concerning security in communications[21] that cannot be solved for the current network trafic, but which could be definitively solved by using quantistic calculators.

Even though quantum-correlation phenomena are objects of experimental investigation, their significance is still being debated. It is well known that some physicists, among them Einstein, are convinced that the entanglement effect is a consequence of some hidden variables, which we are not yet aware of and have not duly been accounted for in quantum theory. However, in 1960 John Steward Bell (1928–1990) demonstrated an important theorem which asserts that the existence of a hidden variable is *incompatible* with conclusive results of quantum mechanics.

We cannot enter into the details of quantum correlation phenomena, but if their interpretation in the light of current formalism is right, we would find ourselves facing a concept for real objects of which no property, not even their existence, could be taken as objectively certain. If this conclusion is not mere mathematical fiction, but is to be assumed as a fundamental truth, we can imagine how many ontological arguments would be implicated among the philosophers.

However, the fact remains that the persistence of correlated quantum states is continuously threatened by every type of perturbation, from interactions with elementary particles to those with electromagnetic fields. The reason for such quantum effects being mainly confined within reduced space-time dimensions is a necessary condition of their intangibility, intended as the absence of any determination, intentional or accidental, caused by their surroundings, including the observer. In this regard we should carefully reflect on what Niels Bohr said:

"We can understand quantum mechanics only if we become aware that science does not consist in the description of what is Nature, but that it rather expresses what we can say about it".

On the other hand, however, Einstein, although he recognises that entanglement of two particles does not allow that these can exist independently one of the other, he remained firmly convinced that physics had to represent an attempt to arrive at a truth independent of the observer.

Questions and perplexing issues never stop being raised to the detriment of quantum physics, and people have tried and are still trying to find alternative theories. However, all attempts to reproduce observed quantum effects using different models have failed up to now. Research and trials continue to take place, but today there is the additional difficulty that models based on substantially different foundations from current ones can be taken into consideration within the established scientific community which, necessarily, acts as a guarantor and caretaker of orthodoxy in science. It must be recognised that, even within official science, some margins of freedom are maintained in order to allow new ideas to be formed, even in contrast with those already established. However, a protracted parallel development of mutually independent or even incompatible theories would in the end entail deleterious

---

[21] IBM has been conducting a research programme on this project whose main purpose is to guarantee that data transmitted across the networks of the financial world be undecipherable to possible criminal interceptors.

consequences, leading to concurrent schools faced with increasing difficulties of mutual communication and contrasting choices in terms of research objectives.

Outside of academic circles it has mainly been objected that certain current physical models are the fruit of an occasional coincidence of discoveries, which, carried on a wave of their first successes, have too rapidly obtained a general consensus at the expense of a necessary and deeper discussion of new hypotheses and principles.

This criticism is in part justified. For instance, one cannot ignore the case of the theory of relativity: it is becoming increasingly obvious that Einstein forced his pace in an attempt to surpass, using not always transparent methods, the results of Hendrik Lorentz and Henri Poincaré, who some years before had developed the formalism of *special* relativity, but interpreted it in a different manner. Conflicting statements made by Einstein in different circumstances have provoked the well-founded suspicion that he might have indeed taken their results and interpreted them in his own way.[22] Subsequent developments enabled Einstein to formulate the theory of *general* relativity in 1916, but what his starting intuition had been became even more obscure and whether his mathematical formalism was the product or the cause of them remained equally unclear. Even in this second phase, Einstein used the results of Hilbert on differential geometry of curved spaces and tensor calculation developed by Ricci Curbastro. which were available at that time. These merely mathematical results may have inspired a definition of *general* relativity. But even in this case it must be said that Einstein possessed to the highest degree an ability to deduce from the mathematical formalism physical significances and correlations.

Today the vast majority of scientific research deviating from the mainstream remains in the end relegated to a grey zone that extends between rare (but always possible) dazzling visions of solitary geniuses and the (frequent) fantasies of presumptuous dilettanti. Only a minimal part of this work emerges via suitable publications, enabling their contents and possible developments to be analysed by experts.

According to most dissenters, quantistic/relativistic effects are described by current theory as a sort of "distillation" of real physical quantities from a fundamentally redundant mathematical context, ensuing from the continuum hypothesis and its analytical objects, which are improperly taken as models of reality. If this criticism were valid, it would be necessary to question the mathematical language of modern

---

[22] It is strange that, in his strikingly innovative article of 1905, Einstein, at that time a simple, twenty-six year-old clerk in a patent office at Bern, though virtually unknown as a physicist, did not cite any bibliographical references, in contrast to the fundamental rules observed for scientific publications. Even stranger is the fact that the article was published without any objections in *Annalen der Physik*, a prestigious magazine edited by Paul Drude. The editor Drude, whose scrupulousness and scientific rigour were well known, was certainly informed of Lorentz's work, but he did not participate in the controversy that in the following years involved Lorentz, Poincaré and Einstein because he committed suicide for unknown reasons in 1906. However, Lorentz thought highly of Einstein, and a few years later in one of his lectures at Columbia University honestly admitted: *"Einstein's general and fundamental principle ... besides the fascinating boldness of its starting point, has a marked advantage over mine"*. Nonetheless, Lorentz still maintained his opinion that there was an aether, referred to which a resting observer is in a position of measuring *"true time"*. This delicate argument is considered by Edmund Whittaker , one of the most eminent historians of physics, in his fundamental work [71].

physics and produce a novel, different one, perhaps less flexible and powerful, but more directly connected to reality as we perceive it. This represents a formidable problem that cannot even be taken into consideration if formulated in these terms. It is instead probable that new mathematical languages can be developed in particular fields, as we have seen above, and from these an idea of a descriptive method of physics would take root, based on convergent or complementary approaches. One could express this idea by the analogy of a microscope the objectives of which can be changed according to the observation field of the object being investigated: an objective of maximum magnification does not allow us to obtain representative statistical information on a sufficiently large sample, while a low-magnification objective does not have enough resolution to put structural details of the object in question in focus. Every objective is appropriate for studying a certain type of detail, but general knowledge of the features of the object consists in creating a model that interprets, unifies and reassumes the images revealed at various levels of magnification.

4.5 The Antinomy Between Simple and Complex

In the preceding sections we have examined instruments of mathematics, which are needed to describe physical phenomena in space-time where the objects and all their attributes are expressed in terms of physical quantities and their measurements. The field of real numbers, to which physical measurements are related, was not the product of a deductive choice, but was rather imposed by the requirements of the mathematical procedures adopted to calculate the combined effects of objects and physical laws.

In the first chapters, however, we have shown that, intuitively, the correspondence between fundamental physical measurements and numbers does not at all require a hypothesis of a *continuum*, which resulted instead from a purely formal context.

This choice seemed to be the only one to successfully yield a complete scientific language. Yet, on the one hand, the more complexity is involved in investigating physical systems, the more the axiom of continuum in space and time becomes significant with regard to the limits of rigorous mathematical procedures; on the other, for extremely simple systems (such as, for instance, elementary particles), a hypothesis of continuum requires a formalism of increasing complexity while the contrary should be expected.

In order to understand the nature of this problem, we must consider its roots in a definition of fundamental physical quantities.

4.5.1 Time

Time, like space, is a fundamental quantity of physics, whose definition has given rise to recurrent and never resolved controversies. The notion of time originated from a subjective perception and memorisation of sequences of changes that were put in relation (more or less exactly) to cyclical occurrences of certain events (for example, the number of heart beats, the rotation of the hand of a clock, the periodicity of an electromagnetic wave, periodical astral cycles, etc.). That subsequent cycles of a standard clock do correspond to equivalent time intervals is a conditional statement. The possible availability of such a clock is justified by a number of observed coincidences, but, in the end, represents an axiom (which is questioned by the theory of relativity). However, independently of the answer to this objection, a fundamental measurement of time consists in counting some cyclical events occurring and manifesting themselves in the *presence* of an observer. The "flow of time", independent of this direct sequential measurement, is a pure abstraction, without which, however, it would be impossible to order our perceptions as a whole in a causal sequence.

The measure of time constituted one of the greatest problems of physics in antiquity. Eudoxus of Cnidus , an eminent mathematician and astronomer born towards the end of the V century B.C., was the first to define time as a _quantity_ ($\mu\acute{\varepsilon}\gamma\varepsilon\vartheta o\varsigma$, *méghetos*) and to assign the property of *continuum* to it. However, a few decades later, Aristotle, dealing with this problem in the last three chapters of the IV Book of *Physics*, seems to have been rather perplexed and appeared, no doubt, very cautious.

In these chapters, he starts from a definition of time based on uniform motion and attacks the formidable antinomy regarding the concept of *instant*: whether this is always the same or changes incessantly. Asserting that it remains the same is equivalent to denying a succession in time, while saying that it changes is equivalent to throwing a bridge across two instants and contradicting the continuity of time. Aristotle tried to solve the dilemma by asserting that the instant was, in one sense, always the same as it was before, while its *attribute* changed. The Latin translation is clear "*quod nunc est idem est quod unquam fuit, quod ipsi est esse, alterum*" (*Phys.* IV 219b 11).[23] However, this distinction is based on the happy syntactic ambiguity of the Greek text, where "$\tau\acute{o}\ \nu\tilde{v}\nu$ (*to nun*)" means "*the instant*", as well as "*that which is now*". Referring, therefore, to the uniform motion of a point on a line, he asserts that time is *not* the *sum* of the instants, if these are defined from a generic immaterial point which, moving, divides the past from the future, but rather from the *magnitude* of the distance covered by the material point, a point that remains *unvaried* during its motion. This definition of time results in being dependent on that of an empirical object (the material point) and the sequence of its positions in space.

The persistence of an object through the flow of time becomes, therefore, a fundamental axiom in the notion of time, contrary to the opinion of the Sophists who asserted that ("*...one is Coriscos at the Lyceum and another is Coriscos at the Agorà.*").[24]

---

[23] "*What it is now is the same as what it ever was, what is different is what is within it*".

[24] Aristotle (*Physica* IV, 11, 219b18) reported the opinion of the sophists who maintained that philosopher Coriscos is, in the two places, two different things. On the contrary, Aristotle's view is that we have a common *substrate* with different *attributes* at both places. The existence of a distinct reality from which change originates is a necessary condition for the intelligibility of change.

The philosopher Joseph Moreau (1900–1988), in his criticism of modern phenomenology, remarked that an objective representation of time as a succession of nows, defined by the motion of an observable object, consisted in a mental operation in which the *continuum* represented what we could call a *procedural limit*. This concept was expressed with great clarity by St. Augustine (*Conf.* XI, 28, 37–38), who placed the principle of continuity of time not in the now, but in the persistent identity of the subject who represents time to himself [6].

Aristotle defines time beginning from the measurement of a uniform motion ($\mu\acute{\epsilon}\tau\rho o\nu$ $\kappa\iota\nu\acute{\eta}\sigma\epsilon\omega\varsigma$, *métron kinéseos*) and, hence, he refers it to a phenomenon, to which he had previously attributed the character of *continuum*. However, he maintains that, since the measure of time is the result of an action of the mind, it is necessarily identified by an "$\ddot{\alpha}\rho\iota\vartheta\mu o\varsigma$, (*árithmos*)" i.e., with a natural number. The question whether time is continuous or discrete is presented here from a mathematical point of view in a different form than that used in previously outlined arguments, which were based on a superposition of ontological and phenomenological arguments. In fact, even admitting that the instant of time can be arbitrarily short, Aristotle repeatedly insists on the *succession* of the instants, that is to say, on the enumerability of time, asserting that it is not possible to simply identify the time instant with a point moving on a straight tine, because, in its realisation, time necessarily contains a "*before*" and an "*after*". The analogy would be, instead, that supplied by an oriented straight line which represents uniform motion, upon which the measurement of time projects a sequence of numbers, while mathematics stipulates that a continuous quantity, represented by real numbers, cannot be ordered in a sequence.

It is thus comprehensible, and also remarkable, that these concepts, which became clearer in modern times, had already been formulated and discussed, albeit in a somewhat confused form, since antiquity, by asserting that: the *point* is an "entity" to which is assigned a position in the space, that the point through its "*flussion* ($\acute{\rho}\acute{\upsilon}\sigma\iota\varsigma$, *rusis*)"[25] generates the *line* and that the motion of the point generates the *instant*.

With these arguments Aristotle undoubtedly intended to refute, on the one hand, the theories of atomistic philosophers, like Democritus and Leucippus , whom he cited in preceding chapters, and, on the other, the radical criticism of motion of the Eleatic school, represented by Zeno and Melissus. He finally specified that time was a quantity we number (*quod numeratur*) and not a reality with its own intrinsic metrics as space has (*quo numeramus*) and, in the last chapter of the same book, he concluded that only motion possessed a "*per se*" reality, whereas time, as "*mensura motus*", would *not* exist if spirit and mind did not exist.

The properties of time continued to represent a fundamental problem in all Aristotelian schools until the late Scholastics. The final answer resulted in a distinction between *imaginary* and *real* time, respectively defined as follows [96]:

"*Tempus imaginarium est <u>extensio</u>* (i.e., generalisation) *ex non interrupta serie <u>successionum motuum possibilium</u> considerata ut mensura omnium successionum*

---

[25] It is interesting that Newton used the word "*flussion*" to indicate the derivative of a function. From this usage we can imagine what importance Aristotelian speculation had on the historical development of infinitesimal analysis.

possibilium".[26] Whereas the definition of *"tempus reale"* is the same, conversely, one refers to *"motuum actualium"* and to *"omnium successionum realium".*

Note that here the adjective possible (*possibilis*) is meant as potential (*in potentia*), in contrast to actual (*in actu*). Real time is thus defined as the set of all real sequences (*successiones*) of motions and is therefore enumerable while imaginary time is seen as a limit notion, which may imply the modern concept of *continuum*. From this view infinite time (*aeternitas*) can be *real* if associated to an absolute absence of motion (or change); and can only be *ideal* if associated to endless motion. In Aristotle's opinion only real time is an object of physics and in the world of change infinite time is never actual, but only possible.

Criticism of Aristotelian concepts of time represents, even today, an inexhaustible source of reflection, mainly because it is based on an interpreting physical phenomena, which encompasses all the categories as we perceive them. We must also remember that even Eudoxus, who had affirmed that time had the character of *continuum*, appeared very concerned about this problem, refusing to define any physical quantity as a ratio of two "$\mu\acute{\varepsilon}\gamma\varepsilon\vartheta o\iota$" of different type (it was for this very reason that the mathematical concept of instantaneous speed was not used in physics until the late Middle Ages).

The attack against Aristotelian physics began at the end of the XVI century in the form of Galileo Galilei who cut the Gordian knot of the nature of time that had substantially hindered the progress of physics by giving a mathematical formulation to the laws of motion. Later on, Cartesius and, in particular, Newton demoted time to the rank of a mathematical parameter represented by a real number.[27] Thus, speed turned out to be the tangent dP/dt to a trajectory of motion in a point P, where, however, the geometric/trigonometric significance of the definition in question was progressively lost.

Newton's reductionism, which culminated in a physical determinism as imagined by Laplace, makes it possible to apply infinitesimal analysis to a description of motion and to associate the concept of causality to that of differentiability of the functions which represent the displacement of a point with time. However, the resulting absolute determinism which connects, without solving continuity, the past to the future, is today questioned by some fundamental aspects of quantum-relativistic physics, where time does not represent a simple parameter, but a co-ordinate of real space where it is fundamentally combined with space co-ordinates. The significance of continuum in this context has taken up, in new forms, the objections of ancient philosophers.

Moreover, the modern theory of differential equations demonstrates that determinism and predictability are independent concepts and that other mathematical

---

[26] *Imaginary time is a generalisation of a non-interrupted series of successions of all possible motions considered as measuring all possible successions.*

[27] Newton asserted that time and space were, respectively, eternity and the immensity of God. Their measure was then only a question of metrics.

concepts must be taken into consideration that describe situations that go under the name of "*deterministic chaos*".[28]

In conclusion, questions on the nature of time are still open and represent a playground of epistemological and philosophical debates whose resolution cannot be obtained using mere instruments of physics and mathematics.

Beyond still unsolved questions on the continuity or enumerability of time, another serious problem exists concerning its direction: One of the most disquieting properties of mechanics—both classical and quantistic—is that equations of motion are *invariant* with respect to reversal of a time axis (from t to $-t$). Therefore, the direction of this axis turns out to be purely conventional, like spatial co-ordinates, where the definitions of top/bottom, left/right and front/back are respectively interchangeable. In other words, an objective criterion does not exist to define the past and the future: every trajectory can be covered in two senses and, if we want, the motion of any system can be inverted. Our common sense says that this is not so, but is it because an objective answer inherent to laws of motion is lacking or because the answer is at least inherent to our interpretation of phenomena? Quantum-relativistic mechanics has not solved this enigma which has emerged as a fundamental difficulty in statistical mechanics, one of the more solidly-founded fields of physics.[29] During the course of our daily lives we are continuously being confronted with objects, whose internal and external motions entail apparently irreversible mutations. To these phenomena we have associated the concept of "becoming" to indicate an evolutionary one-way process. It is, however, clear that these mutations are typical for very complex objects, consisting of a great number of interacting elements. Calculating the behaviour of such objects using dynamics equations is, practically speaking, out of the question; but even from a theoretical point of view, for a system of equations representing their motion it cannot even be demonstrated that a mathematical solution exists. On the contrary, even for systems of modest complexity, equations of motion are known, whose integrals exhibit bifurcation points, in the face of which it is impossible to make a choice of deterministic nature.

The problem of verifying whether a system of a very many particles behaved in agreement with the laws of dynamics or not was faced rather late in the history of physics. The behaviour of such systems had been studied since the XVIII century and categorised as a discipline of thermodynamics. This speciality was soon of great importance, mainly owing to its mathematical rigour; but, until the end of the year 1800s use of thermodynamics was limited to systems in equilibrium. A further development was achieved with its extension to chemistry, operated by Josiah Willard Gibbs (1839–1903), but was still limited to systems in equilibrium.

It is true that defining entropy as the product of the degeneration of mechanical energy and introducing the concept of thermodynamic potential led to formulating the *second thermodynamic principle*, according to which in all irreversible processes the

---

[28] Different from random behaviour, deterministic chaos occurs where the very analysis of the deterministic equations predicts a divergence of uncertainty and non-observance of "strong causality.

[29] We can exempt motions caused by ultra-weak nuclear forces, but this does not eliminate the basic difficulty.

total entropy of a system, including its environment, must always increase, whereas in reversible processes entropy remains constant; however, any kinetic prediction concerning system evolution is at this point impossible because of the absence of a time variable in the theoretical *corpus* of thermodynamics.

Strictly speaking, time does *indirectly* enter in thermodynamics by defining the energy of the system's constituent particles. A fraction of this energy is kinetic, proportional to the square of the speed of the particles and measured, on average, by absolute temperature. However, it is not the absolute speed of the particles, but rather their relative velocity with regard to each other, due to their chaotic movements, which determines the frequency of their mutual collisions. A common group-speed dragging particle swarm does not enter into this average speed. The only kinetic quantity that is allowed for us to know is, in fact, a *scalar* quantity, whereby all information on directions of the real speeds, represented by vectors, is totally absent. The loss of this information is encompassed in the concept of entropy, a merely statistical quantity.

In this context, where a notion of trajectory is absent, an implicit theory of *thermodynamic time* emerges which inflicts a serious blow to the concept of absolute time. We shall see in the following chapter that if thermodynamics is explained by statistical mechanics, a door is opened to a twofold definition of time: one that is related to "*real*", microscopic motions of indistinguishable particles, the other to "*perceived*" changes of the average, *macroscopic* properties of their ensemble with respect to given starting conditions. This latter defines what we may call the *lifetime* and the *age*[30] of the system.

It was only with the work of Ludwig Boltzmann that an explanation of the "dynamics" of complex systems of particles was laid out in one of the most important and far-reaching disciplines of modern physics: statistical mechanics. Boltzmann was able to explain the time development of the properties of thermodynamic systems in states of non-equilibrium, resulting in a corpuscular interpretation of entropy. The fundamental Boltzmann equation is based on defining the statistical distribution of the individual properties of particles and provides a formulation of the variation of this distribution as a function of real time. The inferences and implications of this equation were most important for theoretical physics (they even contributed to the development of quantum mechanics), but did not solve the problem of "the direction" of time, which was only shifted onto a more complex plane.

The thesis that Boltzmann wanted to demonstrate asserted that isolated systems in non-equilibrium conditions evolve *irreversibly* toward equilibrium conditions, as a result of the laws of mechanics.

Let us examine the classic example illustrated in Fig. 4.5: let two different gases of non-interacting molecules be present in two separate closed vessels. After these are able to interact via an opening, the two gases become irreversibly mixed. Classical

---

[30] As in biology, the age of a system can be referred to a process starting from initial conditions (birth) and evolving toward final equilibrium conditions where statistically relevant changes cease to occur (death). The sequence of changes from birth to death defines the scale of the (non-linear) lifetime of a system. For an equilibrium (i.e., dead) system there is no way to relate its properties to conventional physical time.

Fig. 4.4 Henry Poincaré (1854–1912) and Ludwig Boltzmann (1844–1906) in photographs taken in their youth. In the last decade of the 1800s, at the apex of their careers, they were involved in a bitter dispute on the theory of statistical mechanics. Boltzmann had developed a ground-breaking atomistic model, which showed that the laws of thermodynamics, and in particular the Second Principle, corresponded to statistical properties of large sets of particles subject to classic laws of motion. On the other hand, Poincaré, who was superior as a mathematician, had demonstrated that the equations of motion of mechanical systems contradicted Boltzmann's irreversibility principle. In fact, from the point of view of physical argumentation, one can say that he was a prisoner of the rigour of the infinitesimal calculus that he faultlessly mastered. Only at the end of his life did he recognise the immense value of Boltzmann's ideas, which reconciled, with acceptable restrictions, the discrete character of atoms with the properties of continuum as described by thermodynamics

Fig. 4.5 Mixing of two types of gas which were initially separate. Taken singularly, every molecule is subject to *reversible* motion, but the total effect of mixing is *irreversible* in the sense that the probability of returning to initial conditions is in fact zero. In reality, there are cases (for instance, in systems under critical conditions) where this is possible

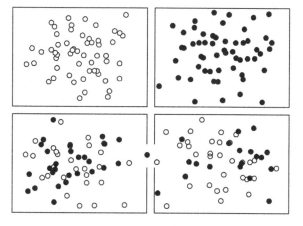

thermodynamics stipulates that, when separate, the gases have a lower entropy than that of the mixture and, therefore, the latter represents a state of equilibrium towards which the systems must evolve. However, in this picture, nothing can be said on the mechanisms and the kinetics of this process. Now, mechanical statistics predicts that:

1. The speed at which the system approaches equilibrium is proportional to the number of collisions per second between molecules, but it decreases with time;
2. Even at equilibrium the fractional concentrations of gases in the two communicating containers fluctuate around the final value;
3. The fluctuations of the concentrations decrease in magnitude with the increasing number of system molecules.

In the event examined, since the molecules are supposed to interact only through elastic collisions, no force fields are responsible for mixing the two gases. The effect is only of probabilistic nature. If, therefore, we assume in a Boltzmann equation an initial molecular distribution as in the example of Fig. 4.5 (i.e., the starting position and speed of each molecule in the two compartments) we can calculate, step by step, the evolution towards the final state of the mixture. The mixing speed will be, *in primis*, an increasing function of the starting molecule's speeds (i.e., of the temperature of the gas), but also of the collision mechanisms (collision cross-sections). If, moreover, intermolecular forces, even extremely weak ones, are present, the speed of variation of the statistical distribution of the gas molecules can change drastically. Therefore, a complex relation exists between the macroscopic time at which we observe a variation of the state of the system and the microscopic time at which we might observe the displacement of the individual molecules. Only this latter is interpretable in the sense provided by classical mechanics. The former, even if formally the same, is practically the product of the fundamental hypothesis of the supposed random probabilistic nature of elementary processes. One of the results of the Boltzmann theory is that a macroscopic quantity exists, described by a distinct function $\Xi(t)$ of the system state, which, for any initial condition and any possibly existent force field, never increases with time. This implies that the time differential, dt, can have only one sign and, hence, that there is only one sense in the direction of a time axis.

Boltzmann's results were the subject of lively discussion in academic circles of his time and objections were numerous. The most important concerns a contradiction using a famous theorem due to Henry Poincaré, which asserted that a closed mechanical system, after a sufficiently long time (called recursion time), had to return to a state as near as one wanted to its initial one. Boltzmann could not deny this serious criticism, but replied that the time in question was so long (for a perfect gas it had to be something like $10^{10000000000000000000}$ years!), so as to render the objection insignificant. But his answer did not convince all critics.

The argument was settled some years later by Marian Smoluchowski (1872–1917), but although there no doubts remained on the validity of the Boltzmann equation to describe the vast majority of physical systems, it remained, however, bound to the condition that a system had to have a sufficiently long recursion time. This was normally the case for thermodynamic systems, yet for most of these, there were particular conditions under which the relation between microscopic and macroscopic time was indeterminate. The meaning of time was, therefore, questioned in the evolution of such systems that exhibit aspects which could not be explained by the determinism of mechanical laws.

In his short—but dense—book on the meaning of time in physics, Stephen Hawking defines three concepts of time: thermodynamic, psychological and cosmological [35]. We have examined the implications that a reversal of the direction of time would cause in the evolution of complex objects. We do not know how mental activity takes place, but if we assume that, by inverting the time, the same holds true for neuronal currents, our memory would then view the future and regress with elapsing time, which would render the existence of intelligent beings impossible. As for cosmological time, Hawking, reporting the results of his models, affirms that, in a Big Bang state, the universe was highly ordered and, consequently, its expansion caused an entropy increase. Moreover, according to his cosmological model, even in a phase of final contraction, the entropy of the universe continued growing, albeit weakly. This corroborates the conclusion that for all three definitions of time, the *arrow* maintains only *one* sense. No other explanation exists why the sense is what we observe and not an opposite one, but that of a *"anthropic cosmological argument"*, which states that, if it were not so, our existence would not have been possible and we would not be here to ask this question.

4.5.2 Form

Nowadays, the success of *mathematical* models of physical reality has led science to progressively abandon *qualitative* models, based on geometric intuitions. Today it is currently believed that a qualitative model represents nothing but a rudimental and imprecise *quantitative* model. Laplace's absolute determinism has, in fact, yielded a conviction that the evolution of physical quantities can be described and predicted with precision by means of mathematical calculations, independently of our ability to visualise the sequence of transformations that constitute natural phenomena. However, a necessary condition enabling us to successfully apply any mathematical model is that the number of variables be limited, that is to say, that the described phenomena be isolated from their surroundings and relegated in a closed box, whose walls are resistant to external perturbations. Therefore, the privileged place for studying phenomena is the laboratory, where instruments and devices are constructed and used to isolate the essential effects under investigation. The characteristic tendency in experimental research to "decompose" complex phenomena and concentrate on their elementary constituents is essentially based on the conviction that, once these latter features are known, it would then be possible to reconstruct any complex phenomenon, provided that suitable measurement instruments and numerical sub-models are available: meaning we are basically faced with a practical problem. Obviously a limit exists for the complexity of calculable phenomena, beyond which a mathematical model is objectively inconceivable; but what the attitude of a modern physicist is when faced with this difficulty is represented by an important consideration by Paul

Dirac in the introduction to his renowned treatise on quantum mechanics [36]. He wrote:

"The main object of a physical system does not consist in the figurative represen-tations that it can supply, but in the formulation of laws that govern the phenomena and in the application of these laws in order to discover new phenomena. If then such a representation exists, it's good, but that it exists or not is of secondary importance".

Dirac was thinking in the first place of abstract concepts of quantum mechanics and of the impossibility of supplying a corresponding geometric representation, but he was also referring to complex systems of classical physics which do not admit other representations but those consisting of collections of formulae and numbers.

The question we must consider is, however, whether such a point of view is sustainable in view of our understanding of reality as we perceive it in our everyday experience. On this subject, the eminent mathematician and topologist René Thom (1923–2002) considered what reality would be like without its geometric intuition and concluded that, in its absence, the observer would find himself in the same condition as "Alice in Wonderland": where everything can happen without knowing why [37].

If intellectualising physical processes in a geometrical way were not possible, Man would have only two ways to escape: to entrust himself to purely intuitive inter-pretations or to fall into a resigned incomprehension and, finally, into a complete indifference, as is typical for animals. The geometric models we are dealing with today are not, obviously, simple graphical images of physical agents, but represen-tations of functions and behaviours, which characterise the morphogenesis and the evolution of the form of objects and their topologic structures.

The perception of an object is tied to its typical space of existence, in which the numerical precision of metrics is often much less important than the geometric and topological properties in question. Let us recall, for example, the innumerable phenomena of morphogenesis in physics and biology, from the macroscopic shapes of solids or fluids to the development of embryos.

It is improbable that their underlying processes can be formulated mathematically starting from the properties of the chemical bonds of the atomic constituents. On the contrary, a topologic model of the object can reveal its essential properties and the laws of its evolutionary process, whereby any alternative mathematical description is inaccessible.

Let us consider at this point how the evolution of a physical system can be consid-ered as a geometrical-topological problem and solved using different methods apart from differential equations of infinitesimal analysis.

In mathematics there are objects, called algebraic varieties, which, in simplified terms, consist of the set of solutions of a collection of polynomial equations. For instance, the set of solutions of the polynomial equation $x^2 + y^2 = 1$ is a variety representing a circle in the plane (x, y) and the introduction of a parameter $a > 0$ in the equation, $x^2 + ay^2 = 1$ (the equation of the ellipse), can be interpreted as the *process* of ovalisation of the circle. For polynomials of higher degrees, the geometrical properties of the corresponding varieties may be very complex. For example, Fig. (4.6) shows the solutions of a fairly simple 3rd degree equation with

Fig. 4.6 Contour plot of algebraic variety defined by the parameterised equation printed at the bottom. The zeros of the equation lie in the white areas of the plot. It can be seen that the combined variations of the integer parameters *a* and *b* produce the evolution of the system morphology passing through different topological structures

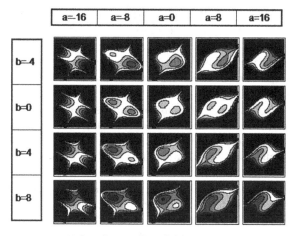

$$4x^3 - ax^2y + 9xy^2 - 9y^3 - 36x + 36y + 10b = 0$$

integer coefficients, two of which are parameterised by multiplying them with two integer variables *a* and *b*. What is interesting about these results is the variation of the topological features of the contours as functions of the variation of *a* and *b*. The general trend of these variations and the differences in configuration may be more important from a physical and morphological point of view than the exact values of the equation's coefficients and the corresponding zero's of the equation. For instance, the equation's coefficients may in some way represent the effect of certain physical quantities, of which we do not need to know the exact value in order to predict the morphological evolution of the system.

The branch of modern mathematics, which combines abstract algebra with geometry, is called *algebraic geometry*. The possibility of addressing scientific problems on a basis of geometrical models has enormous potential in many applications, from physics to chemistry, from biology to phylogenetics, and finally from statistics to economics, etc.

One of the most interesting domains where algebraic geometry is applied is catastrophe analysis where the evolution of a complex system is analysed in the context of force fields which do not fit in traditional formulation.

From this perspective, a system is defined as a general object, which can be characterised by the space, *M*, of its possibly observable states, *m*, and by a vectorial field, *X*, which determines the object's dynamics. Sometimes *M* contains a subset, *K*, in which mathematical discontinuities and/or singularities appear that invalidate the normal procedures of calculation of the "trajectory" of the object in *M*. The set *K* is called *catastrophe set* (the word is used here in its neutral etymological significance). As long as the object's state *m* moves in *M* without touching points of *K*, the object changes with continuity, but in the opposite case it is subject to discontinuous changes, that can only be explained by studying the singularities of *K*. Here one may recall, for example, the formation of a new phase in a chemical-thermodynamic system or the property of a fluid at a critical point, etc.. The cases of catastrophic situations are much more frequent than we are inclined to believe and even occur in simple systems. A case that we would not expect is given by the gravitational orbits of three, mutually interacting bodies, where it has been demonstrated that a great number of catastrophic points are randomly scattered on Keplerian orbits, which, in correspondence to these points, can diverge.

Various types of catastrophes exist, which, in a certain sense, can be defined mathematically. In some cases the main forces in X cancel each other out, whereas in others, secondary (elsewhere infinitesimal) forces, are subject to such amplification that their effect becomes predominant and unforeseeable. Using current terminology one refers to *attractors* and attraction *basins*. An attractor and its basin are sets defined as follows: if a trajectory of the object passes through a point of a basin, the trajectory will have as its final limit the attractor's set.

A catastrophe can be generated by the interaction between various attractors (conflict points) or by one single attractor (bifurcation points) and can also give rise to unstable, oscillating structures between adjacent attractors.

By applying methods of differential topology one can obtain a classification of different types of catastrophe and devise geometrical models of space-time, in which an explanation of the object's behaviour is given. The explanation is essentially qualitative since e.g., behavioural or morphological changes are concerned, but is at the same time rigorous, because it is deduced from the properties of the object analysed and from those of the existing attractors.

A simple example will serve to illustrate these concepts and show the limits of a mechanistic treatment.[31]

Let us then describe the flight of a bug attracted by a light source (Fig. 4.7) and provided with two light-sensitive antennae, the excitation of which is reflected as an increment of the vibration frequency of the wings; we suppose that the bug is confined within a two-dimensional space, where the luminous excitation increases with closing of the distance from a light point-source fixed at the origin of the reference axes. We consider here two types of bugs: For (1) the right-hand-side antenna interacts with the right wing and the left antenna with the left wing, For the other (2) the connections are crossed (right antenna with left wing and *vice versa*). We assume, finally, that the two antennae are sensitive and distant enough so that, given a punctual light source, they perceive different luminous intensities as functions of their variable orientation. According to this simple propulsion scheme, both bugs can only fly with speeds having a positive component in the direction of the two parallel antennae (in simpler words, the bug can only turn but never recede). We then formulated equations of motion for the two bugs. The light source represents a simple attractor, and the flight trajectories can be calculated starting from arbitrarily fixed positions and velocities of the bugs (two trajectories covered in a *short time*

---

[31] Our example was inspired by a cybernetics monograph by Valentino Braitenberg, on what he calls *"experiments of synthetic psychology"* [38]. Though we do have some reservations on his fundamental thesis, we find Braitenberg's ideas enlightening for the complex topics regarding the dependency of cerebral activity on the topological structure of neuronal networks.

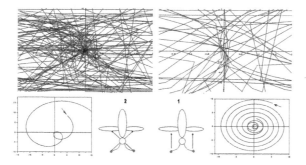

Fig. 4.7 Trajectories in a basin of attraction, represented by a light point-source situated in the origin of the coordinate axes , of two models of flying bugs with, respectively, parallel (bug 1) and crossed (bug 2) connections between their light-stimulated antennae and their wing motors. At the bottom, two typical flight trajectories of the two bugs as calculated by exactly integrating the equations of motion over a short time interval have been plotted. The two diagrams at the top show the trajectories covered over a much longer period of time, as obtained by applying a small statistical noise which prevented calculation arrest at instability points. The lower density of trajectory lines in the right-hand side diagrams is due to the fact that bug 1 has flown for the major part of the time in distant regions, far from the luminous source, while bug 2 exhibits a pronounced tendency to "attack" and hit the source

are plotted in two diagramss at the bottom of Fig. 4.7). However, the integral of the differential equation of motion displays many points of instability in which the smallest perturbations give rise to completely different trajectories. Therefore, we have solved the equations for *long times* applying statistical noise to the excitation of the antennae below which level the bugs in question do not react. On the other hand, the noise intensity concurs in the resolutive algorithm whenever in the effective driving forces become unstable in singular points. The trajectories covered over a long time by the two bugs are respectively shown in the two diagrams of Fig. 4.8 for distances near the attractor. One sees that bug (2) is often found in the vicinity of the light source, with a pronounced tendency to hit it (see a typical trajectory at the bottom left). Bug (1), although also attracted to the source, flies for much of the time in wider regions outside that plotted in the diagrams; only occasionally does it fly close to the source, striking it with circumspection (see a typical trajectory at the bottom right).

In behavioural terms, in the case of bug (2) we could speak of courage/aggression and for bug (1) of fear/flight. These judgements could be quantified using some statistical parameters deduced from the calculated trajectories, but it is clear that these are so irregular and scattered with bifurcation points that deterministic predictions appear useless. We should rather realise that topological properties of the action field (in our case attraction force and the parallelism or crossing of the stimulus/motor junctions) determine to a certain extent the object's behaviour. But this can be foretold using more general methods than that of solving equations of motion (see e.g., [38]).

The behaviour and evolutionary development of complex systems—in particular of living organisms—are regulated by a long succession of elementary catastrophes

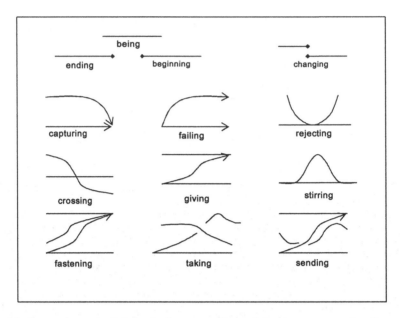

Fig. 4.8 Geometric-topological sketch of some paradigms of elementary catastrophes. Every field can be characterised by an algebraic structure that defines the evolution of the system in which the catastrophe occurs. This event involves a discontinuity in dynamic and morphological properties of the system in question, which cannot be predicted a priori using mechanistic models

that can be referred to a limited number of paradigms (Fig. 4.8). Every paradigm features its own algebraic structure that defines its topological properties. Geometric-topological models are the only ones capable of supplying an intuitive vision of the behaviour of physical (and biological) complex systems.

The attractor/catastrophe investigation provides methods to *explain*, for instance, processes of morphogenesis, but not to *predict* them, as one would expect from a deterministic model since the result of a catastrophe is *a priori* open to a certain number of different outcomes which can be described analogically, but one cannot foresee which one will actually occur. Deterministic models based on differential equations can hardly predict such outcomes unless one were to admit having resorted to unjustifiably arbitrary interventions or questionable simplifications. The final conclusion is that, in nature, mechanistic determinism is often an illusory tool.

4.5.3 Dimension

The variety and complexity of forms in nature is not only due to the great number of degrees of freedom available, but also to the ability of matter to reproduce, following predetermined plans, countless geometric figures often described by complicated mathematical functions, whose physical meaning, however, remains obscure to us. For example, the mechanisms of cohesion and aggregation of molecules can explain—though only partially—the formation of spatially homogenous bodies like plasmas, fluids, amorphous solids and crystals. It is, however, hardly possible, starting from a definition of atomic and molecular orbitals alone to predict the shapes and properties of certain aggregates, whose genesis is controlled by forces and interaction mechanisms which, when confronted with them, theoretical analysis is rendered powerless. Even atomic arrangements in simple dense fluids represent a riddle since their inner state dynamics may lead to complicated structures, sometimes periodic in time and space, for which no explanation exists. But the most important aspect of this variety of shapes is that some of them defy current notions of geometric extension and challenge methods of measurement of some fundamental physical properties.

In a Euclidean space, E^n, objects are primarily classified according to their topological dimension, n, defined as the number of sequences of cuts necessary to identify a representative element of the object: e.g., one sequence for a line, two sequences for a surface, three sequences for a volume (and so on for hyper-volumes in spaces of more than three dimensions). The geometric definition of topological dimensions of a body immediately leads to the intuition of a cellular structure of space, as formed by a compact set of equal elementary cells. It is interesting that if one wants to completely fill the Euclidean space, the cells cannot be perfectly isotropic, i.e., spherical, but, among the regular polyhedra, only the cube can fill completely the space.[32]

Since it is impossible to conceive a compact and isotropic space made of discrete cells of a given shape, a concept of "packing" is immediately entailed, according to an order, symmetry and regularity which atoms must obey when they fill the space. The geometric operations on which the identification of a ordered three-dimensional space structure is based are: translation, rotation around an axis and reflection with respect to a point, a plane or a line.

The symmetrical structures that appear in nature are divided into two classes, called, respectively, periodically regular and non-periodically regular. The latter are typical of biological organisms that develop individually from a nucleus around which peripheral zones grow by widening and modifying the external surface. Among these forms, those of central symmetry are the most interesting since their external surface sometimes assume the shape of regular polyhedra (Fig. 4.9) resulting from a rigorous process of growth determined by the axial symmetry of ideal geometrical figures.

On the other hand, physical periodic structures, typically represented by crystals, assume simpler macroscopic shapes, but cover a wider variety of symmetries. The positions of the points in crystalline lattices are obtained by applying to one point a succession of rotations (C) reflections (D) and translations (T), which finally bring the point back to its initial position in the lattice. These operations are obtained mathematically by applying to a point a sequence of linear transformations of its Cartesian coordinates which form a cyclical group. Although regularity imposes

---

[32] Gauss demonstrated that an assembly of equal spheres can fill the space with a maximum fractional density of $\pi/(3\sqrt{2}) = 0.74$. The cube is actually the only *regular* polyhedron with which the space can be completely filled up. The other possible polyhedra that possess this property are *not regular*.

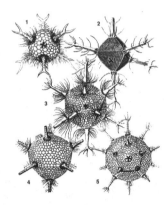

Fig. 4.9 Skeletons of various *Radiolaria*, whose shapes develop according to different symmetry laws. Those indicated with 2, 3 and 5 correspond, respectively, to three Platonic solids: octahedron, icosahedron and dodecahedron. The figure reproduces one of the splendid tables of Ernst Haeckel which were attached to the monograph of Charles Darwin's famous expedition (*Challenger Monograph: Report on the Scientific Results of the Voyage of the H.M.S. Challenger*, vol XVIII, table. 117 (1887))

limits to the order, k, of the rotations C_k (k can only assume the values 1, 2, 3 and 6)[33] there are exactly 230 independent space symmetry groups, subdivided in seven classes, which give rise to a huge variety of crystalline lattices. The possible lattice forms are thus determined *a priori* by symmetry properties or, if we prefer, by the properties of certain mathematical transformations that constitute a defined group. We find in nature crystals corresponding to almost all these groups even if some occur much more frequently than others.

The reason for which a given chemical compound AB forms a well-defined crystalline phase is qualitatively explained by thermodynamics: the crystal nucleates and grows if the reaction of formation of macromolecules: $nAB = A_nB_n$ with $n \to \infty$, if ordered in a distinct stereographic structure, produces a decrease of free energy. This can happen by ordering the mutual positions of atoms A and B, entailing displacement of electrostatic charges or distortion of the valence bonds. In the simplest cases it is possible to simulate the formation of a crystal with numerical models provided that the spatial dependence of the interatomic potentials is sufficiently well reproduced. In these cases the morphogenesis of a crystal can simply be calculated by numerically minimising the interaction energy of the constituent atoms. However, in the vast majority of cases, the factors that identify the crystal lattice of minimum free energy depend on very complex spatial features of valence orbitals, which cannot be predicted with the necessary precision by current theories.[34]

---

[33] Note that in lattices, for purely mathematical reasons, a rotation symmetry of order 5 does not exist, which in nature appears very frequently in non-periodic shapes (for example, in flowers).

[34] Note that the variation of free energy, ΔG, produced in a physical process occurring in a system at temperature T is given by: $\Delta G = \Delta H - T \Delta S$, where ΔH is the variation in mechanical energy and ΔS that of the entropy. Every spontaneous process has as a product a negative value of ΔG.

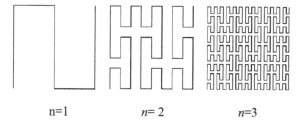

n=1   <i>n</i>= 2   <i>n</i>=3

Fig. 4.10 The figure shows the diagrams of a Peano curve. Constructed by proceeding through iterations ($n = 1, 2, 3 \ldots \infty$), it can be demonstrated that for n tending to infinity the curve passes through every point of the plane in which is defined; the curve is continuous but is *not* differentiable in all its points. Its fractal dimension is $D = 2$, i.e., that of a surface

A crystalline body is, therefore, an object essentially different from the predetermined shapeless mass of a physical *continuum*. Symmetry properties define its relations with other bodies and, in general terms, with its surroundings. The space of a crystal is actually pre-ordered and its three reference axes may be not equivalent, as in an ideal amorphous space, but possess individual properties. The physical interaction of two crystals in direct or indirect contact (for example, if immersed in a saturated solution of their components) involves a real conflict of two independent spaces, which always results in an increase of the crystal with more stable external surfaces at the expense of the other.

One should note that the atoms which make up a crystal are identified by triplets of integer numbers $[h, k, l]$, and, therefore, may be ordered in a linear sequence[35]; the lattice is thus topologically comparable to a folded-up rope. From this similarity arises a crucial difficulty regarding geometric measurements in crystals. In fact, in a continuous body size is defined mathematically as the sum of infinitesimal elements of different topological dimensions (differential of lines, surfaces and volumes) extended to the entire zone encompassed by the body's contour. For a perfect, but limited, crystal the contour should approximately correspond to a polyhedron whose faces are crystallographic planes, normally of low indices. Yet, in real cases, the free surface of a crystal is never in chemical and mechanical equilibrium because one half of the adjacent atoms and their attractive forces are lacking on the external surface). Therefore, the external surfaces may be uneven at a microscopic and sub-microscopic level, displaying convex and concave faceting which may render measuring lengths, surface areas and volumes problematic. In fact, experience has shown that the specific surface (contour area per unit volume) of powders of small crystals of equal average size can differ by several orders of magnitude. This enormous scatter cannot be explained by simple differences in the shape and size of the grains.

Since an ordering process in space produces a negative value of ΔS, it becomes only possible if the temperature, T, is sufficiently low.

[35] The sequence can start from any point, beginning with the polyhedron of its nearest neighbours and then of the second-nearest ones, and so on, by always choosing as the next atom the nearest one.

All matter in the universe is present either in crystal form or in amorphous or fluid phases, which tend to form a perfect single crystal, albeit, in most cases, very slowly, if their temperature is lower than certain thresholds. Therefore, there is a general process of ordering matter, in competition with a tendency to disorder, whose origin is the random motion of atoms. In this picture, the properties of continuity and isotropic, amorphous states correspond to idealised states in which the speed of the atoms largely prevails over the forces of mutual attraction.[36]

Moreover, microscopic and macroscopic structures of matter exist that are not periodic and for which univocal measurements of linear sizes, surfaces and volumes are actually impossible using conventional methods; for instance, spongy bodies, dendritic structures, jets of liquids in states of turbulence, etc.. Even in biology shapes are found whose extension cannot be measured, such as neuronal networks or systems of vascular circulation etc.. The difficulty inherent in measuring these bodies is that their contours cannot be defined with clarity because they are so intricate that they penetrate, on a microscopic scale, the entire body.

Let us take an example:

In one-dimensional space, a classic (and important) problem is measuring the length of a marine coast. This measurement, obtained from standard cartographic surveys, varies with the scale adopted: the more the scale increases, the more the measured coastal length between two fixed points increases.

Now, the pre-determined reference points can get as close as we need them to be, but a practical limit exists: in fact, it would be absurd to measure the coastal line between two points by examining the ground with a magnifying lens. The only criterion of choice of the minimum unit of measure (defined by a necessary approximation) depends on the various uses we can make of the length (for example, in sailing off-shore or along the coast, walking along the shore or evaluating local erosion phenomena). The minimal relevant length in a plurality of possible applications can vary by orders of magnitude and be quite misleading if used in an improper context.

Do concrete objects exist whose size is indefinite? In the example cited above one could answer that it is not the length which is indefinite, but rather the notion of a non-univocal coastal line. However, the answer does not eliminate the difficulty concerning the significance of *precision* in measurement methods. The dilemma may be even more radical from a theoretical point of view since geometric shapes of indefinite measure have been known by mathematicians for more than one century. Initially, they were considered as pure geometric curiosities or monsters. Their analytical representation consists, in fact, of continuous but non-derivable functions, objects on which one cannot operate using calculus and, consequently, were considered in the past to be of little interest.

Their importance grew, however, when one began to consider the surfaces of real bodies as assemblies of atoms, for which, in certain problems, it was impossible

---

[36] The speed of atoms and, in solids, their frequency of vibration around equilibrium positions, supply the foundation for defining *continuum* in thermodynamics, where measurements of quantities are obtained as averages of sufficiently large volumes and long periods of time.

to trace tangent lines and planes and non-derivable functions represented their only possible mathematical description. It is beyond doubt that, in many cases, geometric models expressed by derivable functions may provide a sufficient approximation for measuring lengths, areas and volumes, but the number of pathological cases, for which it must be admitted that this is impossible, has grown in the past few decades in an impressive manner. Today, from cosmology to cybernetics, from biology to hydraulics one encounters at every step of the way shapes that do not belong to a classic geometric-topologic classification. Some of them, called fractals, are characterised by being self-similar, i.e., by appearing the same at any magnification at which they are observed: a property fundamentally incompatible with the definition of a derivative. During the past years, the possibility of graphically reproducing these shapes using recursive numerical calculations has rendered them so popular (see the fascinating monograph by Benoît Mandelbrot [28]), that they are now used as standard instruments in computer graphics.

In practice, these problematic forms reveal the need for unconventional methods for geometric measurements of physical quantities such as, for instance, the effective extension of tortuous lines, wrinkled surfaces, and volumes of involute open porosity.

Actually, this problem had existed since the early 1900s when theory was being developed and was tackled by introducing new definitions of length and topological dimension, more general than the classic ones, based on the concept of an exact superimposition of a number of standard units on a measured object. Among them we report here a definition of measurement by Felix Hausdorff (1868–1942)[37]:

"Let S (the object) be a set in a metric space, E, and D be a number > 0: then the D-dimensional measure of S, called $s(D)$, is the smallest positive number, such as, for every r positive, the set S can be covered by a sequence of closed sets $U(r_i) i = 1 \ldots N$ (for instance, a surface covered by small disks or a volume covered by spherules both of radius r_i with $r_i < r$), such that $\Sigma r_i{}^D < s(D)$ ".

This definition seems complicated, but is easily comprehensible as seen by its application, illustrated in the examples reported below.

The sums $\Sigma r_i{}^D$, measured by covering the given set S, are evidently functions of the radii r_i and of their number N. For example, let us take polygon A of Fig. 4.11. If we measure the perimeter from point to point using a compass, starting with the spread of the compass legs (i.e. of parameter r) larger than the polygon's side, the measured perimeter results in being comparatively small; but if we gradually decrease the spread, the measured perimeter increases until reaching an asymptotic value that no longer depends on the spread, and corresponds to a *real* measure of the perimeter.

Polygons B, C and D represent successive degrees of indenting of polygon A. For these polygons the measure also grows with a decrease in r, but its asymptotic value increase with the indenting. The degree of indentation can be, ideally, endlessly extended, by iterating the mechanism of decomposition of the sides, as illustrated in the figure below. In this case, the curve $s = s(r)$, indicated in the diagrams with X, becomes a straight line, and an asymptotic value of the perimeter length *no longer* exist, but this tends to infinity with r tending to zero. The reason is clear: the

---

[37] The definition has here been somewhat simplified with respect to that developed by Hausdorff.

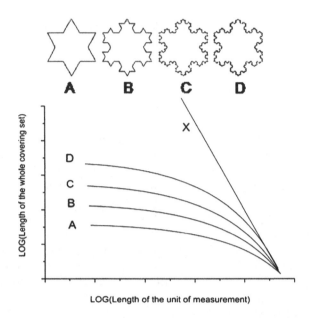

Fig. 4.11 Indented polygons that represent the evolution towards a fractal structure (known as Koch's curve), whose fractional dimension is $D = 1.26$. The curves of the diagrams at the bottom show the variation of the measured length as a function of the length unit used. In the limit case of a fractal (curve labelled X) the length diverges as the measurement unit tends to zero

zigzagging of the perimetric line becomes so dense that the contour tends to assume a *finite thickness* which, at its limit, is "filled" by an *endless* line.

This conceptual difficulty can be overcome by admitting that the contour can assume a *fractional* dimension D, comprised between the topologic dimension of the line ($n = 1$) and that of the surface ($n = 2$).

Various mathematical definitions of the fractional dimension, D, can be given; a simple one is expressed by the formula:

$$D = (\text{Log}[N(r)])/(\text{Log}[1/r]) \text{ for } r \to 0,$$

where N is the minimum number of covering elements $U(r)$ for values of r tending to zero. It can be easily demonstrated that in an Euclidean space E^n of n dimensions the relation holds:

$$\text{Log}[s(r)] = (D - n)\,\text{Log}\,[1/r]$$

That is to say that the logarithm of s as a function of that of $1/r$ is represented by a straight line, whose slope, P, (positive or null) determines the excess from the topologic dimension, n, of the object S:

$$D = n + P$$

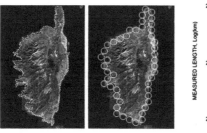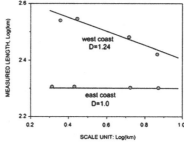

Fig. 4.12 A method to define the fractional dimension of an extremely indented line consists of covering the line with equal, adjacent disks of decreasing diameter and observing the dependence of the length of their chain as a function of their diameter. The slope of this straight line in a bi-logarithmic plot is given by $1 - D$, where D is the line's fractional dimension. The figure on the left-hand side represents the measurement of the coasts of Corsica. The western coast has a dimension of approximately $D = 1.2$ whereas the eastern coast has a dimension $D = 1.0$, i.e. that of a geometrical line. When the fractional dimension is larger than 1, the length of the chain tends to infinity as the diameter of the covering disks tends to zero. Obviously, this happens only in *fractals* that are ideal figures whose tortuosity increases endlessly when the measurement unit decreases. In real structures, the straight line plotted in the figure can sometimes be extended to local details of magnitudes down to the order of fractions of micrometer; but there is a lower physical limit for the unit size set by the structure of matter

Thus we may have in E^1 lines that are quasi-surfaces, in E^2 surfaces that are quasi-volumes and, generally, in E^n quasi-hypervolumes of E^{n+1}.

We are evidently dealing here with geometric shapes that are no less idealised than those represented by differentiable continuous functions. A real object cannot, in fact, have the shape of a fractal, since at dimensions below the order of magnitude of atomic size ($\asymp 10^{-10}$ m), any indenting process ceases to be applicable. the fact remains, however, that, between objects with details of such small magnitude and the largest cosmic bodies, the problem of their geometric measurement dramatically emerges from the need to define the relations between real object morphology and physical effects, which the shape with all its details does *in fact* produce.[38] Numerous problems in physics as well as in biology exist that must be analysed on the basis of these relations. The mathematical laws that govern the formation of fractals can be used in order to study important properties, like, for example, permeability, thermal and electrical conductivity of porous solids, fluid behaviour in hydrodynamics, catalytic reactivity and property of surfaces; moreover, in biology, they are used to explain aspects of the morphogenesis of complex living organisms or the connectivity properties of vascular or neuronal systems.

Thanks to modern computers, models have been developed in these fields where progress in topology, already available for a long time, has been fruitful in sev-

---

[38] For instance, a quantity of primary importance in chemistry is the specific surface, on which the reactivity of a solid reagent depends. Reactivity and catalytic property of a surface depend on this quantity, which can, however, increase by orders of magnitude when the reagent surface approaches fractal structures.

eral applications. However, the mechanisms which govern these processes in nature cannot be deduced from general laws of physics. The forces involved result from interaction phenomena, whose features exceed by far our capacity of analysis. Their origins are often rooted in the combination of space-time symmetries of macroscopic force fields together with local ones of the constituent molecules.

We must conclude that the study of the dependency of morphological and topological forms of physical bodies on the properties and behaviour of their atomic components is definitively leading to two opposite viewpoints, equally vital to understand physical reality and respectively based on opposing perceptions of *continuum* and *discretum*.

One could say, using the language of biology, that the dualism resulting from concepts of genotype and phenotype, between nature and its expression in a predetermined environment, is a reflection, on the one hand, of the order and symmetry associated with *discretum* and, on the other, of the shapeless aggregation of *continuum*. From these two extremes a variety of forms in nature and their countless mutations probably have to originate.

An important, final consequence concerns the impossibility of understanding real forms as products of predictable and expectable elementary mechanisms and the relative value (sometimes ambiguous) of fundamental geometric measures of the bodies, without which, however, any physical laws are indeed useless.

4.6   Ultimate Consequences

From what was discussed in the preceding sections, we must conclude that, in physics, only generalised geometric models are suitable to explain, even if not to exactly predict, the behaviour of complex systems, for which a deterministic description using purely mathematical models is impossible. This assertion regularly provokes bitter criticism from people who see in geometric models an implicit finalism, from which current evolutionist doctrines espoused by these critics fundamentally differ and, what is worse, they see in these models a sort of re-approach to Aristotelian physics and its tenet that the ultimate aim of all physical processes is the attainment of a pre-defined *form*.

Yet, the strength of this type of criticism is definitively declining, together with the belief in a basic simplicity of physical laws and a perfect correspondence between these laws and ordinary mathematical formulae. Understanding nature cannot occur without analogical and geometric models, and probabilistic laws must be an integrating part of any theory of the dynamics of complex systems. For several decades, the work of Ilya Prigogine and his school [29], starting from these considerations, has turned to the development of new methods of analysis open to new horizons.

There is no doubt that an Aristotelian definition of physics as a science of processes and changes has today gained ground over a Newtonian vision, in which the only possible change is that caused by the relation between force and acceleration/displacement which, beyond any apparent complexity, can basically be conducted to a simple, specifiable mathematical formulation. The more or less veiled accusation of Aristotelism, long a synonym for dogmatic ineffectual reasoning, is progressively losing its strength. In his work, René Thom [37] often referred to the ideas and doctrines of Aristotle and even to Pre-Socratic philosophers. In particular, he pointed out the deep similarity of the vortex dynamics imagined by Anaximander with his attractor models. In this vision, small peripheral vortices are generated around a greater central vortex, as envisaged by catastrophe theory, where the variety of forms in the universe appear with self-similarity aspects that are perceived in a specular correspondence between macrocosms and microcosms.

But Thom's speculation proceeded further, asserting that contrasting concepts of love-conflict and justice-injustice that Heraclitus used to describe natural processes, would not be the fruit of primitive confusionism, but deep and valid intuitions that today enable us to apply the same morphogenetic models both to situations of the physical world and to those governing the evolution of man and society. Thom observed that, once we have "geometrised" concepts like information, message and planning, every obstacle to their use in physics is automatically removed, and he went on, saying:

"I have reached the conviction that there are structures that simulate all the external forces of nature in the very heart of the genetic patrimony of our species, at the unattainable depth of the Heraclitean Logos in our spirit, and that these structures are ready to go into action every time it is necessary"

The ever popular idea that Man, the microcosm, reflects the macrocosm, has maintained its charm and strength: he who knows Man, knows the Universe. Thom recognised that for todays's mathematician this pronouncement represents a daydream, but he concluded:

"I accept this qualification, but is it not perhaps a day-dream the perception of the virtual catastrophe that has given rise to the human knowledge"?

Chapter 5
Orthodoxy Versus Heresy

*"Any activity lacking a scope is, for this very fact,
deprived of sense ...Without a scope, science
cannot even elaborate an idea of its own form."*

(I. R. Shafarevich, *"On Certain Tendencies
in the Developments of the Mathematics"*
from the Opening Lecture
held at the University of Göttingen in 1973)

5.1 The Germ of Irrationalism

The cultural climate in the Western World since the end of classical antiquity has been affected by two currents of thought: one of a rationalist stamp, the origin and development of which can be found in the Mediterranean area and the other was of a gnostic nature, which had already established itself, having started in the East, in the known world by the Hellenistic period. This latter way of thinking never gained the upper hand in the West, but neither did it ever completely disappear, on the contrary, it exerted a much greater influence than as one would expect in a civilisation and culture where reason has always been the main guiding reference point for action. While, in fact, science and Christianity tended to consolidate, despite deep contrasts and crises, into systems of orthodox, mutually compatible doctrines, subterranean gnostic currents always constituted an inexhaustible source of heretical thought, which, surfacing from time to time, alternately defied both faith and reason. For example, magical and astrological practices accompanied the development of science through the Middle Ages, and, even more openly, during and after the Renaissance.[1] The

---

[1] By compiling his monumental work on the connexion between magic and scientific experimentation Lynn Thorndike [86] was persuaded that from the Roman Empire through the XVII century social and moral customs in Western civilisation allowed for a peaceful cohabitation of these opposite approaches to investigate the laws of nature. He finally pointed out the great interest in the

C. Ronchi, *The Tree of Knowledge*, DOI: 10.1007/978-3-319-01484-5_5,
© Springer International Publishing Switzerland 2014

European intellectual class was almost entirely infected by this trend. Humanists of great reputation such as Marsilio Ficino and Pico della Mirandola, men of science like Pietro Pomponazzi, mathematicians and physicists like Gerolamo Cardano and Johannes Kepler, and even Isaac Newton, devoted themselves to magical or esoteric practices, to name but a few. In the very century of Enlightenment a variety of more or less secret societies embracing doctrines of mystery were ubiquitous throughout all of Europe. It was, moreover, vaguely believed that magic, through an awareness and investigation of hidden forces, could dominate nature. These were not physical forces acting in the universe. Instead the magical arts were felt to be capable of manipulating psychic forces and other influences that became accessible to the few select initiates of the craft. More recently, these cultural aspects became even more striking. Paradoxically, even in the nineteenth century, in concomitance with the establishment of modern science, one observed an unprecedented proliferation of secret societies and sects (neo-pagan, teo- and anthroposophic, millenaristic, occultistic, spiritistic, etc.) of disparate origins.[2] In the following century, some of these esoteric doctrines not only contributed to propagate abstruse non-scientific conceptions, but also exerted a significant political role in the birth of execrable totalitarian regimes.[3]

Like all gnostic doctrines, their vision of history consisted of a clash between good and evil, light and dark, from which emerged a select group, heroes and wise men, whose mission was to bring about the ascendancy of a hegemonic race and, finally, to shape the destiny of all mankind. It is remarkable that the roots of these doctrines had found particularly fertile ground in France, England and Germany, indeed, the countries where the cultural and social impact of modern science had reached its climax. These movements were born, and grew continuously, and could be found

---

(Footnote 1 continued)

occult which was particularly common at German Universities. We may here add that this cultural climate strongly affected a plethora of deleterious esoteric circles in the nineteenth and twentieth centuries.

[2] One has only to recall the prolific work of Rudolf Steiner (1861–1925) and his proposal of a "Goethian" interpretation of science as an inner perception. His ideas, ranging from medicine to mathematics, as fascinating as they may be, merely represent, from a scientific point of view, examples of fuzzy reasoning—although Steiner himself, who was familiar with basic scientific notions which he had learned in a course of engineering at the Technische Hochschule of Vienna, did often ask his followers to elaborate them in depth.

Following in Steiner's footsteps, Hans Horbiger (1861–1931) developed a fantastic cosmogony of a Gnostic-Manichean nature, which was very well-received by fathers of the Nazis ideology. Adolf Hitler himself was a fervent supporter of Horbiger's ideas since he believed that they provided a scientific basis for a cosmic vision of catastrophes and regenerations, in which cleared the way for the historical mission of a chosen German race.

[3] Among the numerous and variegated esoteric circles which arose in the second half of the nineteenth century, that of Helena Blavatski stands out. Blavatski was the founder of a Theosophical Society which yielded a plethora of neo-pagan movements, whose doctrines focussed on pre-Christians German and Celtic cultures. One of these was *Thule*, a German secret society that, in the first decades of the twentieth century, has woven together mystic elements of Pan-Germanism (*Thule* used the swastika as symbol). In Germany, after the economic and social disaster following World War One, a considerable number of high officers of the *Wehrmacht* found shelter in various *Thule* groups.

everywhere, starting with Masonic lodges of every sort and were well received in the upper social classes. Although of a secret nature, these circles did occasionally let escape some aspects of their esoteric doctrines outside, with the aim of opening the way to a selective proselytism,. By means of publications, which appeared to be scientific, this pseudo-science professed to reveal truths that academic science had ignored or, even worse, had intentionally kept secret. It was, in short, a standard ploy which had always been effective in attracting restless spirits in order to establish a future society based on a new science and governed by an eminent elite, caretakers of both truth and social order. Today, in addition to these movements of Masonic nature, a plethora of sects of Asian origin are flourishing which have experienced broad acceptance throughout the western world. The common denominator of old and new sects is a denial of "official" science, from cosmology through biology, accompanied by proposals of alternative theories, in most cases fully incompatible with orthodox ones.

5.2 The Critique of Rationalism

A number of modern sociologists are inclined to believe that the present mass of doctrines considered heretical by orthodox science might finally contribute in some way to progress, by revealing new perspectives, stimulating comparisons and criticisms and throwing light on unsolved problems, a similar process to what happened with the great religions during the configuration and maturation of their doctrinal basis. It is, however, much more difficult, today to establish whether this is true, or to what extent possibly permanent dissent can be tolerated in such a delicate matter. If, on the one hand, modern science cannot be accepted or refused as a whole, on the other, a comparison, and even an attack on its contents requires acquaintance with its rigorous and complex language, the product of a secular elaboration and of an almost unanimous agreement, but, at the same time, only accessible to a restricted class of specialists. Yet the majority of attacks on academic science have focussed on this very point, attacks which have been carried out by some modern epistemologists. Among these, Paul Feyerabend (1924–1994), a philosopher of irrationalistic tendencies, criticised science not only for having assumed an increasingly dogmatic character over the last three centuries, but also for having named itself arbiter and absolute governor of all human activities, justifying this tyranny by a scale of values which official science has elaborated and imposed on society [72]. This position, he says, puts science out of the reach of all its critics and, what is even more disturbing, tends to raise it to the status of an idol around which a dominant scientocratic class has gathered together in order to defend its dictatorship. Feyerabend claimed total freedom for methods used in gaining knowledge, and was convinced that true human progress could only be possible if this condition were met. With passionate impetuosity, he ended up by defending "alternative" sciences such as oriental and shamanic medicine, alchemy and so on, and including astrology and magic practices. Greatly esteeming Aristotle, he defended the value of common sense, which, apart

from subjective errors of observation, should allow us to perceive the true forms
of real objects. The senses, he asserted, can occasionally be affected by error, but
never be completely distorted, as claimed by Plato and his epigones. Modern science
is obsessed with the reproduction of particularities, where imprecision reigns with
errors of senses, but the laws of nature can be comprehended by healthy reason from
a sensorial perception of objects, which is to say without the monstrous apparatus of
modern science.

These are typical of critical remarks and proposals of the counterculture of the
1960s, with which many modern scientists might partly agree; but, in practice, insurmountable obstacles face such an ideal convergence of views. Science was not
conceived and nourished in an exclusive cultural climate, but has grown bypass-
ing through various civilisations. Through this passage its universal form has been
gradually outlined, by solving problems and removing antinomies, by searching and
researching. Its driving force resides above all in its success in a great number of
applications, from which all of society could benefit. However, the intellectual glamour of science was perceived by a very small number. The fundamentally aristocratic
character of science will always constitute a difficulty at which egalitarian groups
of all colours will bristle. True, a democratic society could be attempted by a scientocracy of platonic ilk, but science and democracy can also form an acceptable
cohabitation. Feyerabend argues that, while State and Religion have agreed on a
reasonable separation in our civilisation, Science has finally usurped all power from
the State. One could reply, however, that a democratic society can solve this sort of
political problem using conventional instruments available to the government. Critics
raise the objection, in addition, that orthodox science swallows up colossal amounts
of funds, out of proportion to its necessity and usefulness. This might be true, but,
in practice, we observe that it is usually public opinion asking for increased research
funding in the hope of benefiting from the reward of its technological applications.
It might be considered reasonable to share public funds on a larger research spectrum, encompassing alternative or even evidently heretical fields, but it is difficult
to predict which the consequences of such politics might be. One thing is certain,
however: in the case of exact sciences, the conflict between orthodoxy and heresy
cannot be considered as a dialectic process, in which, from the contraposition of a
thesis and an antithesis, a novel, more general and valid sort of knowledge will result.
A new science must encompass the old one as a sort of subset, but cannot completely
invalidate it. Neither can it conciliate theses and antitheses that are irreducibly opposite. A continuity of scientific thought might appear as a form of dogmatism, but,
since the very beginning of western civilisation, this continuity was perceived as an
indispensable guarantee against deleterious forms of intellectual anarchism and, for
centuries, constituted, likewise, an equally necessary braking and driving force in
the evolution of science.

5.3 "Normal" Science

The development of science in Western civilisation, after the fall of the Roman Empire and the decline of Hellenism and classical Latinity, lasted approximately 1,500 years: 500 of incubation, 500 of germination and 500 of growth and maturation. During the last two centuries its growth was so rapid that a complete knowledge of all physical laws in nature was considered to have been attained by the end of the 1800s. One was aware, of course, that there still existed things to be discovered and investigated, but one believed that the Great Book of Science not only contained the results of all investigations conducted and the answers to relevant questions, but also the laws and rules to be adhered to in any future research activity: the work of future generations would have flowed along a current of academic science, following a pre-ordained course dictated by a necessary and closed criterion of rationality.

In the early 1900s emerging quantistic and relativistic theories not only took up fundamental questions on the nature of scientific knowledge, thereby opening unexplored fields for experimentation and theoretical speculation, but also gave rise to radical critics who attacked the prestigious and at that point undisputed authority of academic science which was regarded as a perfect, self-containing system of knowledge.

In the 1960s new tendencies in the philosophy of science began to gain popularity among the young generations. With his theory of "scientific revolution" Thomas Kuhn [32] argued that scientific progress was nonlinear. According to his view, when certain conditions have fully developed, science is always faced by a radical crisis that compels us to question the paradigms of what Kuhn called "Normal Science", and to propose completely new ones. These ideas represented the point of departure for a true intellectual revolt that culminated in the extremist doctrines of Paul Feyerabend, who proposed irrationalism as a remedy to the scientific dogmatism that increasingly showed signs of intrinsic sterility.

We may or may not agree with these conclusions, but we must at least admit that the weak point of modern science resides principally in its pretension to proceed according to a rigid formalism founded on an axiomatic basis rather than in its demand for rational rigour. Actually, formal canons should be adapted to increasingly differentiated areas of research and alternative problems. The question is not to accept or reject alternative theories, but to consider the possibility of establishing arguments and procedures according to the types of problems encountered. In fact, the attachment of scientists to a canonical system is often justified only by aesthetic reasons and is sometimes accompanied by a total disregard for the functional nature of research work. These are criticisms that are generally valid for the history and philosophy of science, but have now acquired greater pragmatic weight. The solution to this dilemma is to be found at the end of a difficult course passing between the Scylla reef of a rigid academy and the Charibdis swamp of anarchical creativity.

The crucial question to be considered at present is what we can expect from the future of science? To be sure, interests in newly developed areas of investigation will probably expand: for instance, in advanced optoelectronic processors, micro- and nano-technologies, biological manipulations, exploration of space, etc.. However, these areas of investigations will involve a different kind of discovery than those still being pursued today. In fact, it is inconceivable that the number of physical laws can increase indefinitely. Very likely, within a few centuries, we shall be in possession of scientific models that explain almost everything, and the few explanations lacking will refer to rare peripheral phenomena, whose investigation will become increasingly difficult and expensive until it ceases to arouse our interest.[4] On the other hand, the experimental method on which natural sciences are founded follows an inflexible procedure that can be summarised as follows:

(1) A theory must be able to make precise predictions in a field broader than what had been explored experimentally; otherwise the theory is completely *useless*.
(2) A theory remains valid only if *all* its possible predictions correspond with *all* corresponding experimental results; otherwise the theory is completely *false*.
(3) One single discrepancy arising from *new*, reliable experimental data renders the theory *false*.
(4) Agreement with all existing experimental data does *not* guarantee that the theory is *definitively true*.

As one can see, we are dealing with a dynamic definition of theory, in the sense that its validity must be always corroborated by a continuous expansion of the field of experimental observations. These can be performed with the aim of validating or refuting theoretical predictions or can be used to explore a new field, for which the theory in question is not yet capable of making predictions. Without this continuous incentive to advance in line with experimentation, science is doomed to stagnate, playing only the useless role of explaining to us what we already know.

Under these conditions, our intellectual climate could deteriorate in the sense that scientific thought as we know it today could vanish, a thought which is neither an improbable nor a pessimistic vision. In 1965 Richard Feynman, concluding his series of lectures on the significance of science at Cornell University, expressed himself clearly on this subject:

> *Once we will know all of the physical laws, the vigorous philosophy and the precise attention will gradually disappear. The philosophers, of those who are always standing outside making stupid comments, will enter and begin to explain to us why the physical laws that we have found are right. And we will not be able to object them arguing that their reasoning does not allow science to make any progress, because there will be no more to progress. A degeneration of the ideas will then happen, of the type the great explorers perceive when in a virgin territory begin to arrive the tourists.*

---

[4] Let us consider, for example, the problems of experiments on elementary particles, i.e., on the one hand, the proliferation of unexplained phenomena that grows continuously with increasing collision energy attained and, on the other, the tremendous costs of the accelerators needed to create sub-nuclear particles, which, due to their conditions necessary for their existence and to their short transitory effects, are only interesting to a small circle of specialists.

5.4 Predict and/or Explain

Very likely, Feynman's acidic observation, reported above, would find immediate agreement from those who are directly involved in the arduous work of advancing scientific research. It cannot, however, be denied that the subtle psychological substrate of such a statement consists in a more or less conscious conviction, of a positivistic nature, which subordinates the validity of scientific knowledge to its capacity to predict mathematically the course of phenomena, as represented by the relation and the evolution of numerically defined physical quantities. But is it true? The question is still debated in the philosophy of science and, in particular, in discussions regarding its social implications.

Until a few decades ago the iron law of "the theory of confirmation" was in force, a law which is based on analysing logical-formal relations of propositions in science that contain concepts which are themselves in no way ambiguous. The academic quarters, organised according to various specialties, were acknowledged to be the caretakers of these very disciplines. Indeed, they acquired control of the complex contents of these subjects as well as their respective languages and research methods, on the basis of which they judged the validity or invalidity of new results. The epistemological foundations of the specific contents consisted of a set of experimental observations of indisputable certainty (the so-called *Protokollsätze*), from which theories and models are developed. In this climate, any suspicion of a possible "genetic fallacy" of a doctrine was to be excluded and considered inconceivable.

Starting at the end of the 1960s, increasing interest in interdisciplinary research began to manifest itself in all areas. In this new type of research one could not use the rigorous language of the specialists let alone their complex mathematical models. Instead statistics was applied more and more often to results and information which had already been analysed, both of a theoretical and empirical nature. This sort of statistical analysis took place and was often accompanied by substantial modelling activities, made possible through powerful computer programmes. In this context, rigorous mathematical formalism no longer appears to be essential, even though some details always had to be explained and justified using the tools of competence available in their original specialised domain. Therefore, one comes to a distinction between what Hans Reichenbach [5] calls "context of discovery" and "context of justification" [24]. Within the former context, knowledge is freely pursued under the influence of the most varied historical, social and even psychological vicissitudes. But when, on the basis of certain criteria, one can formulate a new explicit result, *justification* is called to exert its whole critical apparatus of mathematics and logic. From this perspective science displays different essential attributes. In particular, it does not have to be necessarily and exclusively predictive, but can also have a merely explicative value. For instance, the theory of natural evolution is not capable of making predictions on future mutations of a species, but is of great explicative value regarding observations made in disparate fields.

---

[5] Hans Reichenbach (1891–1953), an eminent philosopher, was founder of a current of scientific thought called " *Logical Empiricism*".

However, some modern philosophers of science who, like Reichenbach, agree unconditionally with *logical empiricism* maintain an inflexible positivist tenet, according to which only predictive ability confers scientific value upon knowledge. The famous diatribe between one of these philosophers, Carl Hempel, and Michael Scriven, a theorist of interdisciplinary research, illustrates this point because in order to maintain the supremacy of logical-mathematical methods even in an explicative scenario, Hempel does not hesitate to deny any scientific value to the theory of natural evolution [25].

Who is right? Due to the vastness of the philosophical issues connected with this question, any answer would entail endless objections. But it is to the credit of philosophers that they are more ready than scientists to take a step backward in order to widen the horizon of the speculations. We report here a subtle symbolic consideration due to Stephen Toulmin, a contemporary philosopher of science:

> *The image that comes to my mind if that of a kind of folk dance, which alternates periods of marching and periods of weaving. For a time, the different academic professions march forwards separately but in parallel, each in its own special way; then for a time, they join hands and work together on the general problems arising in the area where their techniques overlap, only to break away once more into separate lines and march along in fresh directions until they are ready to join hands again. This has happened before and, if it happens again, fine: this alternation is probably the only way in which we can preserve our scholarly and scientific concerns from either hardening into a permanent professional scholasticism, or softening into a morass of well-meaning imprecision. That is to say, it is probably the only way in which, in the long run, the academic enterprise can strike a proper balance between the legitimate claims of Truth and Efficiency, on the one hand, Goodness and Justice on the other.*

5.5 Is there a Limit to the Growth of Science?

A great number of complex systems, represented in nature by aggregates of physical and biological objects (or even of psychological and mental states) are subject to self-regulated growth, in the sense that they have a limited dimensional scope and in, most cases, their "size" (as defined in some manner) increases to an asymptotic value, D_∞. The reason is that their source of nutriments in their surroundings are progressively exhausted during the system's growth.

It is worthwhile considering some elementary mathematical models of growth that serve to point out the basic implications.

In the case where growth is *essentially* limited, the dimension $D(t)$ of a system, S, as a function of time obeys a differential equation in which the opposite of the second derivative, $-D''(t)$, (diminishing growth-rate) is proportional to the first derivative, $D'(t)$ (i.e., to the growth speed):[6]

$$-D''(t) = bD'(t)$$

---

[6] For instance, this may represent the case of a car driven at full power if we assume that air resistance increases with the car speed.

The function D of time, t, which satisfies this equation is:

$$D(t) = D_\infty \left(1 - e^{-bt}\right),$$

where b is a parameter representing, in a certain sense, the rapidity of the processes of *dispersion and loss* in the contents of system S. If at time zero the growth speed is c, we have that $D_\infty = c/b$ is the asymptotic limit of the size the system can achieve. The diagram $D = D(t)$ shows that the dimension increases rapidly in an interval of time of the order of $\tau = 1/b$, and then drastically slows down; for instance, at time $t = 3\tau$ the system reaches 95 % of its dimension limit and from this moment until infinite, it grows only of 5 % .

On the other hand, the simplest models of *unstable* growth obey to the equation:

$$D'(t) = cD(t) - pD(t)$$

where c represents the average time of acquisition of new contents and p that of the losses. This equation has as its solution an exponential function:

$$D(t) = Ae^{(c-p)t}$$

that says that a system with starting dimension, A, grows or shrinks, if c is, respectively, larger or smaller than p. The system remains stationary only in the improbable case where it exactly happens that $c=p$. Therefore, the model in most cases tends to predict catastrophic developments of the system (outbreak or collapse of S). However, it is enough to add a non-linear term in D to the above equation in order for the behaviour of S to become stabilised.

Growth equations have a quite general significance and can also be applied to abstract entities, for which a process of increase/decrease through absorption/loss can be imagined, for instance to scientific knowledge if we think of it as a system subject to processes of acquisition and oblivion.

It is difficult to define in terms of numerical parameters what science is to be nourished in order to increase in extent and under which conditions we can speak of losses in contents,[7] in the presence of which this theoretical "nutrient" can turn out to be insufficient to support indefinite growth. However, if we admit that this growth is limited, we can in some way analyse the nature of the possible limits.

We follow an outline proposed by Peter Medawar ,[8] where three possibilities are considered:

---

[7] We must consider that the "size" of our knowledge is necessarily linked to the size of the world of our experiences that defines the knowable space.

[8] Peter Medawar, an eminent and renowned English biologist, Nobel Prize laureate for Physiology in 1960, published works on philosophy of science. One of the most popular is a short, but dense essay entitled " *The Limits of Science* " [39].

(1) The first possibility is that the growth of science itself generates some kind of limit defining a critical dimension, which when science exceeds this, it goes to a standstill or even a dispersion stage.

(2) The second possibility is that the limit is of cognitive type, in the sense that the perception of a new or more complex reality with respect to that already known is definitively failing.

(3) The third one, of logical type, is referred to as the Principle of Conservation of Information that asserts that:

No logical process can increase the contents of information of the axioms, premises and empirical notions, from which the process itself proceeds.

Therefore, in so far as science rests on a definitive, comprehensive logical-mathematical model, it finally results in being intrinsically limited. It can be inquired in which form these limits may be present in our society.

Let us begin with the first case and take as a reference point models of uncontrolled population increase. According to Thomas Malthus [40], the density of population with a surplus of natality (source term positive) increases exponentially with time until it reaches catastrophic proportions. In reality, experience has demonstrated that retroactive factors depending on population density slow down the increase until it has been checked (source term equal zero) to much lower density levels than those of a Malthusian catastrophe. These factors constitute the first type of limit that characterises the simple asymptotic law of increase introduced above. Let us examine what these factors can be in the case of science, starting from Medawar's objections on their effective relevance.

1st Hypothesis: The volume of science has grown to such an extent that for an individual it is no longer possible, within his lifetime, to become acquainted with what is known and aware of what still remains to be known.

Objection: The problem is only of a technical nature. With the aid of powerful electronic processors all necessary information can be transmitted both to the individuals and to the collectivity when needed.

2nd Hypothesis: Science has been specialised and fragmented to such an extent that a total communication gap has been created between its various fields.

Objection: Commonly accessible science has never existed. Specialisation has always been necessary. Communication existed and will exist, if not at the highest levels, then at least as a source of mutual enrichment and interdisciplinary growth.

3rd Hypothesis: The frontiers of science are so advanced that the time necessary to train new generations of recruits who are able to continue the needed research work may no longer be available.

Objection: This would be true if progress depended on one single person. In fact, the progress of science is based on a subdivision of tasks and on a group synergy associated to continuously changing research objectives. Even long ago, single persons were unable to acquire a total understanding and mastery of scientific contents available at that time; nevertheless, individual as well as collective research projects have effectively progressed.

4th Hypothesis: The emergent applications and implications of science are overwhelming the moral stature of man who could destroy himself using knowledge, which should, therefore, remain hidden.

Objection: This argument does not have any logical value: a self-limitation of this kind cannot be a consequence of what precedes it—unless one hypothesises that science has developed to become an enterprise of improbable criminal associations, such as one sees in certain films or novels.

The issues raised in these arguments and counter-arguments are such that further elaborating the ones does not necessarily mean the others will be refuted.

Several historical events seem to indicate that the truth perhaps lies somewhere in the middle. This might be not very enlightening, but it seems to be credible that, under current cultural and anthropological conditions, the growth of science will effectively come to a halt. It is, in fact, undeniable that starting in the latter half of the twentieth century, one can observe, compared with the previous decades, a symptomatic slowing down of the rate of advancement of science—while its technological applications have entered a phase of unprecedented growth that has continued even now. However, even technological progress in various areas has been encountering negative responses and rejections in our society due to the ethical principles elucidated in the fourth point of Medawar's list. This argument, a scarcely inconsequential one, shall be resumed in subsequent chapters.

Continuing the discussion of Medawar's list, one can state that the limits described in the second and third point are of an epistemological nature and have more to do with the ultimate end of science than its historical process of growth. In order to illustrate them Medawar cites the following example:

(a) A biologist is in possession of a microscope with which, using an objective of highest optical resolution, he observes and orders the features and properties of certain bacteria. After classifying them, though not having a higher-resolution objective available, he decides to investigate additional, finer details by merely increasing the ocular's magnification. Consequently, the magnified image appears indistinct and does not reveal additional details. The biologist remains, therefore, in the dark about real sub-microscopic features of bacteria.

(b) An astronomer examines images of luminous bodies in the sky on a photographic plate. To count and catalogue them he uses first a magnifying glass, then he places the plate under a microscope and observes an image that contains a number of points that were invisible under the magnifying glass. Unfortunately, he realises he cannot decide if these tiny spots are images of real bodies or have been produced by impurities in the photographic emulsion.

In both cases, the micrographs represent "models" of reality. In the first one, the biologist is forced to acknowledge the limits of his model and knows that he cannot exceed a given optical resolution until a new, more powerful instrument becomes available. In the second case, the astronomer knows that his model is perfect and does not present obvious precision limits, but some of the observation data it provides may be artefacts.

In the first case we have a cognitive crisis, which relegates the biologist's knowledge within a narrow dimensional space. His instrument is not precise enough. If he had had on hand, for instance, an electron microscope he could have observed important details of bacteria as well as new areas of study.

The astronomer, on the contrary, has an instrument that, for the required purpose, supplies precise answers to the questions for which it had been constructed or purchased (i.e., to reveal the smallest objects), but he is not in a position to estimate the range of its useful application. His case is in many ways insidious, since the model does not give any warning; on the contrary, it provides a guarantee of absolute precision: in this regard there is no place for improvements or extensions. When observation is restricted to sufficiently large sizes where one knows a *priori* that there are no illusory data one would obtain perfect results, but by widening the field of observation, the gain in information *may* itself be illusory and the effects completely misleading.

Today there are many who think that modern science is being confronted with this type of limit. Scientific knowledge is organised and entirely contained in qualified models, which also provide criteria and the means for extending their own application field. However, the models become more and more rigid and new observations and data are encompassed by operating at their periphery and maintaining unchanged the axioms that constitute their central core.

On the one hand, alternative models with "heretical" grounds have no place in today's scientific development unless they go beyond the mainstream. Indeed, scientific communication is hardly available for these types of ideas. Even when it is possible, less qualified circles of scientists are reached who do not have the competence needed for constructive criticism and further development.[9] On the other hand, in academic circles controversies on central topics have become rarer and rarer while marginal ones seem to prevail which are of such limited relevance that one even does not feel the need to respond to them.

Yet, whatever our opinion on these problems may be, we must admit that the extraordinary flourishing of scientific knowledge seen during the last centuries in Western civilisation is presently coming to an end. The stage we are experiencing now has already adapted to a new, different civilisation. The Tree of Knowledge, transplanted in a vast planetary *humus*, might perhaps survive the expected radical changes, but, more probably, will reveal itself to be at last practically sterile—as many aspects of modern Western civilisation may—and produce its last seeds for a new stage of germination of which we presently know nothing.

---

[9] Unfortunately, we are forced to realise that, in the last decades, most of the ideas emerging from these circles, though sometimes yielding high but ephemeral levels of consensus, thanks to the media, become not only heretical, but also completely absurd.

5.6 Ethics and the Final Aim of Knowledge

At this point, we ought to resume the fourth hypothesis examined and countered by Medawar, regarding the dangers of a possible disproportion between advancement of science and ethical foundations of our society. Medawar's opinion is based solely on a logical argument: science does not contain any explicit link that in some way limits how it is used. Rather, the contrary is true: man can acquire knowledge before learning how to manage its applications. But this has nothing to do with its *contents* nor can this entail a negative prejudice of its *progress*.

It is precisely from this point of view that this problem must be examined today. In fact, the acquisition of knowledge and know-how in a global society is fast, much faster than that of moral progress.

As far as we can remember, scientific knowledge has provided society with intellectual faculties and technical instruments to exchange information and control the environment; but the idea that human progress is an increasing function of technical-scientific progress dates back to the philosophies of Illuminism and Positivism, respectively, of the eighteenth and nineteenth centuries. There is no doubt that this assertion has been questioned by past generations, since a great number of examples have revealed that scientific knowledge has, in some cases, lead to creating instruments and applications that are ethically ambivalent. The criteria needed to evaluate progress have changed considerably. The counterculture of the sixties had attacked the power of science and technology due to a belief that their negative aspects were becoming preponderant in the modern world. Theodore Roszak, one of the proponents of this Cultural Revolution was still writing in 2004 on the subject of the tremendous development of information and communication technologies, the crowning achievement among the achievements of the new millennium:

> *Far from destroying Big Brother, computers have given him even more control over our lives. They have been a blessing for snoops, con artists and market manipulators. They have turned global communications into glitchy, virus-plagued networks. Along with some highly valuable resources, the World Wide Web has also washed up a time-wasting flood of trivia, trash, pornography and spam. We have burdened our children with the distractions of becoming computer literate before they are even literate.*

However, despite some doubts and hesitations, and regardless of any ethical appraisal, once scientific knowledge has been acquired, the use of its applications is unavoidable and de facto irreversible. Let us consider, for example, the present condition of the industrialised nations.

Scientific knowledge has relentlessly demolished traditional social organisation by introducing an infrastructure of functions and technologies that support a unprecedented type of civilisation, which would collapse and become an irreparable catastrophe if these supporting features fail to progress further. In fact, developing countries, in particular, often characterised by high demographic growth, are more and more dependent on advanced scientific and technical know-how since they are quickly set to copy life models of industrial countries. However, boundary conditions are changing, since this objective is to be attained in a framework of a planetary society, where, for instance, new and more powerful energy sources must be available in order to cope with increased population mobility and communication, long-range transport of goods, sustainable processes of production, effective control of environmental conditions, etc.: demands which today's science and technology cannot satisfy, and whose fulfilment in a reasonable amount of time will depend on the combined abilities of future planetary populations. Some economists and sociologists doubt that this can happen and, in a rapidly evolving context, they predict periods of serious instabilities. From an ethical point of view, however, it would be difficult to justify limiting the spread of knowledge (even supposing this were possible) in order to slow down development and to maintain a *status quo* that would gratify only advanced countries. A crucial situation could manifest itself in various ways, in the sense that less developed countries might make discoveries and applications considered useful and licit, against which, however, serious objections of ethical character would be raised in more developed ones. The present disparity of cultures would be sufficient to create irreversible trends implying contrary ethical judgements (let us recall, for example, in biology, unlimited experimentation on genetic mutations; in botany, introduction on a large scale of genetically-modified alimentary plants; in chemistry, production of enormous amounts of new substances whose impact on the biosphere is for the most part unknown, in informatics, underhanded acquisition and insidious use of data regarding the citizen's private life; in physics, development of powerful, sophisticated weapons, like neutron bomb, destructive laser beams, electromagnetic guns; in cybernetics, development of killer robots and fully autonomous weapons[10]; etc.). The list could go on and on. The remedies might seem obvious, inasmuch as it would be a question of preventing, if not the discoveries, at least the proliferation of dangerous applications. But in a global civilisation such repressive actions are only conceivable and practicable where there is agreement in judgement and concord in attempts. Once again human civilisation is crossing one of the critical thresholds of its history, at a point where all appears fluid and possible, but the outcome uncertain.

---

[10] Robotic warfare is a recent, dramatic trend toward losing humanity [101]. Neuroscientists are presently devising the "cyborg-soldier", a hybrid man-machine, which can communicate and "interact" with a weapon system via cerebral stimuli. No doubt, these pursued objectives represent the slippery slope towards man's moral suicide.

Chapter 6
The Dissemination of Knowledge

6.1 The Problem of Communication

In every society new scientific knowledge is confined for a certain time to narrow circles and only later tends to become a common heritage; otherwise it is doomed to becoming forgotten during the course of one or two generations. To prevent this loss it is first necessary for the scientific community and society to be ready and able to receive the new message, to validate it and synthesise its contents. These are conditions which are not self-evident; on the contrary, they are often difficult to meet. The following story, a kind of parable, shows what can be the nature of these difficulties. While the example chosen concerns mathematics, the most structured science, the moral of the story and its ensuing considerations can be applied to every scientific discipline.

The hard way to the truth

A. is a capable, young mathematician, member of a recognised research university. For several years he has been working hard to demonstrate an important theorem. On average two or three times a year, he prepares a report on the activities he has carried out and on the progress he has made. In addition, he gives a couple of seminars in his department and a few presentations at international conferences, each lasting half an hour or so. As in many similar cases, although the topic of his theorem is important, nobody is capable of following in any detail the arguments developed by A. and it is because of these very arguments that A. has embarked on a long and difficult journey to demonstrate them. Then, one day, the solution comes to A. With trepidation he goes over again the stages of the procedure he had followed, he re-examines the hypotheses, boundary conditions and ancillary theorems, he checks all passages and, finally, is thoroughly convinced that his demonstration is correct. Full of joy and enthusiasm, he leaves his office and goes next door to the only colleague who is familiar enough with his work. With a pile of notes in his hand, A. tells his friend that he has finally reached his goal. The latter congratulates him on his success, asks some questions concerning certain details of his demonstration and finally advises him to send a letter to an important scientific journal and to write a full-length article as soon as possible. A. is an honest scientist and would like someone to verify his results, but it is not even conceivable that one colleague would have the specific competence and the time necessary to follow, even with the aid of notes, the road that A. has taken during years of hard work. However,

C. Ronchi, *The Tree of Knowledge*, DOI: 10.1007/978-3-319-01484-5_6,
© Springer International Publishing Switzerland 2014

since A. has the reputation of being a good mathematician, his letter is published. After a few months he has also finished his article. As is customary for scientific magazines, the number of pages is limited; A. must, therefore, compress the text, omitting a great number of details and producing a large apparatus of bibliographical references. The editor of the magazine receives the article and passes it to one or two referees for reviewing. These are sufficiently competent to understand the text and analyse the contents, but do not have the time to verify all the details—which would be necessary—hence they merely suggest a few, more or less marginal, improvements to the text. The article is finally accepted and published and A. continues to make his theorem publicly known.

Later on, A. is appointed to a new project and his main interest is now elsewhere. Many years later, towards the end of his professional career, A., while working on demonstrating another theorem, realises that some results contradict the thesis of the now famous theorem he had demonstrated in his youth. In some part, he thinks, there must be some sort of shortcoming regarding some restrictive conditions, or an oversight or even an error somewhere in some part. A. checks every possibility, but cannot find the origin of the discrepancy. Four possible cases are given: 1) the first theorem is true and the second is false, 2) the first theorem is false and the second is true, 3) both theorems are true but the criterion connecting them is false, 4) both theorems are false. In terms of logic, if absolute certainty of the results were a strict requirement, he should opt for the fourth case and, therefore, disallow both theorems. But he might rightly object that, by so doing he would also destroy a possibly valid part of his results, setting to zero the knowledge gained from the subjects investigated. In terms of probability, in fact, the choice should fall on one of the first three possibilities, but by not knowing which, it would be the same as making a partial choice. A. finally decides that the honourable thing to do, in the face of his own conscience and of the scientific community, is to say nothing[1] *and hope that somebody else will detect some kind of incongruence when using one of his two theorems and be able to solve the dilemma.*

If a deontological judgment had to be passed on the behaviour of the protagonist, we would perhaps find an ample spectrum of contrasting opinions. However, if we consider the effects of his choice on the advancement of science, we would be forced to recognise that any individual contribution, whether it be true or false, must necessarily pass through the filter of repeated collective applications before becoming a definitive part of our scientific heritage. In this stage, even an error may turn out to be useful and even fruitful inasmuch as it may help to confirm the truth, simply by considering it from a different standpoint. So that the reader does not think that the example cited is merely hypothetical and highly improbable, we cite here two actual cases:

Some years ago two teams of topologists, Japanese and American, simultaneously announced results concerning a new group called homotopy. Both sets of results turned out to be contradictory, but neither proof could be discredited because both were so complicated that it was impossible to find the error. Only later a third group independently put forth arguments in favour of the American thesis, which was finally accepted, but whether and where the Japanese results had goofed remains unknown.

The second case is even more dramatic.

In group theory, there is a fundamental theorem called "Classification of the Finite Simple Groups" which concerns the existence of a number of groups from which any

---

[1]   Note that the choice made by A. is the less risky one basing analysis of the situation on game theory (zero-sum game).

finite group can be constructed. The theorem has been studied by approximately 500 mathematicians who have separately (in time and space) carried out different parts of the demonstration. This demonstration is partly contained in roughly 20,000 pages published in various scientific reviews and partly written in unpublished documents. There are strong doubts that a single individual, even one of the principal authors, can have a mental grasp of the entire topic and be able to pass judgment unequivocally on the correctness and thoroughness of the demonstration, which is presently considered to be true- unless the opposite is proved (Fig. 6.1).

Today as yesterday, it is sometimes very difficult to recognise the value of new knowledge. The general situation is shocking: only with regard to mathematics, one estimates that every year approximately 200,000 new theorems are published in qualified journals. A good number of them are subsequently contradicted or disallowed, other are questioned with an overwhelming majority of results remaining de facto unimportant. Only a few are appreciated and accepted by a large number of mathematicians and enter into the common heritage.[2]

The processes of appraisal and criticism are often slow and hesitant, independent of the availability of competent judges. Moreover, high quality of underlying individual work is not a sufficient guaranty for the general acceptance and exploitation of the results at a collective level. If we confined ourselves to the field of mathematics, we can see that the road of its history is scattered with credited untruths and rejected truths, with quasi-truths recognised as such, but nevertheless universally accepted. More or less voluntary blindness, incomprehension and misunderstanding represent the ordinary obstacles that new knowledge meets in the process of its dissemination.

A classic case is that of Évariste Galois, a mathematician who lived in Paris in the first half of the nineteenth century.[3] Galois was an *enfant prodige* who attacked problems with a perspicacity and originality granted to geniuses before they are weighed down by excessive academic burdens. Galois had tackled the problem of polynomial equations using a novel revolutionary method that opened the way to a new understanding of the theory of equations. Nevertheless, neither the great Cauchy nor Poisson, who had been asked to review Galois' work, were able to realise the importance of the results of the young mathematician, who died at 20 years of age after having seen his more important articles rejected by the editors of mathematical magazines. It was Joseph Liouville who published them 20 years later without, however, having completely understood the depth and implications of Galois' theory.

---

[2] Discredited or overlooked theorems are not necessarily the work of incompetent scholars. For instance, for more than 350 years Fermat's Last Theorem was repeatedly claimed to have been proved by eminent mathematicians, with demonstrations which later turned out to be faulty (a correct demonstration was first elaborated by Andrew Wiles only in 1995).

[3] Évariste Galois (1811–1832) had a short life, tormented by his passion for mathematics and politics, and died in an absurd, yet fatal duel. On the night before this duel, Galois wrote some pages describing his discovery of important group properties. He eventually asked a friend to submit the manuscript to Gauss and Jacobi, the only ones he thought able to understand the significance of his results. The letter was delivered, but none of the two great mathematicians replied.

Fig. 6.1 The increasingly common use of printed paper produced already by the XVI and XVII centuries such an accumulation of books that one began to perceive even then the problem of handling a considerable number of volumes at the same time. The relation of man to books changed radically. Compared with old manuscripts, often viewed as precious luxury items, books became affordable and were printed in large numbers by skilled typographers working throughout Europe. In the face of the enormous mass of written information, mnemonic learning, which had been a basic skill in the previous centuries, soon became inadequate and the written word replaced memory, becoming an object of necessary, frequent consultation.
The figure shows a machine invented by Augustin Ramelli, an Italian military engineer to the King of France, Henry III, (*Diverse e Artificiose Machine, Paris*, 1588), in which eight small reading desks are applied to a large wheel for as many volumes, whose inclination remains constant while the wheel is turned by the reader, who can, thus, pass from one book to another without wasting time.

The problem of communication and dissemination of original scientific results concerns not only the complexity of their subject, but also the way in which they were obtained in the mind of their discoverers, who sometimes do not possess the necessary linguistic ability to communicate complex concepts to others. This is also true of mathematics, where one might think that a conventional symbolic language should be sufficient to express and demonstrate any proposition. A number of cases show that this is not true. A famous one is cited here:

The Case of Ramanujan

Srinivasa Ramanujan was born in 1887 in India in a small town of Tamil Nadu in a family of Brahmins of modest means. Already as a child in primary school he began to demonstrate a particular interest in mathematics, but was not allowed to attend secondary school. The conventional English schoolbooks that came quite by chance and only occasionally served to impart a few basic ideas on pure mathematics to him. They were, however, sufficient to enable him to conceive and solve complex problems and to formulate theorems pertaining to various advanced fields of mathematics. His activity granted him some local notoriety, but did not allow him to pass the entrance exam to the University of Madras because of his poor literary education. An application for a place as a researcher at the Indian Mathematical Society was also rejected by its Director, Ramachandra Rao, a distinguished mathematician (many years later Rao honestly admitted to have not understood what was written in a greasy sketch book that the young self-taught mathematician had handed to him). Ramanujan ended up by obtaining, with difficulty, employment as a clerk with the postal service, where, however, he could continue his work on mathematics in his free time. Aware of the importance of his results, he managed, thanks to English professors of the University of Madras, to send some of his notes to Cambridge, where these fortunately found a careful and able examiner in the person of Professor Godfrey H. Hardy. The manuscript of Ramanujan contained theorems and propositions that Hardy classified in three categories: 1) important results already known or demonstrable through theorems which Ramanujan was certainly not acquainted with; 2) false results (few in number) or results concerning marginal curiosities; 3) important theorems not demonstrated, but formulated in such a manner that presupposed views on their application field that only a genius could have. Ramanujan was eventually invited to Cambridge where he worked for some years with Hardy, who, in vain, tried to convince him to learn the classical foundations of mathematics and, in particular, the rigorous expositive method of mathematical demonstrations. Every time Hardy introduced a problem, Ramanujan reconsidered it ex novo applying unconventional reasoning which was sometimes incomprehensible to his fellow colleagues. Obviously it turned out to be almost impossible to draw any pedagogical benefit from the methods he was using. Ramanujan died at thirty-two, shortly after returning in India. How he succeeded in formulating his most famous theorems will forever remain a mystery.

For a very long time reports of similar cases (mostly pertaining to exceptional abilities to calculate) have occurred frequently enough to draw our attention to the existence of stable factors within our human genetic heritage responsible for cerebral functions particularly adapted to the treatment of mathematical processes.[4] Paul Davies commented on the presence of these super-abilities in our species [78], observing that, whatever the mutations were that created these genes, their persistence certainly could not be ascribed to adaptations to environment conditions. In order to survive in the jungle we do not have to solve differential equations or know the distribution of the prime numbers. The appraisal and control of our physical actions

---

[4] The most spectacular contemporary case is perhaps that of Shakuntala Devi. Her exceptional calculative abilities had already manifested themselves when she was a small child. Born in 1939 in Bangalore, the child of two circus acrobats, her gifts were put to profit by exhibiting her before a select audience. Later on, Devi made a lucrative profession out of these exhibitions, travelling from one continent to another. In 1977, during a demonstration in Texas, she obtained in 50 s the 23rd root of a number of 201 digits. In 1980, in London, she calculated in 28 s the product of two numbers of 13 digits, randomly chosen at the last minute by an electronic processor (thus excluding a possible fraudulent or telepathic communication with a bystander who knew the result). The time was almost entirely spent in orally reporting the answer, which would indicate that the result appeared to her *immediately*, without any intermediate steps.

Devi never tried to improve, or even acquire, basic notions of pure mathematics. Her abilities did not progress beyond a long series of successful performances in mental calculation of increasing difficulty to audiences of astounded students at high schools and universities. Eventually, in order to augment interest in her exhibitions Devi tried to adorn them with a mystic-philosophical aspect (the numbers would be living, well-meaning entities that introduced themselves spontaneously, and so on), which might be true, but did not enlighten us in the least on her mental state during these mathematical operations.

Shakuntala Devi is, indeed, an intelligent person with a good capacity for introspection, but there are hundreds of autistic subjects with serious mental handicaps who possess similar faculties.

which allow us to jump across a ditch are instinctive and do not happen by calculating the trajectory. What then are the dark ancestral roots of the presence in ours genetic pool of factors that allow for an extraordinary mastering of mathematical properties which, in turn, govern physical laws? And if it is possible to solve mathematical problems without using conventional language, would scientific communication and instruction still be possible ? These are questions to which we do not know the answer. From a practical point of view, however, they call our attention to the extreme consequences of the increasing differentiation of cognitive faculties among members of human society.

The problem of recognition and appraisal of new knowledge dramatically affect all modern science. The flow of information is so intense and multifarious that an immediate filtering and qualification of messages has become more important than could only recently be grasped.

In several quarters the question has been raised if one should at least drastically limit the flow of unqualified information, but the majority feel that complete freedom of communication sufficiently compensates for the negative effects of lack of value, errors, falsifications or even fraudulent manipulation of information. However, with regard to scientific knowledge, the problem is faced is practical one, of a logistic nature, namely how to review, verify and classify the tremendous mass of approximately 8,00,000 scientific articles published every year in major journals. Every one of these articles may be presenting, *a priori,* original contents that could contribute to expanding our knowledge, but the conditions of dissemination are extremely restrictive. Nowadays, when it is published, a standard article is thoroughly read usually only by a few interested specialists even when its contents are valid. Only very rarely is it broadly dispersed through the scientific community[5]; in most cases, within a few months the article is consigned to the archives for future, occasional readers. Only articles frequently cited in bibliographical references of subsequent publications have a reasonable chance of being traced back and re-examined. For the overwhelming majority, the archives, both in paper and electronic form, represent their tomb where they rest like improbable archaeological finds. In practice, the message contained in a publication either is immediately propagated or is almost definitively lost.

In the face of this problem an emergency strategy has been introduced, by attempting if not to assess, at least to safeguard against the total loss of the contents of all published scientific results:

Experimental measurements are collected in data banks managed by dedicated specialists. Contradictory data are, in general, examined and an evaluation is made, using whenever possible, objective criteria. Numerical models are constructed for inter- and extrapolation of data; these models are based on reliable theories and are calibrated and validated by using the best available experimental data. Critical reviews are compiled periodically on the status quo of distinct, important subjects. Implications and consequences of new influential theories are examined and applied in different research fields.

---

[5] Statistical surveys show that one article published in important, international scientific reviews is thoroughly read usually by some tens and, only exceptionally, by some hundreds of readers.

Number of Scientific Articles Published in the Period 1996–2006
(*Science News*, January 2008)

US	3,437,213
UK	9,62,640
Japan	9,83,020
Germany	8,88,287
China	7,58,042
France	6,40,163
Canada	4,73,763

Nevertheless, both production and management of knowledge is today very expensive and, in proportion to costs, hardly effective. Two aspects are at present particularly pernicious: the first one is unnecessary duplication and redundancy of research objectives; the second is the effective, albeit involuntary, penalisation of the quality of work since good papers are often hidden by a proliferation of worthless or useless publications. These faults represent a sort of collateral damage owing to two deontological principles considered as sacrosanct: the first concerns the freedom to choose one's own objectives in scientific research, and the second has to do with the role of public interest and consent, respectively assumed to be criteria of relevance and validity for scientific results. It is hardly a secret that in all research institutions financing of projects has become increasingly dependent on a sort of degenerated marketing, in the sense that objectives are affected by the expectations, often manipulated, of the public, and, consequently, projects are advertised as one does with industrial products. It is not only a question of utilitarian design, but of a perception that science, as a common heritage, must be submitted to a criterion of democratic management and appraisal. Unfortunately every scientific discovery is in and of itself of an aristocratic character. In fact, the result for a discoverer—similar to the final destination for an explorer—is not as important as the road he has taken and covered in order to attain the goal. Eventually, the object discovered may become common knowledge, but this road remains open only to a very few people. The distinction between the value of research and the value of the ensuing result is comprehensible since research consists of a complex mental procedure and only those who are capable of understanding this procedure can also claim to completely understand the object investigated in terms of its properties and connections.

Yet the method of discovery is actually the most intriguing and mysterious aspect in science. For example, in mathematics, to which all theories and models of natural sciences are referred, the main question concerns the source from which formulating the thesis of a new theorem first arises and followed by the methods and procedures needed for its demonstration. Whoever has attended a course in mathematics knows very well that the art of demonstration can be learned but cannot be taught. But how is it learned? There can be no single answer to this question: everyone has to find his own answer: through imitation, analogy, geometric images, aesthetic sense, oneiric visions or even, as in the case of Planck with his *quanta*, through "desperation" in the face of an imperfect current theory.

To give an idea of the psychological background on which mathematicians operate, two anecdotes are briefly described: the first one cited by Meschkowski [14]: A brilliant student of David Hilbert decided, unexpectedly, to switch to the Faculty of Letters. Talking with a colleague, Hilbert commented on the change saying:

"The poor man has joined the poets. For mathematics he didn't have enough fantasy!."

This creative force Hilbert was talking about has nothing to do with fanciful reveries, as in the Italian *"fantasticare"*, rather with disciplined projecting and widening of one's own insight towards completely new horizons. This mental activity, triggered by what we may call "inspiration", gives rise to, beyond any gratification arising from possible valid discoveries, a deeply felt aesthetic experience. Inspiration can originate from seemingly barren observations which then lead the way to illuminating visions when they find a corresponding receptiveness in the intellect in question.

The second anecdote concerns Gottfried Leibnitz, whose intelligence and greatness of culture made him one of the most outstanding European geniuses. He was fascinated by the idea that, in a binary numerical system, all calculations could be carry out by using 1 (the "whole") and 0 (the "nothing"). He saw in this a sign of Creation, in which the best of all possible worlds was modelled by a sequence of "yes" and "no" decided by the Creator. Leibnitz saw his intuition confirmed by his reading of an ancient Chinese treatise of cosmology known as *"Ging"*, where the existence of the two principles of *Yin* and *Yang* was described, which, respectively, represent even and odd numbers, whose innumerable combinations constitute a manifold morphology of the universe.[6]

There are many historical examples for irrational factors directly influencing cerebral processes which eventually led to important scientific discoveries, and, in particular, to solving complex mathematical problems. Why and how this happened remains a mystery for psychologists, all the more since there are testimonials by those having this type of experiences who were of high intellectual level. Jacques Hadamard, a French mathematician, has collected and discussed a number of these cases in an interesting monograph which has recently been reprinted [51].

If we examine the history of mathematics we realise that the mysterious sources of creativity have represented the main motivating force of its progress. If mathematics were simply an application of formal logic, a computer could be programmed to solve *any* problem and even completely develop the entire contents, a statement which is undoubtedly false. If it were true, the young mathematician of the parable reported above, could have translated his demonstrations into the current formal logic language and submitted them for step-by-step verification using a programme which, in the end, would have decided if the proposition of the thesis had the value "TRUE" or "FALSE". But understanding a theorem does not merely consist of checking a long chain of formal logic. This is merely a step in the direction of its comprehension and of complete confidence in its contents. This confidence requires a kind of adjusting the results to one's personal mathematical knowledge and implies gaining insight into the ideas which led to the construction of this particular chain of deductions over all

---

[6] There is here a striking similarity to the theories of Pythagoras mentioned in Chap. 1.

others. Furthermore, a formal logic test cannot be decisive because demonstrations of mathematical theorems are not analytical procedures in the sense that they merely develop information already contained in the hypotheses. The issue of the role of analysis or synthesis in the demonstration of mathematical theorems has historically engaged the best minds: from Kant through Frege to Russell.[7] Not all opinions are in agreement, but discordance lies mainly in defining the terms "analysis" and "synthesis".[8] In fact, everybody finally admits that in a demonstration there are essential intuitions that cannot be completely formalised (for example, the concept of "choice"). If one sticks to a common sense definition of analysis/synthesis, the majority will agree that mathematical theorems, which contain new information, and are not tautologies, are by nature synthetic. However, many maintain that the aims of mathematics are like platonic ideas, pre-existing truths that can be discovered, but not created. We must recognise that, in some cases, it is difficult to contradict this thesis (for instance, in the case of the above-mentioned discovery of the field of complex numbers, which are in perfect accord with the theory of algebraic equations).

The lack of automation of cognitive processes in exact sciences implies a centrality of creativity that eludes any scientific definition. Furthermore, one should not forget that from this perspective intuition is the creative original authority. One of the consequences is denying that scientific research can be programmed as a mass activity, by compensating with number of investigators for a possible lack of individual genius. On the other hand, some aspects of modern research seem to credit the thesis that the increment of research *activity* (for example, by creating new research foundations and increasing financing) is itself sufficient to obtain valid results. The successful use of computers in all research fields—which implies a mere increase in calculation speed[9]—might apparently confirm the automatism of logic and mathematics in pursuing research objectives. But this success is illusory, being based on confusion between *development* and *application*. Discussion of this topic is very widespread in current Western culture, but it is remarkable that one of the most lively and open disputes of this topic took place at the end of the 1970s in the Soviet Union, within an intellectual atmosphere where science occupied a primordial position in the hierarchical scale of social values.

Until that time scientific research in the USSR had always enjoyed unconditional support from the state; the number of active scientists (in mathematics and natural sciences) was[10] of the order of 1,400,000. What results and benefits were to be

---

[7] Russell's *Principia Mathematica* represents the highest achievement of formalists. However, although Russell succeeded in showing that, *in principle*, a demonstration can be reduced to formal symbolic deduction, *in practice* he was able to apply this method only to the simplest cases. If mathematics was to be constructed using this process, we would doubtless not have advanced beyond elementary arithmetic.

[8] The argument has been extensively treated by Ernst Snapper in his article of Ref. [41].

[9] One may recall that the computer's abilities are fundamentally reduced to adding two numbers, to recognise the greater of two numbers and to operate a conditioned choice of the next calculation step. The computer's power resides in the high speed of these operations and in the availability of large memories for storing intermediate data.

[10] Soviet History Archive, Slavic Research Center Library (2000).

expected from this army of investigators engaged in a vast research front was not completely clear. Thus, in a political atmosphere where the concept of productivity began to make its way to the socialist world, these questions went behind the curtain, giving rise to a dispute that was to involve the entire intellectual class of the Soviet Union. The cue was given by four articles published in the authoritative *Literaturnaya Gazeta* in issues from October 1979 through January 1980. The articles directly concerned themselves with research in mathematics, but raised a problem that concerned scientific research in general terms. In these articles opposite points of view were supported indicating that the discussion was taking place in political spheres looming over the individual research institutions. We report some of the more provocative affirmations:

– The modern age is facing a divorce between natural sciences and mathematics, whereby the computer represents the third party.
– Mathematical demonstrations of intuitively reasonable results merely represent a waste of time.
– Today all application problems can be solved by a computer; consequently, applied mathematics is superfluous.
– Modern pure mathematics is reduced to a form of art and, as such, only people who possess great talent should be encouraged to practice it.
– Teaching mathematics in schools must be taken out of the hands of mathematicians, and be entrusted to computer experts since students will never be asked to solve problems following a rigorous mathematical formalism, but rather by using computers.

The article continued by citing some "incontrovertible facts" (*sic*):

– In the last 50 years none of the discoveries in mathematics has found an application in natural sciences.
– It is improbable that in the future somebody will benefit from new advances in mathematics since the already difficult communication with mathematicians will become absolutely impracticable.
– In the next 50 years modern mathematics will be completely forgotten by 99 % .

The mathematicians, directly addressed and concerned with the matter being discussed, countered critics by citing examples of recent theories that have found ample applications in physics (*e.g.*, operator theory, groups, topology of non-Euclidean spaces, etc.). Nevertheless, except for some exaggeration in the level of accusations, the impact of the blow was hard and convinced them to formulate a defence based on three points, by which one could cast some light on future developments of research in this field:

(1) Reform of mathematical education in schools, whose methods, in the opinion of many specialists, have not advanced beyond levels of the seventeenth century: it is necessary to reorganise instruction by placing emphasis on methods of formulation of problems, construction of models and numerical solutions of problems using computers.

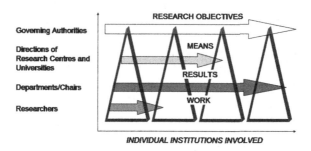

Fig. 6.2 The *graph* illustrates the quality of scientific communication between modern research units at different hierarchical levels within a project-oriented organisation. The length of the *arrows* indicates the magnitude of the communication area and the *shade* of *grey* the amount/quality of its scientific content. One can see that scientific knowledge is concentrated in hierarchically intermediate units, where competence is available together with greater possibilities of information exchanges

(2) Restriction of research activities in mathematics to particularly talented and creative people.[11]

(3) Maintenance of a broad scientific *"middle class"* dedicated exclusively to applicative problems, in particular to modelling and numerical calculation. Freeing this class of any obligation to publish (the ill-famed principle of "publish-or-perish" which currently holds sway), for this obligation represents the principal cause for the proliferation of articles lacking any intrinsic value (what is usually called junk literature) and, instead, being obligated to continuously widen individual competencies.

These kinds of questions were finally extended to scientific research in general and, though expressed in less draconian terms, criticism presently involve research in all countries of the globe.

As for the specific question regarding what is trustworthy and what is valid in the continuous flow of new results of scientific research, before expecting an answer one has to find an adequate interlocutor. The pyramidal hierarchical structures, which in the past reflected a concentration of competencies towards the apex of research institutions, are gradually being replaced by managerial structures in which the highest hierarchical levels are exclusively dedicated to what we may call marketing of objectives and results, where decision-making criteria are no different to those in a free play of demand and supply (Fig. 6.2).

Research, both public and private, is almost totally oriented and channelled into "projects" whose objectives consist of practical achievements, mainly of a techno-

---

[11] A policy which could truly be put in place in the USSR at that time. Yet, in democratic countries where the right to education and complete freedom in the choice of one's own profession are guaranteed, this condition would hardly be possible. Actually, in democratic systems there is an inherent tendency toward cultural homogeneity (and mediocrity), even though the average intellectual level tends to grow; in this context, all kinds of qualitative selection, of both people and type of education, eventually become spontaneous, but are slow and expensive.

logical character. The basic work for the progress and consolidation of science is entrusted to executive cadres organised into small units (university chairs, laboratories, departments of research institutes), whose power resides in their mutual contacts and ability to operate at interdisciplinary levels. It is through information exchange that allows the processes of criticism, validation and absorption of innovative knowledge to take place. In most cases, an appreciation of new results does not occur through any sort of rigorous monitoring, but rather by examining the consequences of their implementation in the context of disparate applications and comparisons.

Thus a new idea, which first seemed of little interest, may turn out to be fertile for new developments in fields which are far from those where the new development had been initially conceived. The criteria needed to appraise these developments are often based not on rigorous mathematical demonstrations, but on ascertaining whether the hypotheses are reasonable and the consequences plausible which themselves possess a demonstrative capacity. We are considering here a criterion founded on a powerful conviction that natural laws are of general validity and that, consequently, a mistaken idea is in the end doomed to be confuted by an experiment. This is something more constructive than Karl Popper's statement that a theory must be considered valid until its predictions are contradicted by an experiment. From a strategic point of view, new theories, with all their corollaries, should be applied in all possible fields and their predictions confronted with results of diverse types of observations.

Therefore, a social aspect to scientific progress exists in modern civilisation, which in the past had only played a marginal role. Genius, intuition and imagination, on the one hand, and interest and competence, on the other, contribute together to the germination and growth of new knowledge until this is eventually integrated in a common heritage.

6.2   Knowledge and Society

What the parameters are that initiate and control the progress of knowledge in our society is a question which modern historians and anthropologists have spent much effort attempting to answer. It is enough to turn our attention to the history of science in order to realise that progress has been slow in some civilisations and, for thousands of years peoples only developed rudimentary knowledge and techniques barely sufficient to maintain an almost unchanged intellectual level and standard of living. On the other hand, other civilisations flourished and matured in the shortest of times, of an order of a few hundreds of years. What caused this difference? Under what conditions does the potential of the human brain unfold in a wonderful variety of cognitive elaborations?

To be sure, there are innate cerebral structures that allow us to organise our perceptions on a basic level of connections which, in turn, defines *a priori* categories such as space, time and causality as well as unity, variety, and equality of the objects perceived, etc. In the millenarian history of gnosiology, different theories have been developed by various philosophical schools, trying to explain the bases of the evolution of knowledge, but none can explain how and when knowledge acquisition was realised.

From the point of view of individuals, the stimulus to elaborate and acquire knowledge varies from case to case. It is generally accepted that the average quotient of intelligence in sufficiently large groups does not depend very much on race, but this does not imply that, with intellectual *potential* being equal, an equivalent *developmental* as well as a high intelligence coefficient in a young individual will necessarily mean that he will acquire exceptional knowledge. In fact, a number of emotional and irrational factors determine the development of individual knowledge.

Two external agents initiate and sustain the process of elaboration of knowledge. The first one is bound to the stimuli of the person's surroundings where the individual is confronted with dangers and difficulties which can be avoided or overcome only by care and due application of talent. If the problems encountered can actually be solved and are proportionate to the means available, a continuous progress of knowledge takes place in the individual within the context of his interactions with the environment. Numerous examples exist in history that indicate that the most advanced civilisations flourished in geographic areas with moderately unfavourable geologic or climatic conditions, where, for example, agriculture demanded complex irrigation systems (Mesopotamia and Egypt) or optimal exploitation of the seasonal cycles (China, Indochina, India, Peru, and Mexico); or where the territory was fairly barren and could scarcely be cultivated and supply of primary goods was necessarily bound to sea travel (Phoenicia, Crete, Greece).

On the contrary, civilisations which developed in naturally fertile areas or in places where resources were constantly available often hardly seemed to have improved beyond the hunter-gatherer stage (North America, parts of South America, Australia, central Africa).

On the other hand, where environmental conditions were too unfavourable, knowledge and talent could turn out to be insufficient to solve problems of adaptation which were simply too challenging and civilisation very rarely succeeded in developing beyond what was necessary to survive (arctic regions, Siberia, Patagonia).

The second aspect is of social nature and concerns the interaction of the individual with the heritage of knowledge of the group in which he lives or which is available to him. The production of new knowledge through the elaboration of what was already acquired in a group is a complex process, especially in well-developed civilisations, where an individual can only be confronted with a small part of thecollective cognitive patrimony. In such a context it is obvious that the most important factor is communication: starting from instruction of new generations to evaluating and disseminating newly acquired notions.

When we consider the progress of science in the past centuries, we are often astonished at how effectively new notions were discovered, accepted and divulgated by a restricted group of "man of learning"—corresponding to the current larger "scientific community"—a scattered group (in total perhaps a few hundred contemporary individuals), who were working separately, at great distances from one another, communicating through rare written documents, sometimes imperfectly translated from a variety of languages. Nevertheless, this method of advancement of science has worked wonderfully well for more than two millennia and, despite some bizarre and fantastical appendices, has supplied the base on which countless generations have been instructed and educated.

In the last century, however, the situation changed radically . By turning research into a public enterprise, the scientific community expanded to become an international class of millions of specialists. In fact, we may speak of industrialisation of the process of knowledge production, where, however, the definition of the quality and the quantity of the product cannot be established and managed using the standard means available to industrial policy. In fact, as in the past, the great discoveries in science still take place within most restricted groups of individuals whose intelligence and creativity are unquestionably above average within the international scientific community.

A classic example is supplied by the development of quantum physics.

The birth of quantum mechanics had a long period of gestation in the second half of the nineteenth century when statistical thermodynamics was applied to the properties of particle systems, and Ludwig Boltzmann, one of the more brilliant minds of that century, related energetic states of particles to their statistical distribution. When Planck published his theory of electromagnetic quanta in 1900, he meant to solve the serious contradiction encountered by classic theory in calculating the energy of the electromagnetic radiation emitted by a black body. A few years later, the hypothesis of Einstein that light was composed of quanta, called photons, opened the road to a new interpretation of reality. Soon after, in the tormented period between the two world wars, a select group of scientists of exceptional brilliance—mainly Albert Einstein, Werner Heisenberg, Max Planck, Louis de Broglie, Niels Bohr, Erwin Schrödinger, Max Born, John von Neumann, Paul Dirac, Wolfgang Pauli, Richard Feynman and a few others—practically rewrote modern physics opening the way to a period of unprecedented scientific and technological progress.

One interesting aspect of the progress of modern physics was the prompt reception of these new ideas by the scientific community, made up mainly of university faculties. These supplied the *soil* that received these new seeds and leading to a germination of innumerable plants, as it were. Therefore, the existence of a class of honourable specialists able to understand and elaborate new ideas has been the key to an explosion of scientific progress in the Western World. After World War Two, this community grew and consolidated in the wide, open structure of institutionalised research and in a large number of projects for technological applications.[12]

---

[12] In the last three centuries, scientific research has been carried out in an academic milieu of universities and has been considered to be an integral part of student instruction. In the years after World War Two, public and private research institutes, each with their own specific missions, benefited from much more substantial financing which was needed to support the exploding costs of advanced instrumentation and equipment, and soon became centres of excellence for experimental research. Universities, however, maintained their leadership in studies of a theoretical nature, but the indiscriminate increase of the student population in the last few decades has unfortunately created, especially in Europe, serious problems with regard to managing all kinds of academic research activities, and, consequently, the educational level of the students involved has been lowering at a remarkable rate.

Therefore, a new aspect of research was represented by its direct link to engineering for a full exploitation of scientific discoveries. Thus, in view of possible economic profits, investments of public and private capital were made to support research advancement in all fields and at all levels.

In the preceding chapter we have emphasised that the rudimentary nature of scientific and technical research in primitive societies tended to assume an inverted pyramidal structure with the apex solidly implanted in the ground of public demand. Today the situation is reversed: the apex of the pyramid is situated in inaccessibly lofty heights and the collectivity is reduced to the lowest step, i.e., at a level of mere enjoyment of what has been achieved. From these conditions it turns out that in a compact and homogenous society the development of new knowledge depends on:

- *A congenital factor expressed by the frequency at which individuals are born, whose creativity and genius fall in the upper tail of statistical distribution of the intelligence coefficient of the population.*

There is no doubt that scientific knowledge would never have reached its current state without the contribution of a very few individuals of exceptional quality, whose capacity for abstraction, analysis and synthesis produced results which would have otherwise been impossible even with the combined efforts of many "normal" people. From considering historical documents we can conclude that genius is *not* a product of civilisation, in the sense that it can only flourish within a pre-existent context of advanced knowledge. If we consider the work of great mathematicians in antiquity, such as Euclid or Archimedes, we see that their ability to abstract and reason was not less than those of modern geniuses, like Gauss or Einstein. On the other hand, psychologists and psychiatrists have raised important questions concerning the character of a genius that may manifest itself in logical, mathematical, mnemonic, musical, artistic, or, simply, manual abilities of an exceptional degree. But what genius consists of exactly still remains a mystery.[13] Recently some psychiatrists, in analysing the public and private life of various geniuses of the past, have concluded that most of them were affected by behavioural troubles, symptomatic for an illness called "Asperger's syndrome", a mild form of autism (see M. Fitzgerald [42, 43]). Although this diagnosis cannot be generalised, it still holds true that a large part of our scientific and, in general, cultural heritage is the fruit of mental states, which we may consider atypical, if not pathological. If, therefore, we reverse the argument, we have to admit that the price of "normality" is an absence of genius, and it must be concluded that progress of our civilisation has been guided by a small group of individuals with genetic characteristics deviating from the norm.[14]

---

[13] In the life of geniuses creativity has often shown a fluctuating course with unimaginable peaks. In the case of Einstein,, for example, the year 1905 (since then called "*Annus Mirabilis*") was truly productive in a miraculous way. March: quantum theory of light; April and May: proof of the existence of atoms through Brownian motion; June: theory of special relativity; September: explanation of mass-energy equivalence.

[14] We do not intend here to embrace the theses of certain historians who have tried to explain the success of the greatest producers of scientific discoveries using "psycho-biographic" methods . These radical approaches started with the work of Frank Manuel on Isaac Newton, [44] in which he claimed that the *Principia* and the Law of Gravitation are, from a certain perspective, the product

- A factor representing the average ability in a group to further elaborate acquired knowledge.

What is mainly meant here is the average quotient of intelligence (IQ) that can be quantified using sufficiently objective methods. It cannot be denied that differences in average IQ exist between different populations. For ethical reasons, in the last few decades, comparative analyses have not been conducted based on differences of race. It seems, however, undeniable that IQ does not depend only on hereditary genetic qualities, but also—and perhaps to a greater extent—on environmental and social conditions. We must at any rate be realistic enough to admit that the average intelligence quotient in a compact and sufficiently numerous population is a variable parameter that can influence cultural development.

- A factor that reflects the ability of the group to perceive and grasp external knowledge through contacts with other groups.

Here we are talking about a particular quality which, in different contexts, biologists and anthropologists call capacity of *mimesis*. It consists in learning and taking control of newly created behaviours or advanced concepts, without passing through the steps that produced them. Owing to this quality, available knowledge can be rapidly acquired and transplanted in the soil of another culture where it is more fruitful than in its original ground. This knowledge acquisition process is highly remunerative as it only demands an ability to observe as well as intellectual mobility.

In conclusion, the process of acquiring and elaborating knowledge is a social phenomenon, where individual qualities are in synergy with communication whose effectiveness primarily depends on the social organisation at hand.

We have made these general remarks because one of the most serious questions in our present society concerns the capacity of science to solve at a *global* level problems of increasing size and complexity. These problems do not concern, as in the past, only the frontiers of science, but rather problems regarding its capacity to sustain a future civilisation, problems—it must be said—which might defy being solved.

Unfortunately, a disgraceful opinion was propagated in the messianic positivistic culture stating that whenever a problem is well- formulated a solution always exists; in other words, science makes everything possible, provided one struggles enough for it.

Now, the above considerations should have made clear that society cannot expect science alone to solve all its problems because science's achievements are propor-

---

of a Freudian neurosis. It could be observed that this vision of the history of science represents the counterpart to historical-materialistic interpretations of science as a product of human economic needs and activities. In fact, one should note that in 1931 the Soviet historian Boris Hessen in a famous and hotly disputed article [45] intended to demonstrate that the roots of Newton's work were of a socio-economic nature. These ideas have given rise to doctrines of Marxist inspiration like sociology of science and scientific revolutions. This argument will be taken up in the next chapter.

tionate both to the intelligence of specialists and to the wisdom and prudence of society as a whole, where a balance between prognosis and providence must always be maintained. Whether our present global society possesses these qualities is, unfortunately, highly questionable.

Chapter 7
Beyond Mankind

"I do not feel like an alien in the Universe"

(Freeman Dyson, *Disturbing the Universe [76]*)

"We are truly meant to be here"

(Paul Davies, *The Mind of God [78]*)

7.1 The Collective Memory of Man

Knowledge is the product of activities of individual minds, activities whose subjects and investigative methods are—at least in theory—free; in reality, however, they are rooted for the most part in the social base where man is educated. In the modern Western world, this base consists of an established system of knowledge collected and transmitted to subsequent generations through a well-organised school system. For every branch of science, therefore, it holds true that progress takes place at continuously advancing frontier lines, thanks to which the process of learning in a new generation can benefit from pre-established cognitive structures, which, on the one hand, allow for an enormous economy of thought and, on the other, offer the possibility and the instruments to attack problems of increasing complexity. Scientific progress is, therefore, a permanent legacy in advanced, stable societies.

From the quaternary to the modern age, individual cerebral abilities have essentially remained unchanged, but they have been employed in such a way so as to increase with time a collective patrimony of knowledge, owing to a tradition involved in every aspect of knowledge and maintained in an exact fashion as well as owing to an astonishingly synergistic amplification. This form of education and learning constitutes a very defining aspect of human society as opposed to other species, whose generations start from the same point and cyclically and exactly follows an immutable learning itinerary. Therefore, within rigorously organised animal societies, complex

C. Ronchi, *The Tree of Knowledge*, DOI: 10.1007/978-3-319-01484-5_7,
© Springer International Publishing Switzerland 2014

activities can sometimes be perfectly executed, which, however, invariably, yield the same results, without any possibility of progress or deviation. [1]

Progress is, in fact, possible, because knowledge is not inherited genetically by more or less complex coding, but, in a certain way, is continuously being codified and retained in some kind of *external* memory. This collective intellectual heritage is transmitted through communication, which presupposes a unity, a super-individual state, a communion (*"cum unione"*) of cognitive processes. Human civilisation is essentially bound to uninterrupted, multifarious traditions, oral, written, depicted, which encompass all intellectual activities. Already in prehistoric societies the individual was facing a patrimony of knowledge, which exceeded his learning capabilities, but he accepted it in compliance with an authority principle, whose alleged origins may even date back to a mythical past. The idea of progress, as conceived in our times, is completely alien to this traditional vision of knowledge that survived until historical ages. In all ancient civilisations, speculation was always made with a view to the past, believing that knowledge was something previously lost by man, and to be discovered again. The myth of a Golden Age of Mankind has always recurred in foundation myths of new civilisations, and its echo is still found in all great religions. This fundamental belief was of practical importance since the focus on the past supplied the psychological stimulus needed to preserve memory as a common root and source of knowledge. It is an instinctive behaviour stemming from the belief that a state of perfection cannot *spontaneously* develop from an imperfect one, while the contrary is likely.[2]

Whether it is true or not, if one rejects this principle, which is based on a belief in an initial Design, what remains is explaining progress mechanisms as a sequence of events starting from random change and resulting, respectively, in their promotion or elimination, according to their effect. This sort of concept is not easily accepted in the absence of arguments of modern natural evolution theories (even though these ideas circulated in the past among ancient atomistic philosophers). But what is more important for us, from a pragmatic point of view is this conclusion: if in antiquity

---

[1] A famous example, studied by several mathematicians, is that of the honeycomb. Charles Darwin wrote that bee architecture is the most striking product of an instinct, which, through natural selection, leads to the highest degree of geometrical perfection. In reality the role of instinct in the bee is much simpler: it has been demonstrated that the close-packed assembly of hexagonal prismatic cells and the non-trivial shape of the closure at their base indicate that the honeycomb structure is dictated by the principle of minimum use of wax. This rule is automatically established if all insects are of equal size and performance and if they start working side by side. The wedges of the cells are initially rounded, but the surface tension perfects them by transforming the wedges into prismatic shapes. Once the general principle of matter minimisation is established, the ruling mechanism leading to the final shape of the honeycomb is analogous to that of crystal formation governed by a minimisation of free energy, a process which is certainly not due to the "instinct" of atoms.

[2] One should recall that in *Timaeus* Plato affirms that animals are the product of man's decay through the transmigration of the soul. This belief is explicitly present in ancient Hindu philosophy and religion.

man had pursued a strategy to attain progress which was consciously based on the principle of natural evolution, questioning the entire tradition every time, human civilisation would probably have been dissipated into anarchical individualism from the very beginning.

In all great historical civilisations creation myths are bound to a concept of knowledge as memory, but, in reality, mythical accounts are products of human intuition and speculation in a framework of the knowledge man possessed. Therefore, mythical accounts not only represent *in nuce* the first philosophical systems, but exerted a fundamental influence on the birth of science. In his monumental work on comparative mythology Joseph Campbell [57] lists four essential functions of myth, without which science would never have originated:

- The first function was to provide a foundation of a cosmological vision that supplied an image of the Universe and an explanation for its progress and development.
- The second was to focus attention on metaphysical problems, concerning the existence of man and the meaning of his presence in the Universe.
- In third place, mythology provided criteria of confirmation of the established social order.
- Finally, it allowed man to psychologically centre and harmonise himself in his environment.

In modern society these functions of myth, still indispensable, have mainly been taken over by history and science, where this latter is proposed as a guarantor of three additional functions:

- To make reliable predictions.
- To collect and organise new knowledge.
- To continuously exercise self-criticism, submitting his own predictions to the ultimate authority of experimental confirmation.

For this reason the tradition/transmission of science now seems to be one of the most important tasks of our civilisation, which, therefore, is urgently called upon to develop adequate methods for recording and storage.

The care needed to preserve collective knowledge induced primitive societies to develop a consistent oral tradition, employing various materials and mnemonic devices. The art of improving an individual's capacity with mnemonic aids, the *ars reminiscendi*, is very old and, from the Middle Ages through the late Renaissance, a rich source of literature can be found on this topic. The ability of individuals in the past to learn by heart even lengthy and complex texts was much more developed than ours. The absence of this ability meant that any form of progress in intellectual activity would have been nearly impossible.

With the spread of printed books, collective memory surpassed individual memory to such an extent that, in view of ensuring a comprehensive tradition, techniques of memorisation have completely lost theirs significance. On the other hand, the gradual disappearance of direct communication and oral tradition created new problems and raised questions that are today of vital importance.

Ancient Libraries

The idea of constructing archives for documents, situated in protected surroundings, such as temples or rulers' palaces, dates back to the first Mesopotamian civilisations. The most ancient library found in archaeological excavations is that of Ugarit (thirteenth century B.C.), probably belonging to a private person. That of Nineveh (seventh century B.C.), constructed by Ashurbanipal, is the first to be organised in a systematic manner. Famous is of course the library of the Attalids (Pergamon, third century B.C.), which contained Hellenistic and Persian manuscripts mostly written on sheepskin (*pergamenae*, parchments). The introduction and widespread use of more compact and cheaper paper-like materials increased the capacity of great libraries.

Among the largest and most illustrious is that of Alexandria. It is reported that, in order to increase its size, king Ptolomaeus Philadelphus (309–246 B.C.) sent emissaries all over the world to acquire volumes on all topics of interest, even those written in different languages. His passion for thoroughness was such that those travelling to Alexandria with books in their possession were obliged to hand them over to the Library who then kept the originals, returning copies of them to their owners. According to actual witnesses, the library of Alexandria continued to expand for nine centuries and contained something like 700,000 volumes. This mammoth enterprise was not due to a mere antiquarian passion or in order to gain prestige on the part of the kings and of the governing class. In fact, a school was associated with the Great Library, directed by famous scholars, which continued its activity until the destruction of the library upon the Arab conquest. The school at Alexandria was the centre of intellectual activity in the history of the Mediterranean civilisations for almost a millennium.

We must not forget the libraries of the Roman Empire, the most of which famous were found in Rome, the Ulpian Library and Forum Library, containing Greek and Latin manuscripts as well as numerous libraries of wealthy private citizens. It is remarkable that the volumes of the latter group were the only ones to escape barbaric destruction and, having been transferred to Christian monasteries in a daring fashion, substantial part of classical literature could thus be preserved.

Although the aims of those promoting and maintaining these vast undertakings of collecting, cataloguing and conserving the cultural patrimony of their civilisation might have been varied and different, all were certainly aware that time inexorably and quickly erased every trace of material and intellectual activity of men, regardless of their historical vicissitudes. Thus, concerning literature of classical Greece in the fourth century B.C., we know that little remained of all that had been produced prior to the fifth century B.C., in spite of an exceptional period of cultural and political expansion which had begun in the seventh century B.C. For instance, although a tradition of philosophical and poetic works flourished in the Ionian colonies, only the Homeric poems and the works of Hesiod were almost entirely preserved, apart from which only scattered fragments cited by posterior authors survived. While it is true that the overwhelming majority of this literary production was not primarily meant for reading, but rather destined for the occasional recital (theatre, assemblies, lectures, and convivial meetings), it should still be emphasised that this vast body of material was almost completely dispersed in less than three centuries, despite having been collected and handed down in written form. The most regrettable loss was that of philosophical works, but its loss was lamented by scholars more for archaeological reasons than for necessity of study. In reality, the vital essence contained in the great works of the past had been already extracted and assimilated and, from this aspect, they had exhausted their historical and cultural function. Particularly scientific treatises, all designated in the same way with "$\pi\varepsilon\rho\acute{\iota}\ \Phi\acute{\upsilon}\sigma\varepsilon\omega\varsigma$, *perí Physeos*" (on Nature), had been condensed and reformulated in subsequent schools. Already by the Hellenistic Age, they constituted the first steps of an ascending staircase that, though representing basic elements, meant that one no longer wished to start from the very first step at the bottom.

From this point of view, one should emphasise the actual supremacy of the intellectual activity of living generations and the secondary role of historical memory. In reality, there are situations, in which this judgement must be more cautiously made. A typical case is often when a hegemonic civilisation collapses as a result of natural catastrophes, military reverses or deep political or economical crises. The unexpected disappearance of the intellectual elite of a generation can in fact cancel out the collective cultural heritage as well as requiring knowledge to be reconstructed, starting once again from the bottom of the staircase. In this phase, recovering historical memory ceases to have a merely archaeological character and becomes a catalyst for a reconstruction process. Historical examples are, for instance, the transition from Hellenism to Latinity, from the Persian to Muslim culture, from Latinity to Roman-Germanic culture and from the European Middle Ages to a National Renaissance.

From this perspective, within a fully developed civilisation, a collective perception of rapid change has always stimulated plans for organising and conserving what was then considered to be precious knowledge. Often, however, the results of these enterprises were not commensurate with their expectations, the actual course of history being often different from that imagined. Thus, in the eighteenth century the realisation of a French Encyclopédie (the etymon indicates its educational purpose) should have favoured, in the view of its editors, the fall of the old society and the birth of a new social and cultural order, based on scientific progress. Now, this work, though admirable, had a minimal practical impact: separated by a distance of a few decades the progress of science and technology had rendered it totally obsolete. The reason was an extraordinary advancement of science in the following century which itself was actually the result of a *uninterrupted* tradition involving generations of scientists who had flourished in the sixteenth and seventeenth centuries. Owing to the wide distribution of their original works, their contribution to this heritage survived without breaking its continuity despite dramatic political changes that shook Europe at the end of the eighteenth century. The many, more or less monumental series of encyclopaediae, which have been produced in various countries until today have basically suffered the same fate.

However, even conserving original works in libraries is becoming increasingly problematic. Today in France the Bibliothèque Nationale contains 13 million printed books, in French alone, a number which increases annually by approximately 35,000 new titles. The Library of Congress in Washington, the largest in the world, with a collection of 32 million catalogued books, must deal with an annual increment of approximately 1 million titles, with the tendency being one of constant increase. To these impressive figures must be added the mass of electronic documents in the form of ROM and those accessible via the *web*, whose number is attaining hyperbolic figures.[3]

In the face of this colossal mass of information a dramatic question must be raised which concerns its global and specific value or, at least, its practical usefulness. The answer can only be found in two fundamental reasons which were already known in antiquity, *i.e.*, safeguarding historical memory and instructing future generations. Today one is also aware of imminent epochal changes and of the fragility of our civilisation when confronted with possible geopolitical changes and natural catastrophes. The faster changes appear to be, the more obvious it becomes that memory must be saved from absolute oblivion if it is to have any part in a future civilisation. The same feeling arose during important historical mutations of the past; the global picture of the present has not substantially changed and, whether consciously or not, the questions that we raise remain the same:

What will be the nature of future generations? To what historical memory will they be linked? Will it be possible to maintain continuity of the criteria used to evaluate its contents? Will a unification of languages take place and on what basis [4] ?

If we consider the present state of Western culture, we must recognise that the collective memory which can be handed down to the next generations, is so vast that it can scarcely be identified, which makes safeguarding the value of its contents almost impossible.

The striking increase in power and ability of electronic processors will soon play a vital role in recording all information produced in a future global civilisation, but its intelligent selection and organisation will require a power exceeding by orders of magnitude the joint efforts of all contemporary people. On the other hand, the traces of any human activity, even if stored in the most durable artificial memory, remain of limited interest and thus of equally limited relevance and existence. For a short time these traces continue to exert a direct influence on society; their contents may be used as models, as sources of inspiration or be a source of criticism and reflection; those of greatest interest are absorbed by other more relevant subjects. In the end, the original remains are left to rest in some external memory from where, with only a few exceptions, may be taken up again from time to time until they fall into a permanent state of oblivion. The periods of vital relevance, of possible influence and of mere archaeological value of these products of human activities can be estimated from our own historical past. Two aspects immediately appear: the first one is that the effective lifetime of intellectual constructions is at most of the order of four to five generations; the second is that in exact sciences like mathematics and natural

---

[3] One should remark that even digital archives present serious maintenance problems. This is shown by the following example: The various scientific measurements collected during the NASA Apollo missions were stored on tens of magnetic tapes that were placed at the disposal of select American scientists. Yet after only a few years reading them became increasingly problematic for a variety of reasons so that the tapes were eventually declassified and offered to anyone who felt they could extract something useful from them. Today almost all this data is lost forever.

[4] There are presently approximately 500 languages, including dead ones, which have their own (mostly) literary production.

sciences as well as technology, new knowledge is assimilated by the ruling culture in such a short time that a need for the original memory gradually disappears within a few decades with even its archaeological interest completely vanishing within two or three centuries.

We provide two examples that illustrate how lifetime and intrinsic value of a literary work are more or less connected with its original form, but are always doomed to disappear:

- The famous publications of Albert Einstein on the theory of relativity were studied, and discussed for approximately forty years, during which time they exerted a fundamental influence on modern physics. Today they are read only for historiographic purposes. Within a few decades they will only be of antiquarian worth and their hypothetical disappearance would not entail any loss, neither for modern physics nor for our cultural patrimony.
- A great literary achievement such as Joyce's *"Ulysses"* has been read, studied and discussed for approximately forty years exerting an important influence on language and style in modern English literature. Today its role as a model has been exhausted, but the novel is still enjoyable and its disappearance would involve a great loss for literature as long as English remains a living language.

We can, thus, appreciate through these brief examples that mathematics and physics and, in general, all scientific disciplines have the ability to digest within a short time new results, and, at least in theory, they don't need any historical memory.

In the preceding chapter, we briefly presented the state of modern scientific literature. Although every new discovery or result is very well documented, only a minimal part of the total available information has been used and has effectively contributed to the progress of science. The difficulty in exploiting any residual potential is probably insurmountable. Even by inspecting distinct and limited fields, the articles published worldwide are contained in hundreds of journals and can only be located through precise bibliographical references. Every journal publishes annual indices of its contents and authors. Every article is usually preceded by an abstract and by keywords that place the article in various scientific contexts in which it may be relevant. Moreover, for some disciplines collections of the summaries of all pertinent articles are published. But even these are mammoth-sized for a person merely wanting to get a feel for the status quo in distinct research areas.

On the other hand, libraries, the only institutions able to purchase expensive complete collections of books and journals, are overflowing with volume and operating costs, as they desperately strive to transfer all contents onto electronic files, believing that keeping this mass of paper will eventually become impossible. An increasing number of reviews are published today in electronic form and online. During the last few years this practice has facilitated bibliographical searches, but, at the same time, the user is submerged by a mass of information from which he can draw some profit only if he is sufficiently competent and quick-thinking, but soon no one will be able to sift through this mass because it will become impossible. Furthermore, this apparatus represents only a starting point for a process of learning and acquiring the complete information contained in the potentially interesting documents selected. How could it then be possible to proceed with hundreds, if not thousands of pages? This depends essentially on the qualities of the user in question and on his strategic decisions. In the end, one realises that automatic selection of information may provide an effective didactic aid to reaching the solution of a specific problem, but only very rarely can it furnish a cue for important innovative developments.

Thus, even the work of the greatest genius is first studied and then sectioned, squeezed, filtered and condensed in order to distil the essential contents and, finally, put aside for purposes to which science is indifferent. Science, therefore, exhibits the alarming aspect of a gigantic organism that swallows and digests the intellectual production of man and seems to possess its own alien life form.

On the other hand, we are facing the exterminated memory of other types of human production, which can hardly be catalogued and has barely been organised in distinct structures. Yet, it is precisely from this disorganised mass of information that science emerged. Furthermore, without this very background, it would not be possible—or even conceivable—to sustain any progress, including of a scientific nature.

In the last few decades, philosophers, scientists, historians, sociologists and anthropologists have increasingly been emphasising the question concerning how the future of the human species is connected to preserving its collective memory. The projected focal point, which extends from a near future to cosmic times, crossing the threshold of meta-historical horizons, remains centred on the problem of the relation between man and a machine endowed of artificial intelligence, where only this latter will ineluctably contain and manage the contents of a history and of a knowledge originally possessed by mankind.

One cannot imagine these transformations without being filled with anxiety and anguish, even if there are people who see significant progress being realised in these anthropological changes being forecast, asserting that man will be able to use and dominate these intelligent machines, and to construct his future from his living memory, always alert and able to provide the necessary criteria for appraising and judging his choices.

Literature on this subject is vast and current opinions disparate. However, it is very difficult today to imagine long-term objectives of artificial intelligence programmes implemented in electronic processors, whose power and efficiency continue to improve with unexpected rapidity.

The problem has recently been addressed by the principal groups of research and development of computer operating systems. Among these, the IBM research centre of Yorktown Heights, in the USA, published a collective work of about twenty volumes [67] in the 1908s, in which conclusions were reached that still remain completely valid. We report them here as they are summarised in the last volume of the series.

Human mental activities can be divided into two general categories:

– The first concerns *analytical thought* and comprises typical activities of a subject, of which one can measure their Intelligence Quotient (IQ). These activities are characterised by their step by step linear procession as a function of time, with a complete perception of information possessed and of the operations in place. These can be deductive reasoning, logical or mathematical operations or explicit plans of problem solving. Every step can be described and accurately taught to other subjects.

– The second is called *intuitive thought*. This does not proceed according to a pre-defined plan, but tends to include manoeuvres that imply simultaneously diverse perceptions of the problem as a whole. The subject can arrive at a valid conclusion without clearly knowing how he did it. Transmission or instruction regarding this method of progression are almost impossible or take place through inexplicable ways.

Today neurologists identify the first type as an activity of the *centre of intelligence* located on left-hand-side of the cerebral hemisphere, and the second as an activity of the *centre of creativity* located on the right-hand side. Suggestions have been made for a deeper analysis of these activities, which in humans seem to be anatomically *correlated*. This entails extremely complex intermixing flows of rational and emotional processes in our brain, where imaginative and material stimuli meet together to determine mental activity. In this scenario it very difficult to conceive of a computer being able to completely simulate a human mind.[5] In fact, since their first construction computers have surpassed man in arithmetic problems and strategies of games with fixed rules. Generally, they are superior to man in applying standard methods of problems solving, but exceedingly weak in discovering new methods [68].

Today, the vision of a future role of Artificial Intelligence is open to opposite horizons. On the one hand, there are the very critical positions of philosophers, who assert the intrinsic impossibility of a machine being able to think and understand, if one takes a common sense approach to these two words; on the other hand, teams of specialists in computer science are working with a view that a computer simulation of human brain activity is a reachable goal.[6]

---

[5] Computer programmes simulating (numerically) simple neuron circuits are presently available and are fast enough for some practical uses. They, however, represent rudimentary applications in comparison with the real potential of developing computer programmes of artificial intelligence; however, all reasonably conceived objectives are far from simulating cerebral elaborations. It is also true that some people think that the great computers of the next generation will be in a position to make great progress in data processing strategies. Nonetheless these opinions are still premature, if not unrealistic, ; even by starting with a simplified model of biological neuron circuits, the problem is not to simply upgrade memory and data processing speed, but to develop algebraic theories and methods of numerical calculation within a new discipline called *computation geometry*, an interdisciplinary branch of mathematics, cybernetics and computer science which is only now starting to take shape [46, 47].

[6] The dispute in the 1980s involving John Searle [50], a philosopher of science at the Berkeley University of California and a research team at Yale [49] became well-known. The Yale team had developed one of the first programmes able to read simple texts and to answer questions regarding their contents. Searle asserted that the machine could give reasonable answers despite a complete lack of understanding of the message contained in the input text. He argued that, by mechanically following step by step the same procedure as the machine does according to its

The problem must, however, be viewed from a wider perspective than that involving the possible codification and elaboration of human memory in an intelligent machine.

In a series of interesting articles collected in one volume [58], Jacques le Goff, examined the relation between history and memory, contending that it would be ingenuous to assign supremacy to the latter, considered alive and real, in contrast with the former, seen as a product of intellectual manipulations. He continues, saying that without history, which is incessantly projecting itself into the future, it is impossible to guarantee continuity in a civilisation which would eventually fall prey to catastrophic millenarianistic prophecies or utopian revolutionary visions. *Homo civilis* has always possessed an historical consciousness, which has guided him in his operations, furnishing matters for reflection and meditation, exciting passions and affections, a sense of familiar and social responsibility and, first of all, has allowed him to conceive his own life as being part of a grand plan, which is realised through an order of events in which he participates. It is the subtle difference between ability to plan and ability to realise where one finds the difference between agent and instrument. As powerful as the instrument may be, the first position on a hierarchical scale will always be maintained by man so long as he remains capable of forming his personal vision of the future from his own memory. But how long will this be possible? This dramatic scene is described by a famous citation of I. J. Good [7] :

> *Suppose that an ultra intelligent machine can be defined so that it surpasses all intellectual activities of the most intelligent man. Since the plan of the machine is one of these intellectual activities, an ultra-intelligent machine will be able to plan even better machines; that would certainly lead to an "outbreak of intelligence" and intelligence of man would be left behind of many lengths. Therefore, the ultra-intelligent machine would be the last invention that man would have needed to make, provided that the machine will be always obedient enough to tell him how he can keep itself under control.*

Today we can only meditate on these sorts of problems, keeping in mind that the improvement and advancement of artificial intelligence has by now gone on an unstoppable course. The only way of preserving the supremacy of man is to place him ideally in a world of the spirit, freeing him from a deep and ancestral persuasion

---

(Footnote 6 continued)

programme instructions, he could report on texts written in Chinese although he hadn't the slightest acquaintance with this language.

Actually, the question to be answered is: "In which sense does a machine understand an input datum in comparison with how the human brain does?". Strictly speaking, it must first be asked whether semantics is reducible to mathematical logic or whether , more generally, a specific field of psychology is. The problems connected with the interpretation of language are so complex that an answer based on a unified theory appears to be absolutely impossible. The most famous supporter of psychological theory is Noam Chomsky [53]. Conversely, Richard Montague [55] claimed that syntax, semantics and pragmatics are branches of mathematics. Finally, some philosophers have pointed out that the structure of language cannot be reconstructed like *objects*, but rather as rational *practices*, which are *not* entirely convertible into to formal logic [56].

[7] Irving John Good, an eclectic English mathematician, was very interested in Artificial Intelligence since the very beginning of its applications. The quotation is taken from one of his articles on this subject [59].

that the only thing worth fighting for is a victory against adverse forces of nature. But the vision of the future now includes horizons that we can only indistinctly perceive.

7.2   Traces of Man in the Universe

A fair number of futurologists and cosmologists, regardless of their philosophical school, have considered the question of the destiny of mankind and have tried to find answers in accordance with current models of origin and development of the Universe.

Today we know that, depending on some basic assumptions, the Universe could be unique or plural, open or closed, stationary or moving, in expansion or contraction. While the first models were constructed following classical thermodynamic schemes, the most recent ones have synthesised quantum mechanics with general relativity leading to mathematical descriptions of the Universe that cannot be represented by a simple geometric intuition of an expanding fluid. From a modern perspective, the evolution of the Universe should take place in a space-time abstract with unattainable horizons or scattered singularities that, like vortices, may swallow the Universe to regenerate it after a break in the causality chain.

In a picture where energy and matter reciprocally transform one into the other, where all elementary particles annihilate or aggregate, where stars and planets are born and die, where galaxies are formed from cosmic dust and then collapse to form bodies of unimaginable density, there have been attempts to calculate if "man" as a physical phenomenon had, *a priori*, a reasonable probability of appearing on a planet like the earth and, in case other similar planets exist, if other beings can also exist which, even if not human-like, have characteristics so as to be considered as "observers" of the Universe (called in specialist jargon ETI: Extra-Terrestrial Intelligence).

The majority of cosmologists think that, even if that were possible, any interaction between us and them is exceedingly improbable. One of the arguments was produced by Enrico Fermi, who argued as follows:

"Since the age of the Universe is approximately 14 billion years and since there are in our galaxy 250 billion stars, inasmuch as *homo sapiens* needed only a little more than 200,000 years from his apparition on the earth to progress until being able to formulate the Big Bang theory, if there was even the smallest probability that some ETI's were born on other planets, they would have had all the time in the world to arrive here among us".

One calls this argument "Fermi's Paradox", though it isn't at all a paradox. There are cosmologists who have tried to express mathematically the probability of the existence of ETI, but their arguments are always affected by the *petitio principii* that intelligent beings appear wherever ambient conditions are compatible with life as we know it on Earth. [8]

On the contrary, the probable uniqueness of the "human phenomenon" is affirmed by cosmological models based on the so-called *Anthropic Cosmological Principle*. This principle starts from the unquestionable argument that there is a very strict correspondence between the values of the universal physical constants (*e.g.*, gravitational constant, speed of light, *etc.*) and the present state of the Universe: therefore, if these values had been different, even very slightly, the conditions necessary for the appearance of life would never have been met. This claim is obviously based on current cosmological models that describe the expansion of the Universe from an initial "Big Bang", followed by conversion of energy into matter and transformation of its states of aggregation. In spite of the great complexity of the problem and of the uncertainty of the hypotheses, cosmological models have to provide a non-obvious result of predicting the present state of the Universe and the distribution of matter—from elementary particles to the galaxies—in agreement with existing observations. The equations of modern cosmological models are formulated in terms of a small number of universal physical constants and a common characteristic is their high sensitivity to hypothetical variations of these input constants. In most cases even small deviations from the *right* values lead to predictions of the present state of the Universe very far from the real one. The formation of planets in which chemical and thermodynamic conditions are compatible with creation of life is, therefore, closely bound to strict values of a few physical constants of which we know neither their nature nor their origin.

Even from the simple point of view of thermodynamics we are faced with formidable problems. We have seen in the preceding chapter that the second law of thermodynamics asserts that irreversible processes exist, in consequence of which the entropy of the Universe can only grow. This law implies that time, considered as a numerical variable, has a defined direction (the "arrow of time").[9] Therefore, for example, a chemically non-reactive gas at a super-critical temperature always tends to expand since its entropy grows with its volume; conversely, dispersed matter, subject to gravitational attraction, tends to agglomerate when temperature decreases towards absolute zero and the impact energy of the colliding atoms does not prevail on the attraction potential of their masses.

---

[8] In reality, rather than a probability calculation one is dealing with a decomposition of the probability of existence of ETI as the product of a sequence of probabilities regarding the possible realisation of physical and chemical conditions necessary for living organisms, some of which can be reasonably estimated, while others are only the result of vague conjectures.

[9] In *reversible* microscopic processes, which characterise the thermodynamic equilibrium of a system, the sign of the time variable can be inverted without changing the system's properties because statistical random fluctuations of the variables can produce either a positive or negative deviation from the average value. In fact, in describing a thermodynamics systems in equilibrium time does not appear as an explicit variable. In the Second Principle time enters *indirectly* when the passage of a system from one state to another is supposed to occur in a finite time.

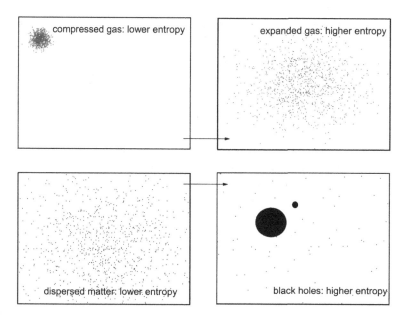

Fig. 7.1 The two pairs of diagrams in the figure represent the evolution of two different systems from an initial low-entropy to a final high-entropy state. In the first case (*top*) we have an inert gas, whose particles do *not* interact with attracting or repulsing forces, but only collide elastically, and gas entropy, which is proportional to the logarithm of the number of possible positions of the atoms, increases with the expansion of gas. In the second case (*bottom*), atoms are mutually attracted (*e.g.*, by gravitation). Metastable equilibrium configurations are those for which atoms are isotropically attracted by their neighbours, whose forces are thus counterbalanced. When, however, attraction processes prevail, precipitation of matter into a distinct dense phase takes place. If this phase is a black hole, the number of possible positions of atoms in the gas and in the condensed phase increases and, again, entropy diverges. In the two cases the behaviour of entropy is opposite with respect to expansion/contraction. The objection could be raised that atoms of the gas also possess mass and are thus subject to gravitational attraction; however, in agreement with the definition of a gas phase, their speeds must be high enough to provoke elastic collisions, making agglomeration impossible. Yet, at temperatures sufficiently close to absolute zero, atom speeds can become so low that the gas behaves like a system of the second example

Matter aggregates, in which internal thermal pressure is too low in order to counterbalance the forces of gravitational attraction, tend to shrink and to attract and capture additional matter, thus increasing their internal forces of cohesion and their density. *Black holes* are the product of such processes of aggregation where the highest densities are attained.[10] Now it has been demonstrated that, contrary to gases, the entropy of these very dense bodies increases with the square of their mass. The two processes of expansion and contraction of matter are, therefore, concomitant in the evolution of the Universe, and their extent depends on the initial state of the Big

---

[10] These large black holes are formed by stars that have exhausted their thermonuclear fuel and have used up the kinetic energy of their mass through continuous emission of electromagnetic radiation.

Bang, which, however, cannot be calculated *a priori*. One has attempted to estimate the probability that the parameters describing the initial state of the Universe had values compatible with its actual current state. In the space of phases the initial conditions needed are restricted to a volume whose fractional dimension is so small that it defies our imagination (its *logarithm* is of the order of magnitude of -10^{120}). How did these conditions come to be? Why were the initial co-ordinates of the Universe only found in this infinitesimal element in the space of phases?

If one attempts to answer, starting from statistic considerations, one arrives at two possible reasonable conclusions: either there is an infinite number of universes (called *multiverse*) one of which is ours, or we must admit that some *choice* occurred in view of a pre-established [11] design.

In addition, there are numerous unanswered questions regarding the evolution of the Universe: *e.g.*, its size and finiteness, its isotropy or anisotropy, the reason for the observed preponderance of matter on anti-matter, the significance of cosmic space and of fields of force, and so on. Even to grasp the nature of these problems, we may consider the enigma of the inexplicable difference between gravitational and electrical forces. For instance, the ratio between the force of electrostatic attraction of a proton and an electron is 10^{39} times larger than the force of gravitational attraction of their respective masses. What does this enormous number mean? What quantity can have determined it? What property can a Universe have which is governed by two important forces so similar in their mathematical expression and so different in intensity? These questions have intrigued generations of physicists. Some think that the only quantity that can be related to this large number is the total number, N, of particles in the Universe [12] , or its age expressed in sub-atomic units of time (*e.g.*, the transit time of light in an atom). While the first hypothesis is not verifiable experimentally, the second one, due to the work of Paul Dirac, entails a dependency of the constant of gravity, on time and should be, at least in principle, experimentally verifiable. Although, at first glance, this seems to contradict available observations, the hypothesis of the dependency of physical constants on time has given rise in cosmology to a school where new interesting approaches to the theory of gravitation have been considered.

The physical laws, as we know them today, indicate that the evolution of the Universe has followed a complex course which, when described by mathematical equations, has an outcome strongly dependent on the experimental values of a few physical constants. As asserted by the Anthropic Principle, there is a strong connection between these values and the variety of forms in which matter has developed, so as to render the appearance of man possible.

---

[11] Note that here the prefix "*pre-*" doesn't imply creationist hypotheses only if one considers a cyclical evolution of the Universe. In this case the answer refers to ontological aspects of the initial question.

[12] Various estimates of the number of particles in the Universe exist, all converging towards a value that is situated around $N = 10^{80}$.

It has been said that the Anthropic Principle implies (or suggests) the presence of a "fine tuning", of an intelligent design, as if by a cosmic architect, or, without directly referring to teleological doctrines, the existence of an infinity of different Universes, and in one of these the conditions required for the appearance of man were effectively satisfied ... and there we are.

These ideas are vehemently opposed by evolutionists of mechanistic tendencies, whose cosmic theories of evolution are mainly based on statistical models. It is clearly impossible to demonstrate the validity of a cosmological model using scientific criteria since verifying its predictive extrapolations is impossible. In addition, the ability to reproduce the few available data can be deceptive since these data can assume different meanings, depending on the point of view of the observer (see, *e.g.*, N. Bostrom [60]).

It could be objected that, from an epistemological point of view, the present state of the art of cosmology is not essentially unquestionable and, from this point of view, not different from that of cosmogonic doctrines of the past, and that every controversy on this argument is, in a certain sense, ineffectual. In our context, however, it is interesting to examine the image of Man based on new theories that refuse the reductionism of models based on a general theory of evolution.

If the time horizon of the anthropic principle is considered, what meaning would the final destiny of man have in a Universe *ab-initio* oriented towards the birth of intelligent life, if this very life, after have existed for a very short time, completely disappears without leaving behind any trace? The religious and eschatological question that man had always asked regarding the purpose of his being represents today the assertion of a cosmic force, inherent in life and consciously exerted by intelligent beings, a force which is opposed to the phenomenological and mechanistic forces of thermodynamics, which incessantly enlarge the entropy of the Universe, pushing it towards a shapeless end.

Hence, since intelligent life first appeared, a real fight for its cosmic survival has taken place where *homo sapiens* knows that he represents a sort of concrete awareness of the elapse of time, of the present and the future; but he must also realise he is the intermediary protagonist in a process that outstrips him in time and space.

What will man then be able to transmit to a future Universe where his body certainly cannot possibly exist? The answer, from a cosmological perspective, supports an anthropic point of view and is unequivocal: "The entire contents of mind and memory of all mankind, recorded on devices or, still better, on thinking systems much more robust than the human brain".

From this premise, visions of progress have been created, the outlines of which can already be discerned in our lifetime.

"Universal Machines" equipped with processors programmed with artificial intelligence and memories of enormous capacity will be constructed to preserve human knowledge and to extend this knowledge by applying human criteria, but using instruments of an incomparable magnitude of logical and analogical activities. These machines will be equipped with limbs (in a general sense) of any dimension, organs and sensors for every type of activity and perception. Moreover, these machines would be able to replicate themselves, since the construction of a machine containing plans of its own creation is conceivable and might even be feasible within a few hundreds of years. In a distant future its physical dimensions could grow until it reached the size of a city or even a planet. The machine could design and fabricate space vehicles and begin a process of colonising the Universe, settling and replicating itself on new planets, even where life, as developed on earth, would be impossible. The machine would represent a new ETI and would continuously improve its own structure in order to resist increasingly adverse conditions, maintaining, however, its entire corpus of knowledge developed from an original body of knowledge initially stored in its prototype.

Though presenting science fiction features, an evolutionary process of this sort is not at all unrealistic or absurd. Its conception is due to astrophysicists and cosmologists of renowned competence. Thirty years ago, J. D. Barrow and F. J. Tipler analysed these ideas extensively in a well-known book [61]; we can briefly summarise here the steps involved in this hypothetical cosmic enterprise:

1. Man gradually transfers all his knowledge to computers which are resistant to conditions increasingly adverse to biological life.
2. Man creates a kind of Universal Machine driven by Artificial Intelligence, capable of replicating itself. The entire contents of Man's memory are stored in the machine.
3. Colonisation of the Universe begins.
4. *Homo sapiens-sapiens* comes to an end, perhaps together with the Earth.
5. In some new planetary colonies the Machine continues to replicate itself, forming aggregates of increasing mass and power and establishing mutual contacts.
6. The number of machines evolves toward colossal systems in which all inner energy is used to increase, organise and maintain information and memory.
7. All the energy in the Universe is gradually converted into information until a "Point Ω" is reached as the apex and ultimate state of the Universe.

Echoes of the philosophical thought of Pierre Teilhard de Chardin [62] are evident, in particular of his idea of a *"Noosphere"*, a thinking network of cosmic dimension and of the *Point Omega* as the apex of a process of spiritualisation of matter. It is, however, on this aspects that materialists attack this kind of possible implications of the Anthropic Principle. But also the Christian speculation sees in this principle one of the manifold reincarnations of the ancient Gnosis. Teilhard, who was a splendid writer, spent a good part of his life trying to explain and to justify these theories [63], but he never succeeded to completely convince the ecclesiastical authority of their orthodoxy. We try here to briefly see some of the criticisms.

Thought, in all its manifestations, is *indissolubly* bound to both the thinking subject and to the reality in which he is living. The contents of the mind do not possess any ontological justification in the absence of a thinking human subject with his limitations and his struggle to acquire new knowledge; therefore, any kind of information, from a mathematical theorem through a literary work, which has somehow been coded and recorded in a permanent memory system, but surrounded by a space definitively emptied of human presence, ceases itself to exist as such. What does, therefore, an eschatological vision of a network of cosmic "machines" mean, machines acting as caretakers of a mass of information comprising the history and an explanation of the Universe? What information will these beings be able to exchange with one another if not messages of the sort : "Here I am" or "Are you there ?"

The ancient Gnosis was well aware of this objection and its definitive answer is that the ultimate aim, the *eschaton*, is not a victory of Mind, but the destruction of Matter and the annihilation of the Mind in an immobile All, freed of a Universe, which was created by an evil Demiurge. Therefore, in order to give meaning to an anthropic final perspective, its principle must eventually be reversed, turning around the original design of a Universe which, from an initial outbreak from which it originated, is oriented toward Man and consists of order and complexity. This argument is, after all, the weak point of all pantheistic doctrines: although their vision of the universe is mostly based on cosmic cycles, they recognise their absurdity and assert that the final jump into absolute Nothing represents the only possible final liberation of conscious beings.

Such a conception of the aim of human knowledge appears to us no less disquieting than that of a Universe where intelligent beings appear at a certain time, progress until they are capable of "explaining" the Universe, and then disappear. Refusing to accept this kind of non-sense does not arise, as somebody might assert, from a kind of "human chauvinism", but from the struggle to give significance to the continuous activity of trying and searching on the part of Man, pushed by a mysterious mental force which, in some way, precedes him.

7.3 In Search of *ETI*

The stimulus to devise and solve problems is innate to man and represents the source of his instinctive exertions for knowledge. New knowledge is, therefore, continuously acquired during individual man's life through conscious or unconscious mechanisms, though only part of this *subjective* knowledge becomes *objective*, that is to say something which is definitively formulated in some language and suitable for external communication. When, therefore, we speak in general terms of knowledge, we refer to both subjective knowledge present in the totality of living beings—something that develops in them and dies with them—and objective knowledge, whose contents can be written and recorded on an external memory.[13] Though not definable in the absence of man, objective knowledge can exist as a sort of footprint of the collective human mind. In our day, the majority of this knowledge is stored on an extinct paper collection and on a variety of advanced artificial memories. Therefore, at least in principle, one could imagine that the objective knowledge of the human species can be transmitted to any "observer" that might be present in the Universe and be capable of deciphering it.

Concrete steps in this direction have been already made in the form of "messages" sent to space. In 1974 an American team of the radio-telescope at Arecibo, in Puerto

---

[13] Despite an extensive debate between Karl Popper and John Eccles on the functions of the brain [64], they both define three Worlds of Mind to which knowledge is to be referred. The first one is the physical world, the second is the world of mental states (both conscious and unconscious), the third is that of the products of the human mind.

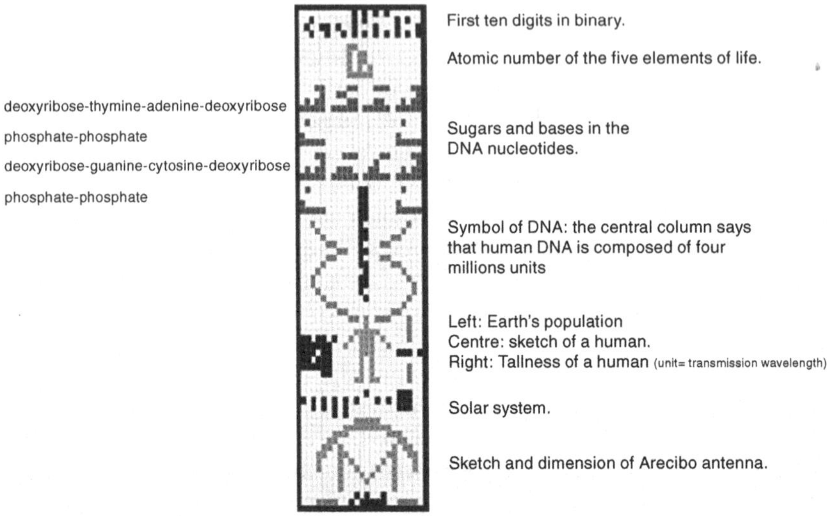

First ten digits in binary.

Atomic number of the five elements of life.

deoxyribose-thymine-adenine-deoxyribose

phosphate-phosphate

deoxyribose-guanine-cytosine-deoxyribose

phosphate-phosphate

Sugars and bases in the
DNA nucleotides.

Symbol of DNA: the central column says
that human DNA is composed of four
millions units

Left: Earth's population
Centre: sketch of a human.
Right: Tallness of a human (unit= transmission wavelength)

Solar system.

Sketch and dimension of Arecibo antenna.

Fig. 7.2 Graphical representation of the binary coded message repeatedly sent out to space from the observatory at Arecibo, in an attempt to impart some fundamental aspect of our scientific knowledge to possible "extra-terrestrial intelligences". The message contains mathematical, physical and biological data. The signal was transmitted sequentially, but one supposes that an intelligent being should be able to understand that the package contains a number of sequential impulses that can be ordered in a two-dimension map, consisting of a rectangle whose side lengths are exactly equal to two prime numbers, whose product is the number of *bits* of the message. In this case the contents of the message were elaborated after serious considerations. More recently, several other messages were sent containing various items, from pictures to music

Rico, the largest in the world, elaborated a message addressed to possible extra-terrestrial intelligent beings (ETI). [14] A coded signal sequence of 1 Megawatt power was sent towards a dense star cluster some 25,000 light years away. The message was compiled under the direction of Frank Donald Drake, a renowned astrophysicist of SETI (*Search for Extra-Terrestrial Intelligence*), an international scientific organisation. The message is reproduced in a graphical form in Fig. 7.2. Transmitted sequentially in binary digits, the diagram can be reconstructed as a simple mental exercise. However, a basic requirement is that the receiver should first of all be capable of conceiving natural numbers, reported from 1 to 10 in binary form in the first four lines of the message. After this information follow the atomic numbers of the elements H, C, O, N and P (the constituents of our DNA) and the formulae in which they appear in the relevant nucleotides. The double spiral of our DNA is written in subsequent lines. Next there is an image of man with his dimensions in units of the

---

[14] Since then, a number of "messages" to hypothetical ETI have been sent to space containing less compact information, including musical pieces and other items. In this regard it must be said that the number of believers or people just interested in the presence of aliens in regions not too far from—or even near—the earth is growing, and includes some influential astrophysicists.

wavelength of the message being transmitted as well as an outline of the solar system and, finally the features of the radio telescope from which the message was sent.

If the message is intercepted in some way, only an anthropomorphic ETI could manage decoding it within a reasonable amount of time. Should this hypothetical ETI exist and be equipped, for example, with receptive organs different from ours, decoding might demand exceedingly advanced abstractive abilities; as a consequence, it would hardly be likely that our simple message would stimulate these beings into developing a completely new concept of number and measure, for example. We could, therefore, expect an answer only from beings similar enough to us and having attained a level of civilisation much higher than ours.

Obviously, we do not have any serious proof of the existence of super-intelligent ETI's, so opinions on this matter can only be based on personal feelings; nonetheless a group of enthusiastic ETI believers is constantly claiming and producing the most fantastic "evidence". However, there is also a number of influential astrophysicists who are persuaded of their existence and who have succeeded in using radio telescopes and even space mission equipment to send all kinds of "cosmic messages" in the hope of obtaining an answer (even NASA has beamed Beatles' song "Across the Universe" !).[15]

Very likely, within one or two centuries from now our heritage of objective knowledge will be completely stored in electronic processors and could be sent in the form of coded messages into space. However, if the messages reach planets inhabited by ETIs, these will be found in one of the following three possible conditions: (1) their degree of intellectual development is lower than ours and the message could not be understood or even detected as such; (2) their degree of development is comparable to ours (which is the least probable case) and decoding would be very difficult and of little use; (3) their degree of development is much higher than ours and the message would only have the attraction of an archaeological curiosity.[16]

---

[15] The International Astronomical Union's Central Bureau for Astronomical Telegrams, strongly recommends (8th Principle of the Declaration Concerning Activities Following the Detection of Extraterrestrial Intelligence) that *no response* to any signal or other evidence of extraterrestrial intelligence should be sent until appropriate internal consultations have taken place (safety first!). In fact, believing or not believing in *ETI*, if an unfortunate encounter were to take place, most astrophysicists, among them Stephen Hawking (see an interview of his broadcast by Discovery Channel, May 9, 2010), are convinced that it will certainly no be like an encounter between two gentlemen and, most probably, we would find ourselves facing "trigger-happy" aliens, with consequences more similar to those described in famous movies like *War of the Worlds, Independence Day* or *Mars Attacks* than to what most "*ufologists*" dream of. Actually, this pessimistic view simply results from attributing to *ETI*'s the kind of behaviour humans have historically shown whenever they have met people of different races.

[16] It could be objected that in the last two cases the discovery of an interlocutor would be in itself an important event. This is true, but once the value of the contents of the message is lost, the message would be reduced to the type of interaction cited in the preceding section ("Here I am - Are you there?"). Apart from modern fantastic tales of extra-earthlings that in the past came to found our civilisation, we must conclude that only forms of intelligence very similar to ours could, albeit with great difficultly, *communicate* with us—if one takes here a benign significance of this verb is intended.

The first consideration to come to mind spontaneous is that the transmission of our knowledge to hypothetical extra-terrestrial intelligent beings does not have any sense, inasmuch as objective knowledge is only of intrinsic value in relation to the intelligence that has produced it.

The same consideration can be applied in analogous terms to the process of transmission of knowledge from man to machines equipped with artificial intelligence of increasing complexity: the three cases are given, which are respectively characterised by: (1) complete subordination of a machine to man, (2) difficulties in mutual communication and (3) complete subordination of man to machine.

The following arguments serve to better illustrate the possible situations.

Perhaps in a not too distant future the database of our total objective knowledge will be organised so as to be used by programmes of artificial intelligence for solving scientific, technical and practical problems. In this situation, the computer will have a far superior ability to access database than humans will, though the purpose and the method of its use will initially remain under the control of man.[17]

The problem becomes more serious when the machine is programmed, through a feedback procedure, to modify, cancel, correct and expand the original database, and can generate, in some way, new information and knowledge, of whose reliability the machine is its own guarantor. Commercial programmes of this type are today rather common. Obviously, before any programme is executed, the original database is preserved in an archive copy, so that, in cases of obvious error, the programme can be reset to its starting condition via a suitable instruction. When, however, the results of programme self-correction are considered valid, the archive copy tends to decrease in importance, in proportion with the reliability of the new results supplied by the computer itself. After repeated feedback procedures, the operating database undergoes evolutionary changes, whose path is difficult to retrace , while the initial copy is eventually devoid of any practical use.

Complex cases exist, in which the operator is absolutely *not* in a position to judge if the results of the programme are valid, and the terms of the dilemma of accepting or not accepting them are reduced to the only possible and simple answers of "yes" or "no". There are then exactly two opposite criteria on the basis of which can be decided when there is disagreement: our confidence in a programme of a machine or a reluctance to abdicate our own judgement.[18] At this point an absolute incommunicability between man and machine is produced.

---

[17] One of the first and most famous AI programmes is BACON.4 and was developed by Pat Langley and his collaborators in the 1970s at the Carnegie Mellon University in Pittsburgh, USA. This programme was used to "discover" some theorems of Euclidean geometry starting from classical postulates as well as certain laws of physics and chemistry starting from general and hierarchically organised experimental data. AI applications and connected problems are described in an interesting collection of articles [48].

[18] Note that whatever problem needs to be solved, the human operator is never neutral when confronted with a solution worked out by a machine. His bias for expecting a certain type of answer may be dictated by guesses, feelings and even fears or wishes. Sometimes this bias may be weak, but it is *always* present.

With the progress of machine performance, a time will come, perhaps in a near future, when man will decide to give, in cases of doubt, preference to the answers provided by a machine (a tendency already perceived in present times). In many cases, the automatism of a procedure produced by machines seems, in fact, to offer, on the one hand, a sort of guarantee of objectivity and fairness in decisions and, on the other, an exclusion of risk due to possible incompetence or lack of reliability of a human operator. Though these criteria today pertain to isolated problems and draw our attention mainly to the instrumental aspect of the choice, an awareness of the real dimension of this subtle and insidious problem is making its way in modern society. The role and the importance assumed by machines in human civilisation mainly depends on the size and urgency of the problems to be solved. The evolution of the relationship between man and machine was till now governed by the limited capabilities and selected functions of the latter, but both are in a phase of rapid increase of these two characteristics.

Probably the fundamental question concerning the authority credited to machines will remain unanswered for a certain time, during which time, however, the practical applications of microprocessors will continue to be extended to a point at which the problems in our society, associated with their use, will increase to such an extent that the issue of the primacy of man *vs.* machine will be defined in completely different terms. Nevertheless, in the present situation, the objections to futurologists of a "mechanistic" school, supporting a final dominance of machines, can be examined in the light of some considerations regarding the form and the elaboration of memory. The arguments that follow are laid out following a thread of controversial opinions, which are collected in a vast body of literature involving philosophers and scientists. The prevailing ones will serve as milestones marking the way of the future of mankind.

7.4 Two Paths for the Future of Our Civilisation

By considering the present state of our society and looking at its future, two different scenarios are outlined at the horizon.

The first one is that of a *quasi-stationary* civilisation, characterised by a stable equilibrium and harmonic, controlled progress. The resultant society is generally confronted with problems whose solution can be attained by available means; for example, problems concerning food and industrial production, urban development and maintenance of *habitat*. Under these conditions the progress of knowledge maintains a fundamentally exploratory character and applications are focused on *quality*.

Contrary to this scenario is a *transitory* context, marked by uncontrolled demographic increase with marked differences in the standard of living within the world's population, as well as by disparate intellectual and technological resources, concentrated in distinct geopolitical areas. Under these conditions progress assumes a character of urgency; recourse must be made to innovative technologies, even the most dangerous and hazardous sort, and the times needed for their implementation become shorter and shorter. The horizon of research is restricted by the extent of applicative aspects, which are absolutely preponderant, and *quantity* is the primary objective of all productive activities.

These are indeed the conditions under which automatism unfolds all its advantages: speed in processing data and controlling variables, strict determinism in decisions, a tendency to concentrate and to develop knowledge within a limited number of key sectors.

For more than one century our modern civilisation has found itself in this latter stage. Total confidence in progress of any kind, whose mechanistic nature had already been defined in the eighteenth century, has represented, in all political and economic projects, the driving force of a fast and indiscriminate industrialisation realised in the past hundred years. This has triggered irreversible processes of growth in all activities that carried the seed of an epochal crisis.

If, however, we examine the current situation in the face of the dramatic problems that characterise this crisis, we can discern a general tendency on the part of man to reassess the criteria needed to evaluate progress and to reclaim decision-making authority over all his activities, freeing his choices from any kind of automatism.

The evolution of modern global civilisation has produced dramatic problems which, after all, are neither new nor insoluble, but whose true difficulty is of quantitative nature.

Rapid achievement of knowledge and development of technical applications shall represent in the next future the primary objective, whereby the strategic choices will be imposed by the limits of our capability to put them into effect within a reasonable time.

But another important aspect must be considered. Whether good or evil, knowledge is today a heritage belonging to all mankind. Even in the most disparate cultures common standards have gradually been established concerning a quality of life which can reasonable be attained as well as social relationships including ethical behaviour. It is a slow process based on ethnic interpenetration and crescent economic and cultural interdependence of different countries, which will finally lead to a geographically homogenous planetary population. Under these conditions, conservative tendencies will emerge creating global society to evolve towards near stationary scenarios. We have seen in the previous chapters that advanced historical civilisations of a stationary nature risk being eventually swept away and supplanted by "younger" civilisations in a state of transition. The main cause resides in their cultural heritage, which is the richest, stagnating until it crystallised and was conserved by a restricted social class and expressed in forms which can hardly be assimilated by new, alternative elite classes.

Now if global civilisation did not have in future external contenders or enemies, its evolution could likely be arrested and remain in a state of collective feeling of well-being made possible by adequate scientific and technical knowledge and by a stable socio-political order (fair production and distribution of goods, peaceful resolution of conflicts, control of birth rate, *etc.*). In this context, lacking any Promethean ambition and completely engaged in safeguarding *Homo Sapiens* species and his *habitat*, all attempts to develop and implement artificial intelligence and intelligent machines would be limited to pre-established functions. The conception of an intelligent machine as a trans-human being, heir and perpetuator of man, would be totally alien to this scenario in which questions concerning the destiny of our earth and the Universe would always be relegated to a distant historical horizon and the problems associated with the survival of man are dealt with step by step, depending on the situation at hand.

Conversely, the most serious problems that man will have to solve from the perspective of a distant future are, in order of priority: climatic, geologic and astronomic ones. The following example shows what might be the magnitude of these problems and what the consequences could be to their solutions.

Changing Geography

Colossal projects aimed at modifying our environment on a planetary scale do not meet with great agreement in our times; however, in the first decades of the past century there were a considerable number of such plans, nourished by new achievements of science and technology.

One of the most ambitious was developed between the two wars by Herman Soergel, a German architect, who proposed blocking the strait of Gibraltar with a 15 km long dam that would have allowed the level of the Mediterranean Sea to be lowered by hundreds of metres. The resulting land would have created a connection between Europe and Africa, forming a new geopolitical unit called Atlantropa. The plan would have been completed by means of gigantic dams of the central African rivers enlarging Lake Chad by one order of magnitude, allowing irrigation of the North African deserts. The immense costs would have been offset by a series of hydroelectric power plants constructed on the Gibraltar dam, supplying a power of the order of 50,000 MW, at that time an astonishingly large figure (today the same power could be produced by fifty nuclear reactors of current type). The Atlantropa project enjoyed wide international approval and was supported by the League of Nations. Less enthusiastic were the Mediterranean countries, especially Italy, which saw with horror their coastal towns separated from the sea by tens or (as in the case of Venice, where the North Adriatic Sea would have been been entirely reclaimed) hundreds of kilometres. Promoters of the project, however, did not surrender and countered by envisaging futuristic urban solutions devised by architects of great repute (among them Le Corbusier). For instance, Genoa with its important harbour would have been reconstructed on the resultant new coast, leaving the old town on the edge of an internal artificial lagoon (Fig. 7.3). For Venice, which would have found itself at a distance of 450 km from the sea, one would have created an artificial sea contained by a dam 30 km away from the town.

The Second World War put an end to these speculations, but during the years of the reconstruction the plan was resumed and again succeeded in obtaining a financial support by UNESCO. It was eventually the development of nuclear reactors that reshuffled the claimed unique economic advantages of the dam power plants and the Atlantropa Comp., purposely created to prepare the details of the project, was finally dissolved in the 1950s.

Less ambitious projects, but of equally significant ecological impact, are still being considered even now. For instance, consider Israeli plans to create a waterfall channel from the Mediterranean to the basin of the Dead Sea that would feed a number of power plants, satisfying electrical energy demands of the whole country.

In recent years, plans intended to modify our surroundings on a large scale are being considered with growing caution, influenced by a better understanding of the possible catastrophic effects of a climatic and geological nature. However, in a more

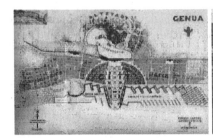

Fig. 7.3 In the first decades of the past century one began to conceive projects of man-made geo-planetary modifications. In one of these, one proposed to lower the level of the Mediterranean Sea in order to join Africa and Europe into a single continent. If put into effect, this project would entail reconstructing *ex novo* a number of harbour towns. *Left:* The map of new Genoa developed and published in the 1920s by a famous study of architects in Munich, Germany. *Right:* A recent graphical representation of the project by Miromar Entertainment AG. On the background of the artist reproduction one can see the real Gulf of Genoa with the old town

distant future and on the basis of more reliable predictions as well as control of their consequences, significant changes to our environment will not only be feasible, they will even become necessary for our descendants, with regard to solving problems related to controlling the purity and composition of the atmosphere and of fresh and salt water and, in general, to the fixation cycle of carbon in the earth's biosphere. The underlying chemical processes are the product of combined industrial and biological activities[19] and, even today, the problem of controlling and maintaining them in a steady state can be formulated and analysed in realistic terms.

The next step will be to prepare a plan to adapt our civilisation to significant climatic variations and to foresee unavoidable adverse events in time. From this perspective analysing probable planetary catastrophes and how to solve them will become relevant.

It is not necessary to review the abundant fiction which has been written on the end of the world, Doomsday, in order to realise that the phenomena with which we will be confronted sooner or later may have gigantic dimensions. Moreover, human resources are not so feeble as to exclude any mitigating human intervention. Let us examine, for example, one of the most feared events: that of a possible collision of a great celestial body with the earth: the event of a planetary impact.

Threat from the sky

There are thousands of asteroids and comets that orbit around the sun and periodically come dangerously near to Earth. Since 1993 the Minor Planet Center, located in the US and supported by NASA under the auspices of the Near-Earth Object Program, has been officially entrusted with systematically observing the sky to identify, follow and report the trajectories of celestial objects of a size of at least one hundred metres, the smallest of which would destroy an area comparable to that of a big city, should it collide with the earth. Since there are approximately 4,000 near-earth objects, one is constantly employed in methodical

---

[19] The atmosphere which is to be expected on an earth devoid of life would be quite similar to that of Venus: approximately 97 % CO_2 and 3 % N_2.

and systematic work demanding continuous corrections of their calculated trajectories based on frequent astronomical observations. Today the forecast of a possible impact can precede the event by some decades. Presently, the only object that has provoked concern is Apophis, an asteroid of approximately 300 m diameter discovered in 2004. Since its 323 day orbit crosses twice the trajectory of the Earth, initial calculations seemed to indicate that that Apophis could come dangerously close to the earth in 2036. More accurate calculations of its trajectory now show that there is no threat of collision, at least in this century. In any event, for an object of this size it would be possible to envisage a last minute intervention strategy to deviate or destroy the asteroid.

Nevertheless, the threat described above would be fatal if one day an object of such dimensions so as to completely destroy life on our planet through its impact were detected on a collision course with the Earth. In fact, an event of this magnitude can today be forecast, but certainly much too late to undertake any action to sufficiently modify the trajectory of the object (Fig. 7.4).

But if astronomical observations of the object were precise enough to allow us to determine its exact position at a sufficiently far distance, its trajectory could be changed with a thermonuclear explosion producing an angular deviation that, though small, might be sufficient to avert a direct collision with the Earth. In order to achieve this result, it would be necessary to operate a network of terrestrial and spatial astronomical observatories and continuously and without interruption scan the firmament with instruments of increasingly high resolution: a massive enterprise, both in terms of technology required and costs involved.

Let us consider a hypothetical example: the case where, at a distance of the order of some light years, a planetoid of the size of the moon was seen to be moving along a collision orbit with the Earth at a speed of 100,000 km/h. To sufficiently deviate this body (say, so that its shortest approach distance to the earth be 100 times its size) with a thermonuclear explosion produced by the most powerful bomb (today 100 Megatons TNT = 10^{17} J), it would be necessary to hit the body at a distance from the earth not less than about 0.8 light-years (7×10^{12} km). The missile containing the bomb as a payload would have to be launched as soon as possible and its trajectory continuously corrected on the basis of more and more precise observations of the trajectory of the planetoid. After this body had been hit, the expected effects of the explosion could only be confirmed approximately 8,000 years later. Distances and times could be reduced by a magnitude of two to three orders by using bigger thermonuclear devices. In the framework of the CYCLOPS project[20] it has been estimated that energies of the order of magnitude of 10^{23} J could be produced by feasible H-bombs. However, to produce the necessary quantity of deuterium (about 200,000 tons) all resources of the earth (presently we produce 200 tons deuterium per year) yielded at maximum levels would need to be exploited for at least 100 years.

These cases, though currently falling beyond the horizon of our reasonable fears, are here meant to show that scientific knowledge and technological development in conjunction with socio-political conditions making it possible to manage projects of very long duration, can allow man—even within the limits of the present state of advancement of science—to effectively manipulate objects of planetary size and at enormous distances.

---

[20] The CYCLOPS project (NASA Report CR 11445, 1971) was launched in 1971 by NASA in collaboration with the University of Stanford and research centres of the USA and other countries in order to study possible contacts with extra-terrestrial forms of intelligence through messages sent—or possibly received—by networks of radio telescopes. The project was later abandoned out of budgetary considerations.

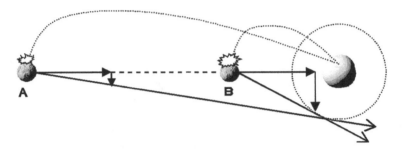

Fig. 7.4 A celestial body on a collision course with the earth can be sufficiently deviated by a thermonuclear explosion favourable detonated by a bomb launched from the earth (position A). However, if the body is first detected at a shorter distance from the earth (position B) the explosion would have to be much more powerful in order to produce the same effect as in position (**A**). Realistically, even celestial bodies of great mass can be sufficiently turned aside from their collision trajectory with conventional thermonuclear bombs if they are hit at great distances (and if the calculations of the trajectories are precise enough). Such projects would entail enterprises lasting a long time: centuries or even millennia. One has observed approximately 5,000 large asteroids and comets moving around the sun at distances of the order of 200 million km. Of these, approximately 1,000 could, in a relatively near future, come dangerously close to the Earth. The data of their orbits are being continuously collected at the Minor Planet Center, which works in collaboration with an astrophysical observatory in Cambridge, Massachusetts

But current research in this area is pursuing different aims. Space exploration projects, the most expensive and, at first glance, least productive ones, still enjoy great favour in the public eye. However, their financial support by governments of advanced countries is justified, to a great extent, by possible military applications or by resulting development of spin-off technologies that do often entail huge commercial benefits.

On the other hand, the objectives of a good part of scientists working in space research are much more ambitious and concern, more or less consciously, a vision of the future in which space missions are undertaken where "intelligent machines" are predestined to replace man or, in any case, to surpass him in a meta-historical struggle to take possession of the Universe.

It is difficult to have an opinion on these positions, but we can at least say that present objectives of scientific research in the most advanced fields are either too short-sighted or too far-sighted, or, in ethical terms, too egoistic or too sterile. This is perhaps the consequence of the fact that our modern society does not have a clear vision of its near future and hence fails to perceive the necessity of steering research towards a global project of common progress.

Chapter 8
Artificial Intelligence and Post-Humanity

"The idea of non-human devices of great ability to carry through a policy, and of their dangers, is not new. All that is new is that we now possess effective devices of this kind."

(Norbert Wiener, *Cybernetics*, 1949)

"We believe computers are "smart", so smart that we cast ourselves as "dummies" in their presence We have begun to believe that the mind, the defining feature of human nature, is a somewhat inferior information-processing machine."

(Theodore Roszak,
Los Angeles Times, January 28, 2004)

"Die Geister spielen Mechanik, doch nur für eine gewisse Zeit."

(Pavel Florenski, (1882–1937), [99])

It seems paradoxical that in an age in which conquests of science and technology have been realised beyond our every expectation, there is a sort of apprehension in our society caused by a conscious psychological fragility of man in the face of his work.

Fear and anxiety reveal a profound insecurity in individuals who live in increasingly man-made surroundings, whose equilibrium is extremely precarious and whose stability depends on human decisions and power. While, in the past, the survival of a fragile individual was dependent on his ability to fight against potent forces of nature whenever they appeared to be adverse and harmful, today these very forces, which, after all, have somehow sustained the stability of our world for millennia are starting to look fragile and precarious.

Man's creative thought, continuously realised in new, great enterprises, whose dimensions in space and time have exceeded limits, which used to be considered impassable, obliges us to look at the progress of knowledge as something that cannot

C. Ronchi, *The Tree of Knowledge*, DOI: 10.1007/978-3-319-01484-5_8,
© Springer International Publishing Switzerland 2014

be stopped and must follow its course by changing the world beyond long-established frontiers where the survival of man and his species is being dramatically challenged. The humours that nurture these visions are fickle, sometimes of opposite tendencies. Some envision a victorious final war of man against biological death; others see the necessity of creating an advanced species: successor and heir of man. In both cases man is called upon to put at risk his own nature and modify it, activating a sort of feedback genetic manipulations, by which he would take full possession of the natural mechanism of biological evolution.

Today these ideas find resonance in materialistic circles as well as in others permeated by mystic-pantheistic currents. Yet it would be wrong to consider them as daydreams that have found a place in the deserted realm of Myth and to argue that they are nothing but more or less bizarre forms of fiction. As in myths and fairy tales, it is not the specific contents which are important, but the spirit that breathes through them. It is always necessary to recognise this spirit and to discern its influences, examining their potentially positive and negative consequences.

8.1 Who Will Come After Homo Sapiens?

The (verifiable) history of the experiments aimed at manipulating human genes in order to create a new species using scientific or pseudoscientific methods is almost one century old. The first experiments were conducted in the 1920's by Soviet veterinary Ilya Ivanov [70], who tried to couple men and primates via artificial insemination of female orang-utans (complementary experiments with human females were not carried out solely due to a technical accident). It is well known that man possesses one chromosome less than apes, a state which doesn't prevent possible hybridisation. In fact, Ivanov had previously carried out a number of experiments on different species of animals. The Soviet Materialistic Biology Society, associated with the Academy of Communism, financed this project with US $ 10,000, corresponding nowadays to approximately one million dollars, a considerable sum for the USSR at a time when it had been financially devastated by the revolution. The scope of the experiments was mainly ideological since one intended to demonstrate a biological continuity between man and animals. It seems, however, that Stalin was later interested in these experiments because he hoped that a new species with superhuman force and resistance and subhuman brain could be eventually "manufactured" to be employed in the Red Army.

One would be inclined to think that such ideas could be conceived only within totalitarian systems lacking any ethical scruples. In reality, Ivanov carried out the majority of his experiments in the French laboratories of the Pasteur Institute, situated in the colony of Guinea. Moreover, similar experiments were performed—more or less secretly—in China, USA and Great Britain; apparently without producing any positive results.

With the progress made by science these crude projects lost whatever interest they may have evoked among biologists, who eventually saw in *in vitro* fertilisation a cleaner and more effective method for genetic manipulations. The declared objective of the majority of modern projects involving genetic manipulation is to find a relation between our genetic heritage and possible diseases of congenital origin, and to develop techniques for "repairing" the DNA: a very important and ambitious objective open to possibly troublesome discoveries and applications. Nonetheless, although significant progress has recently been made in this area, our knowledge on molecular genetics is still in an infant stage and much development is needed before the above-mentioned objectives can be reached.

However, behind this rationale, considerations on eugenic applications have been increasingly gaining ground. From this different perspective, the scope of genetic manipulations would not be to ensure the development of an embryo towards a fundamentally healthy body and long life, but rather to select a genetic heritage according to a pre-determined typology of what an adult should be. Cloning, intended to perpetuate the genetic qualities of an individual whose features had been previously evaluated according to pre-established criteria - called "default fitness function" (for example, degree of intelligence, creativity, force, agility, *etc.*) - has become a point of departure for a type of research, which is paving the way for the construction of *ad hoc* individuals, where "*hoc*" represents the optimisation of human behaviour according to a model of society which is based on abstract ideological speculations. An anthropological reversal of the man-to-environment relationship has quite evidently come to pass:

The *Homo Faber*, who, contrary to other animal species, is characterised by his ability to modify and adapt his surroundings to his aims, would now himself be finally adapted *a-priori* to a pre-determined society. The step from accepting this point of view to planning research in order to create a post-human or trans-human species is a small one—even though our present knowledge doesn't yet allow us to operate with certainty and precision on DNA components.

The field of eugenics is much more insidious than it might appear at first glance.

The definition alone of a "genetic improvement" entails considerable problems and lends itself to ambiguous answers. Already the enormous differences which exist between human individuals create a problem when it comes to defining a "*normal*" personality, in the sense of stability, equilibrium and behavioural limits within a populous and variegated community. On the other hand, the range of normality in a population corresponds to the centre of a statistical distribution at whose tail-ends a variety of cases is present which are not always classifiable. Abnormality, in general terms, is considered as a negative feature, even though individuals of intellectual or physical qualities of an *exceptional* nature in a positive sense must also be included in this category. The negative judgement of abnormality is evidently of a sociological character, inasmuch as abnormal individuals create a crisis in behavioural and relational schemes on which social life is based.

Returning to the argument mentioned in Chap. 6, one should remember that genius is often affected by troubles of an autistic type known as Asperger's syndrome [1]. Famous personalities of art or science are known (curiously enough, this syndrome is almost exclusively observed in male subjects), who have paid a high price for their genius, in the form of continuous depression, immaturity and serious deficits in their social relations, so that psychologists have reversed the conclusion asserting that "the price of normality is a lack of geniality". This is not a tautology in as much as normality refers to a satisfactory mental equilibrium. We must, therefore, conclude that the best of all possible worlds is certainly not that inhabited by a society exclusively made up of geniuses. On the contrary, a superman society, envisaged by eugenicists, could likely turn into a sort of hell from which its members can only escape through a door of regressing back to old "*normality*"; a consideration that is not new and that sometimes one is tempted to apply to post-modern man.

The vision of a future genetic mutation, influenced, if not fully controlled, by man, has assumed new aspects in recent years, some of which have caused great perplexity and distress. Obviously we are dealing with ideas cultivated in restricted quarters of specialists, who anticipate unlimited progress for some modern scientific fields, such as cybernetics, neurology and genetics. In this scenarios, however, we are confronted with the paradoxical situation of an optimistic appraisal of human abilities in a mid-term time frame (and mid-term means here centuries or at most a millennium) being opposed to the conviction that man is in the end doomed to disappear, after having created an advanced species that will succeed him on Earth and in the Universe. The research objectives inherent to these predictions can roughly be summarised as follows:

1. Man will be genetically modified in a laboratory to be endowed with sufficient intelligence in order to achieve the scientific and technological progress necessary to realise the following objective 2
2. Systems of artificial intelligence will be developed and implemented in future compact and powerful processors and installed in intelligent machines, which will continue to grow in quality and efficiency until they have far surpassed human intelligence, and will be capable of realising objective 3
3. The last generations of Men will create a new, superior species that, in symbiosis with intelligent machines, will head for immortality.

For some years now, these ideas have enjoyed considerable attention especially, in the United Kingdom and in some Americans Universities. The related topics are introduced with professional competence and the objections countered with vehemence in publications and conferences. An example of this is a monograph written in 1970 by Frank H. George, a specialist in psychology and cybernetics, who discusses these theses in a well-balanced manner [79].

---

[1] A popular essay by Michael Fitzgerald on the relation between autism and creativity [42] provides a good introduction to a vast literature on this topic, ranging from clinical to socio- psychological studies.

It is important to note that arguments in defence of the thesis of the ineluctable decline of man and of the necessary realisation of a new, advanced species are based on a generalisation of the law of natural evolution, a slippery ground on which it is exceedingly difficult to conduct an open debate without incurring the immediate wrath of evolutionists of every school. On the other hand, we cannot help observing that what is now happening to the law of natural evolution has already occurred with with the Hegelian dialectics of the late 19th century, whereby all problems were formulated and solved by reducing their terms to an omnivalent resolutive mechanism of a "thesis-antithesis-synthesis" triad [2]. This kind of dialectic was eventually used by innumerable (and discordant) Hegel's epigones in order to explain everything and the opposite of everything, from philosophical systems to political ideologies.

For natural evolution, the same is happening with concepts of "mutation-adaptation-selection". To the objection, previously examined, that it is very difficult to explain how a man-constructed "intelligence" can surpass its constructor in extended reasoning ability, one answers that, if this were not possible, natural evolution would not have been able to take its course. In other words, one assumes that a processor, equipped with a suitable learning/feedback programme, will eventually overcome any obstacle it faces, thanks to repeated executions and adapting changes within its programme. In short, what in the evolution of a species is represented by a selective filter of accidental genetic mutations becomes, in our case, the guarantee that repeated applications of programme variants do eventually lead to the solution of whatever problem is faced. Yet, seeing as it becomes increasingly difficult to explain the mechanism of any improvement in organisms of increasing complexity as a product of natural evolution, so must the likelihood diminish of an artificial intelligence programme with a finite memory continuously modifying itself without, in the end, falling apart. One may object that artificial intelligence will not only be equipped with programmes of defence/aggression procedures, but also with criteria to carefully evaluate economy, aesthetics and even ethics, thus rendering the programme resistant to deleterious evolutionary changes and uncertain responses to ambient perturbations. But these considerations are only circumlocutions concerning the definition of the degree of freedom of a crucial feedback procedure. Practically, only two cases are given: either the system is partially safeguarded, and its evolution is actuated *a priori* or it is allowed to gamble its entire contents, exposing itself to a high risk of incorporating germs, thus leading to a final lethal outcome. Furthermore, if we consider the case of several AI-units operating in different environments and connected in a network and the possible variances among operating systems, a catastrophe would be bound to happen, no matter how well the machines are able to communicate with one another.

On the other hand, there is no doubt that even now intelligent machines are superior to man in all activities based on calculation procedures, in finding search strategies

---

[2] Consider the use of evolutionism in the propaganda of militant atheism as in the case of Richard Dawkins, or its role in the thoughts of David Deutsch [81], where evolutionism as formulated by Dawkins is assumed to be one of the four pillars of human knowledge (the three others, according to Deutsch, are quantum physics with a belief in a plurality of parallel universes, Popper's epistemology and Turing's universal computing theory).

and data treatment, in displaying consequences of postulates and so on. Moreover, robotics is today at the threshold of applications that will significantly affect our lifestyle. Robots endowed with artificial intelligence, interacting individually and learning from man will become an extension of his limbs, his body and his memory, encompassing his world of images, voices and writings. At the time of writing, robots are being constructed of such complexity that they are able to learn, listen and converse as well as acting independently. Technology will allow us, sooner rather than later, to construct androids as described in past science fiction stories. However, they are and will be machines whose guiding authority will remain completely in the hands of man. No matter how exceptional their abilities may be, the functions that robots exert have human decisions as their input. To paraphrase, it can be said that, even when somebody regularly loses in playing chess with a robot, he still does not lose his pleasure in a game that he has invented and has decided to play, being just as concerned about the superior ability of the computer as about his car beating him in speed and power, but which he uses when and how he likes.

The relative value of man compared with a machine must be principally based on this assumption, which is merely of ontological origin. From a materialistic point of view the one could come to the opposite conclusion: bestowing supremacy on the robot. But in order to make this judgement, there is no need to open the window of our imagination to future scenarios. In fact, even in the past, there have been people who considered certain inanimate objects more valuable than their fellow men.

Today a constant threat has always come from the mechanistic conception of man, accompanied by a mystical Titanism. We are not dealing here with a Faustian impulse, but with an anxiety of giving meaning to the Universe which, deprived of the presence of Man, makes us despair of the present and feel horror at the future. More or less consciously, certain types of research in advanced fields of genetics and neurology are motivated by this cultural background. Obviously the risks of genetically manipulating man are perceived, but these are placed in a context of insufficient knowledge and know how. We cite a remark by F.H. George taken from the above-mentioned reference (p. 51) :

> *...we shall have to face sooner or later the problem of immortality, and that in turn is followed by the problem of being overtaken by the more highly developed species; ironically, this could be a species we ourselves manufacture in the laboratory. The imminence of such results should serve as a warning that we must carefully examine the implications of what scientists do right now, in the short term, let alone in the immediate future. Any failure to anticipate sufficiently or to understand enough, and so far we are nowhere near satisfying either criterion, is bound to spell disaster..*

Actually, the greatest risk, if we proceed toward the future with the sole purpose of acquiring new and more knowledge, regardless of the cost, is that we lose along the way our most precious baggage without any equivalent compensation. The efforts needed to survive and to progress, even under adverse situations, can serve to polarise and upgrade the faculties of man in selected functions and, within this context, feedback consisting of gratification would be reduced to nothing more than merely corroborating the *status quo* in the case of success or to some kind of penalty in the opposite case. Actually, in a civilisation governed by this basic rule it would be difficult to assert the supremacy of man over intelligent machines, inasmuch as the former would be devoid of all values alien to the machine's operating logic.

Actual danger is perhaps to be expected in a not too distant future. In the preceding sections, we have repeatedly emphasised that civilisations can decline and disappear because of a disproportionate increase in selected fields of knowledge to the detriment of others. This crisis is typical of societies suffering from excessive organisational concerns, obsessive interests, difficulties in social relations and problems of communication. We are dealing, at a collective level, with symptoms of an autistic type (the same symptoms that often accompany genius). For this reason civilisations at this stage of development disintegrate when clashing with other less evolved ones, but which are richer in their variety of interests and resources.

These considerations remind us of the crisis faced by current Western society. In the face of events and changes generally called "globalisation", problems undoubtedly arise, whose solution depends on our scientific resources and technical know how. It would be, however, a mistake to subordinate the efforts of all the peoples of the earth to the progress of science and technology. This progress only possesses in fact a value if related to human progress and hence in the absence of man any definition of "value" cannot exist [3]. In other words, knowledge must first be judged not only in the context of where it is *produced*, but also where it is *accessible*. For example, which value can have the ability to conceive and construct powerful weapons in a society dominated by violence and intolerance? Or to be able to create and maintain sophisticated artificial environments for the use of a few at the expense of the of the rest of the Earth becoming uninhabitable? Or to create a species of supermen in a global civilisation where the great majority of the population lives under sub-human conditions?

If we examine the present situation of our society, we must realise that the destiny of humanity is mainly bound to the effects of primitive, conflicting emotions such as egoism and altruism, aversion and sympathy, hatred and love, aggressiveness and passivity, intolerance and tolerance, confidence and suspicion rather than to the progress of science. These mental states are innate both in animals and men, but in humans are governed by the power of will and reason, and, therefore, assume positive or negative "values". In mediaeval psychology they were called *vitia* and *virtutes*, and the ethical behaviour of man was tied to their sometimes very intricate, interactions. Nowadays some biologists are persuaded that, through a genetic manipulation of the human species, it will be possible to suppress "defects" and upgrade "virtues", using the same criteria by which they believe one can artificially improve intellectual capacity. Once again we find ourselves facing a sad sort of reductionism. In fact, motivation and value, both behavioural and ethical, of man's action cannot be obtained by his genetic imprinting, but rather by the particular resolutive process he adopts, inter alia, in situations where he faces contradictory tendencies and the necessity of choice. It is from this process of psychological enrichment that the char-

---

[3] The root of the word "value" comes from the Latin verb *valeo*, which means "to have force, possession and dominion". Therefore, in its etymological significance, the "value" of an argument is linked to a subject who seizes it as an instrument or as a weapon.

acter of an individual is formed with all the connections that render him responsible to his society and, according to Christian theology, to the final judgement of God.

Today mankind still has a long way to go on its own before arriving at a decision concerning its genetic survival and the possibility of bequeathing its whole knowledge to a post-human species that probably will not be in a position to comprehend its "value".

It is more likely that, within a future planetary civilisation, hopefully characterised by an efficient organisation and satisfactory social relations, those traits of the human race that today are considered negative - like dissatisfaction, love of risk and unpredictability - may become the driving force in progress in a future history.

8.2 Is There any Mission Left for Post-Humanity?

Visions that used to be popular only in the world of science fiction literature became objects of scientific speculation and the source of extensive, publicly financed undertakings after the first successes of astronautics in the 1960's. The amazing progress of rocket technology and the swift and brilliant development of electronic microprocessors in the years to follow exceeded even the wildest expectations. The consequence was that members of circles of astrophysicists and astronomers began to fear to be left behind in the rearguard of a technological progress that involved disparate branches of science and engineering at the frontline. In fact, after the first initial enthusiasm, some voices could be heard criticising space research programmes, arguing that progress in astronomy and astrophysics did not justify the enormous cost of certain colossal projects which were being planned [4].

With the launching of research projects like CYCLOPS and DAEDALUS in the 1970's these criticisms were answered, where a theoretical basis for possible interstellar exploration was set in motion, implicitly indicating new, much more ambitious objectives than those considered in astronomy and astrophysics of that time. Beginning from this time, practically all specialists in these fields have been more or less directly involved in what is now called "colonisation plan of the Universe".

Independent of the various faiths and currents of thought, the initial justification for this kind of research is that one day all natural resources on the earth will become insufficient to sustain human civilisation and that a new place to live will have to be found in the Universe. The time needed to colonise a suitable extra-planet will be very lengthy; therefore, one has to plan and prepare how to get there as soon as possible . We cannot say *when*, but there are detailed plans on *how* this will happen; in fact, various models have been published on this subject in the scientific literature of the past few decades.

---

[4] From the very beginning the main interest in spatial research lay in the development of spacecraft technology, whose military strategic importance requires no comment. Even the launch of Space Lab was intended to develop permanently manned orbital stations. The majority of the experiments carried out so far under gravity-free conditions have produced results of modest scientific importance.

The world from which the colonisation of the Universe shall start is still one dominated by man, but by a man who has completely used up his biological and intellectual potential. Therefore, even the enterprise of founding a new human civilisation on another planet might be only an intermediary solution. Very likely, man will eventually have to realise that all possible new environments will be increasingly adverse to biological life and that he will be absolutely unable to follow any path leading to survival; his only chance will be then to yield his place to an intelligent machine that will have to travel this path in his stead.

Detailed designs of this machine have been sufficiently studied by computer and robotics specialists who have declared that the construction of a rudimentary prototype might be possible within a few hundreds of years. The brain of the machine is to consist of the fastest central processor with the greatest memory, programmed for artificial intelligence and containing initial instructions for constructing the machine itself; it must also be able to learn and improve itself, even on planets far from the earth. The machine's peripheries consist of robots and modules to execute all operations necessary for the duplication of the machine, starting from materials supplied by the environment (a similar objective doesn't appear so extraordinary, since, even now, fabrication of certain electronic devices is completely automated and the same is true of an increasing number of mechanical machineries).

The ultimate mission of man would be to repeatedly launch intelligent machines of this kind into space, for exploratory voyages that could last thousands of years or longer. Having found a suitable planet, the machine would receive fresh instructions from earth. Under favourable planetary environments the machine could create sizeable colonies, increasing its own dimensions and improving its abilities, until it was itself able to construct and launch space vehicles that would allow them to take further steps towards even more distant worlds. In short, it would be a question of starting a chain reaction which one could, in the beginning, control to a certain extent, but that, after a sufficiently long period of time (perhaps millions of years or even more) had elapsed, complete self-sustenance would have been achieved. In the improbable case that the earth still existed and were inhabited, the messages that man could send to the machine would in the end become superfluous and worthless to the supercomputers that the machine would have meanwhile produced in distant regions of the universe.

It is doubtful that such a perspective could arouse general enthusiasm in society of the 21th century. In fact, this enthusiasm is currently restricted to small circles of specialists, in some cases esteemed, highly cultured persons. It is, however, amazing that in these studies which have been the focus of much reflection and thought one considers, on the one hand, the ultimate aim of human history to be the conversion of all human knowledge into a sequence of impulses recorded on some form of artificial memory and to its blind cosmic proliferation. On the other hand, in improbable models where man still appears as a protagonist in the colonisation of the universe, he is supposed to be reduced to an *ovulum* frozen in space vehicles to be developed in an artificial uterus, once the goal of a new habitable planet has been reached. Some people argue that the *ovulum* could be synthesised upon arrival, by following instructions sent via radio from the earth (the memory of the machine might be insufficient to contain the necessary information). The machine could, finally, generate children on distant planets, nourish them, rear them, instruct them and assist them in forming a new human population. This could be possible, at least in theory: Indeed, these new men would probably find themselves in conditions much more favourable for proliferation than those met by Adam and Eve (after the fall). Moreover, the time which man has needed in order to develop an advanced civilisation (approximately 40 000 years) is very short when compared with the time in which a universal colonisation process is conceived to take place.

At first glance, such projects may seem to correspond to the legitimate will of man wanting to preserve the inheritance of his own species, but a deeper analysis reveals an alarming side to these ideas.

A thinking and self-replicating machine does not necessarily represent a monster to be exorcised from the vision of our future. Serious concerns, however, arise when this machine is intended as an equivalent or superior substitute for man and it is maintained that, through a sequence of manipulations, internal physical states can be created in the machine's processor that are perfectly equivalent to mental states imprinted on the human brain, in the same way as an operating system is loaded on an a newly fabricated computer. The analogy which has often been used and misused between "hardware/body" and "software/soul" reveals here its limitation. A human mind is much more than a programmed computer. Its memory and its states are the product of unrepeatable individual lives consisting of relations with other human beings, of interactions with the surroundings, of free decisions and non programmable creative or playful activities, of artistic production, of ethic and aesthetic motivations, and, finally, of the permanent influence of various passions. All this cannot be dealt with at the level of noise in the framework of mechanistic behaviour, as some people claim.

In reality, in the strategies of exploration and colonisation of space, independent of various, predetermined scenarios, the protagonist is not Man, but a being generically called Intelligent Observer, whose only proper and suitable attributes are promptness in elaborating data, a vast memory, longevity and an ability to survive and to operate under adverse conditions. These qualities can indeed be found to the highest degree in sturdy machines equipped with an artificial intelligence processor. Yet their progress is exclusively oriented towards one sole objective: to attain greater and greater expansion of themselves in the Universe and to reproduce all the while maintaining, for a limited time, contacts with their place of origin. From this perspective, most of what represents our cultural patrimony, both objective and subjective, turns out to be an hindrance, a useless encumbrance in this process of cosmic propagation of these "observers", whose behaviour would be more similar to that of a virus than to that of a human being. This remark will perhaps not surprise some biologists and anthropologists, but it must still be admitted that, from an evolutionary point of view, this kind of survival and reproduction has an essentially regressive character, since every reduction in the variety of vital functions when confronted with increasingly difficult environmental conditions represents a regression towards the order of inorganic matter, as dictated by the inexorable law of minimising any energy required for interaction.

Chapter 9
The Beginning and the "End"

"Multum adhuc restat operis multumque restabit,
nec ulli nato post mille saecula,
precludetur occasio aliquid adhuc adiciendi."[1]

(L. A. Seneca, *ad Lucilium VII, II*)

This rapid survey of the vast material from the preceding chapters has led us to consider the history of science as a universal indication of Man's path from his beginnings to modern civilisation. Moreover, this survey has caused us to view future horizons with circumspection beyond which point man's role as protagonist is uncertain, for one perceives his substantial dependency on the process of cosmic evolution of which he is a labile, miraculous product.

We have reached a point where we can summarise the sequence of considerations made so far and try to devise historical co-ordinates for our own civilisation as well as to guess at the direction in which science seems to be moving. Our purpose, however, is not to attempt to foresee the future of science. We have seen that this is hardly possible, even for exact sciences. Actually, whenever we try to pre-establish the path of scientific progress we must be prepared to expose ourselves to the risk of going down a blind alley or, perhaps, of committing fatal errors. Nevertheless, the only compass we can effectively use in our navigation of the future is our knowledge and the rudder is how we use it. Even amid contrary winds and stormy waters they remain the only instruments that allow us to change the course whenever we deem it necessary and allow us to stay on to the new trajectory. But we cannot live in a dream whereby a course can be directly given or influenced by science alone.

---

[1] *"Much work still remains to be done and much will remain, and the opportunity of adding still more will not be denied to anyone born in a thousand centuries."*

C. Ronchi, *The Tree of Knowledge*, DOI: 10.1007/978-3-319-01484-5_9,
© Springer International Publishing Switzerland 2014

9.1 Doomed to Progress

Strangely enough, the issue of the destiny of mankind in a far off future has always been the object of religious, but rarely of scientific concern. Conversely, however, none of the great religions considered the history and progress of human civilisation in a cosmic backdrop, as an integrating part of the end of days of creation, the "*éschaton*". The individual who renews himself from generation to generation represents the only precarious, but central being whose ultimate destiny religion tries to elucidate. From this perspective, the acquisition of knowledge takes on a relative value for man when confronted with the responsibility of his ethical choices. Nevertheless, even within this religious vision, knowledge is still part of the fundamental reality represented by the individual. Since time beyond our reckoning, men have perceived, albeit incoherently, a connection between knowledge and acquaintance with good and evil. In the Prologue to this book we mentioned that Biblical references are clear on this point, but even the ancient peoples of Indo-European origin had felt it necessary to distinguish between the twin mental states defined by the Latin "*sapere*" and "*cognoscere*", with their correspondent abstract concepts of wisdom and knowledge. [2]

"*Sapere*" indicates a contemplative state, characterised by ending the conflict between perception and reflection and always implies knowledge in a positive sense, as a fundamental value in relation to the perfection of a human being.

On the other hand, the verb "*cognoscere*" indicates an evolving mental state, an action provoked by a perturbation caused by a variety of feelings, such as curiosity, expectation, desire or fear, *etc.*, primordial impulses that continuously arise in man, as well as in other highly evolved living beings. Although knowledge has an instrumental character, which can be discerned for the most part by its capacity to soothe these perturbations, its acquisition does not necessarily entail attainment of well-being and happiness or, in an ethical sense, of goodness. The primary effect of knowing is to expand the time distance between the moment being lived and the horizon of consciousness, freeing the individual from the constant threat of unexpected events that are incumbent, without, however, saving him from recurrent conflicts with his surroundings. For millennia peoples of every culture have constantly been obsessed by wanting to know the future, whatever it may hold for them. To this end, they cultivated and practiced the arts of divination in its various forms, while at the same time they used magical arts to try to influence causal chains of events of which

---

[2] These Latin terms have retained their original significance in all European languages, but evolving differently. Latin *sapere* (to have taste, discernment) was maintained in the Romance languages, but in German changed to *wissen* (Germ. *Weißheit*, Engl. *wisdom*) deriving from an Indo-European root that means "to see". The root of Lat. *cognoscere* is found in Engl. "*know*" and Germ. "*kennen*", *etc.* Between the Latin *sapere* and *cognoscere*, *scire* can be found (and hence *scientia*), a word derived from an Indo-European root, which means "to saw, to divide". Science , therefore, referred to an act of criticism (Gr. κρίνειν: to divide) and discrimination.

It is worthwhile noting that in English culture, solidly founded on empiricisms, the word "*sapience*" (Lat.*sapientia*) has lost in spoken English its original significance and has instead taken on an ironical nuance.

they were not in a position to know their individual nexuses, but they perceived the fortuitousness and precariousness of these ties, and hoped to be able to manipulate them. Divination and magical arts were employed as ambivalent instruments for good or evil, according to the agent's will.

We have seen that a particular form of exploration and observation sprouted from this *humus* and from these beginnings modern science was distilled, retaining, however, its original ethical ambivalence. Until natural philosophy achieved an awareness of its own methods and autonomy, knowledge and divination, science and magic walked side by side, even in evolved civilisations, without conflict or contradiction, since exercising these arts was perceived to be a common attempt to inform the individual and protect him, as much as possible, from ominous future incognitos or to provide him with destructive forces to use against enemies or threatening events.

From this dual perspective and within a social body of common consolidated interests, knowledge assumed an important stabilising function and, therefore, its acquisition continued uncontested in all intellectual and practical activities of man. However, in early human history, knowledge did not constitute a common patrimony from a modern point of view, but rather belonged to consortia or restricted classes, where it was used as an instrument of domination or as a symbol of social prestige. On the one hand, its transmission was often kept secret and its practical application restricted to narrow circles, on the other hand, research and experimentation in all possible forms was generally promoted, or at least not hindered, by the whole community.

The situation changed with the historical development of Western civilisation. Starting from the Middle Ages, knowledge became a privileged resource of *clerici*, who still constituted a caste, albeit variegated, recruited from all social levels and rapidly assuming an international flavour. This class was highly influential on a cultural level, but political power lay in other hands. Nonetheless, although mediaeval universities were subject to ecclesiastical jurisdiction, owing to their political disassociation, disparate schools were able to flourish and, even despite rancorous mutual conflicts, prospered alongside one another. In the newborn universities natural sciences did not enjoy the status granted to theology, philosophy or exact sciences, such as arithmetic and geometry, but they benefited from almost complete independence.[3] After centuries of stagnation, the trend towards empirical sciences and technological developments started increasing again: astrologists searched the celestial bodies mainly to discover their influence on man, alchemists experimented on both inert and living matter without limit or scruples concerning the scope of their experiments, physicians had free access to human bodies for investigation, chemists

---

[3] That there were almost no impediments in the choice of research instruments and objectives is also demonstrated by extreme and sometimes monstrous examples. Some horrifying cases are known about the experiments conducted by the King of Sicily Federick II of Hohenstaufen (the founder of the University of Naples), who drowned men in sealed water-filled casks in order to extract their souls and bottle them through a hole; or when he confined children since their birth to completely isolated cells in order to establish which language they spoke when they reached the appropriate age (a thing which never happened).

produced and applied every type of remedy and poison, engineers were called on to design new machines of every type and architects to construct cathedrals whose boldness remained unequalled in the centuries to come.

Whatever the social fall-out of these intense experimental and speculative activities was, they very rarely were opposed effectively and, when this happened, the negative effects were always ephemeral.

There are indeed situations where new developments in science or technology appear to be in conflict with the views or interests of authorities or large population groups. Some cases of public repression of scientific or technical activities have been reported and analysed in detail.

The most notorious is that of Galilei, whose sentence didn't, however, influence or delay the progress of physics.[4]

Examples of opposition to technical progress in historical ages are relatively few. One of the first occurred in the eleventh century when European noblemen revolted against the use of the crossbow with a steel arm, whose arrows could pierce a knight's thick armour. The same opposition took place 200 years later against the use of fire weapons. In both cases, the protests had no effect; on the contrary, during the successive centuries the progress of steel metallurgy was mainly due to the fabrication of guns of increasing power, at least until World War One.

Other cases of opposition even to beneficial technical innovations have been reported, when the interests of certain professional categories were endangered. We may cite for example the canalisation of the Rhine river realised in 1817 by Johann Gottfried von Tulla and fought against by boatmen using guerrilla warfare but their opposition was harshly repressed. In the same years, we may recall the revolt of Parisian tailors against the increasing implementation of sewing machines that terminated in 1841 with the complete destruction of the Thimmonier factory.

Finally, we may also mention political movements which have opposed atomic armaments, movements which, in spite of their great international resonance, could

---

[4] For more than three centuries Galileo's case has been used by an army of controversialists to vituperate the Catholic Church which was accused of having stymied scientific progress. Even in 1980, in a BBC programme disclosing scientific "truths" by J .Brunowski [77] we can find the emphatic statement: that "the sentence (of Galilei) put an end for centuries to the scientific tradition in the Mediterranean countries". To judge the weight of this verbal nonsense it should be enough to consider the case of the great astronomer Giandomenico Cassini. In 1648, only a few decades after the ecclesiastical censure of the heliocentric theory, Cassini was appointed director of the observatory at Panzano (Bologna), which was situated in the Papal States, where he further developed the ideas of Galilei, applying them to his excellent astronomical measurements that contributed substantially to the progress of modern astronomy. Furthermore, when between 1655 and 1659 Newton's master Isaac Barrow was living in Italy for a lengthy stay, he was strongly impressed by the scientific culture and the refinement he found there and, returning to Cambridge, he could pass on to Newton many ideas and notions he had learned there. Finally, one should remember that the *editio princeps* of the work of Newton was published in 1726, together with a scholarly *Commentarium Perpetuum* by two French mathematicians Thomas le Seur and François Jacquier, both Franciscan friars. Evidently the dispute on geocentric vs. heliocentric models had very soon become uninteresting and theological irrelevant, despite Galilei's formal sentence having only been revoked in 1757 by Pope Benedict XIV.

not prevent large nuclear arsenals in USA, England, France, Russia, China, Israel, Pakistan, India and North Korea from being constructed. Very likely, the same will happen with the opposition to the pacific use of nuclear energy, which is presently vehemently disputed by groups of environmentalists.

One can doubtless assert that the progress of knowledge in the history of mankind and its applications has been always relentless and, one could even say, inexorable. In his famous analysis of the development of science, Robert K. Merton [82] observed quite rightly that the great discoveries first mature in the cultural substrate of a given society and then appear almost at the same time, in diverse, independent forms.

On some historical occasions, it can be clearly seen that scientific and technological progress even occurs when it does not imply any immediate benefit for all. For instance, in Europe the first industrial revolution at the end of the ninenteenth century was triggered by new scientific discoveries and supported by positivistic philosophical currents that preached the messianic role of science. The social changes that took place at that time were of great significance. Among them we can number the rapid spread of urbanisation and the concomitant gradual disappearance of agricultural civilisation. If we consider from a modern perspective some aspects of these changes, for example the workers quarters in certain English towns or the living conditions in industrial plants and mines, we are filled with horror and outrage against the supporters of this revolution. The meagre advantages that it offered to the masses were accompanied by an unprecedented degeneration of the environment and of the standard of living. The industrial *habitat* was a true hell compared with that of the country. Nevertheless nothing could stop the spontaneous migratory flow towards the cities. There are at least two reasons for this: the first was an objective consciousness on the part of the rural population of being deprived of any possible promotion because the peasant's life was bound to a restricted territory whose limits had been effectively unbreakable for centuries. The second was the persuasion of wanting to be part of the winning side, a side which possessed knowledge and produced goods and money in ever-increasing quantities. The labourers did not look at their current grey surroundings but at the shining Sun of the Future, a vision the peasants could not picture for themselves, since for them the sun rose and set every day, always the same within an alternating sequence of the same few joys and many pains. With the passage of time—decades and not centuries—the choice turned out to be the right one, inasmuch as the standard of living of labourers increased more than proportionally with respect to that of the wealthy classes. However, the success of the winning party had its price: to be able to maintain its advantage on activities of peasant's and artisan's classes, production and productivity in industry could not stop growing. That was made possible by new scientific knowledge and further technological applications, but the final consequence was that all the resources of food, materials and energy available on the earth had to be implicated. This process was so comprehensive and all-pervasive that concepts like exploitation, optimisation, integration, recycling, conditioning, dismantling, disposal, obsessively present in modern technical-scientific jargon, can be found more and more frequently in corresponding forms in economic, social and psychological languages.

It is a matter of fact that, starting from the period following World War One, the success and achievements of any enterprise depends, in a first stage, on its rate of growth (starting impetus) and, in a second stage, on its ability to compete with similar competing enterprises (acceleration). For this purpose, whatever the nature of any enterprise, every available knowledge is used with a view towards improving production and introducing innovation. The result with which we find ourselves confronted is a great and ever-growing inequality between societies that promote and control rapid progress and those irremediably inadequate to sustain its impact.

Nowadays, unlike the past, underdeveloped and highly civilised societies are not necessarily separated geographically, but are often intimately connected through massive and increasing immigration and interpenetration, which continuously creates a much deeper and variegated social stratigraphy, especially in urban areas, than was the case in the past.

Therefore, when one speaks of promotion of scientific progress and freedom in research, it is necessary to consider not only the speculative value of the knowledge acquired, but also its influence and effects on the current global society, characterised by radical inner contrasts and strains, which threaten to tear it apart.

In our times, scientific and technological progress necessarily entails a parallel, substantial increase in the flow of information within the whole planetary population. This fact has actually not only destabilised totalitarian political systems in countries where a rigid control is exerted over all aspects of local culture, but has also made difficult governance of democratic countries. This information flow has been emphatically hailed by many people as the confirmation of two equations "knowledge = democracy" and "ignorance = totalitarianism". In reality, the matter is much more complex and, hence, more problematic. Even if it is true that it is much more difficult to exert absolute authority on a well-educated society than on one composed of ignorant and uneducated people, it is equally true that the former is strongly conditioned by a mass of information and commonplaces, which are strictly accepted and effectively considered beyond discussion; in this regard, one should not forget that the word "instruction" has the etymological significance of "prepared, manipulated",[5] a significance which still remains attached to one of the aspects of instruction.

Furthermore, until about two centuries ago, the foundation and growth of cities had always produced cultures and lifestyles, which were sustained by a harmonious development of composite socio-geographic surroundings. Today, science and technology have accelerated and amplified the growth of human mega-agglomerates, whose only purpose is to carry out an increasing number of technical functions, like communication, transport, production, trade, *etc.*, which are absolutely indispensable for their daily survival. The spirit of the inhabitants, their feelings and aspirations, are relegated to a social subsoil, from which they can hardly emerge. Lewis Mumford, a distinguished historian of science and technology, renowned for his studies on urbanism wrote in 1937 [80]:

---

[5] Oswald Spengler in his analysis of various forms of civilisation repeatedly evoked the ghost of a culturally uprooted class, even though not necessarily of a low status, called *"Fellacher's"*, which is deprived, *de facto*, of any political and decision-making role [65].

"Our capacity for effective physical organisation has enormously increased, but our ability to create a harmonious counterpoise to these external linkages by means of co-operative and civic associations on both a regional and a world-wide basis, like the Christian Church in the Middle Ages, has not kept pace with these mechanical triumphs."

When Mumford was writing these considerations on "social mechanics", he meant all mega-machineries, sometimes uncontrollable, which consist not only of material objects, but also of social organisms, such as dominating inflexible bureaucracy, inaccessible political and industrial hierarchies, tightly framed professions, oppressive fiscal apparatus, monetary and banking systems, and so on (today we could add and place it at the top of the list, automatic systems of surveillance, control and documentation of a citizen's life).

Society is in rapid evolution—concluded Mumford—and, if it is true that only the most adapted individuals will be able to survive, we see no reason for believing that these will be our ideal types. On the contrary, those who will survive will likely present characteristics which are at present repugnant to us. The example of the complete success of a rigorously well-organised social system was at that time provided by German Nazism. In the beginning, there was nothing, in the theory of social evolution, to indicate the evil nature of this system and could prefigure its final historical sentence. The moral judgement that might have been passed in this regard was, in fact, to be found in a scale of values completely foreign to the universe of science and technology.

On the other hand, even modern democratic States are in a state of deep crisis. Increasingly conditioned by economic problems, the power of governments and of the parliament has been left with little room to manoeuvre. States, organised as a great business enterprise, are subject to profit dynamics, with all its concomitant problems: financial regulation, productivity and competitiveness. On this level, modern States are confronted with multinational industrial and financial enterprises, whose activities support substantial parts of social infrastructures and whose economic power is large enough to influence and even determine political choices.

It would be, however, a mistake to think that new industrial elites are in a position to pursue their own pre-established policy. Even entrepreneurial and managerial classes are transitory and short-lived. Subject to the iron logic of success and profit, in an atmosphere essentially characterised by permanent internal and external challenges, rivalry and conflict, it must act pursuing short- or medium-term objectives.

In this scenario, even the acquisition of knowledge is considered as a productive activity ruled by market laws so that the character of the scientific research objectives being pursued has to fit this role. While in the past, scientific research was always intended to pursue general or, when possible, definitive answers to precise questions, something is changing dramatically in our lifetime.

When Plato introduced his views on the political and social function of knowledge in his *"Republic"*, he proposed an outline that could be summarised as follows (Fig. 9.1):

From the incentives of the world of the senses (A), the human mind rises into the world of ideas (B) where it formulates questions, whose answers, (C), obtained via

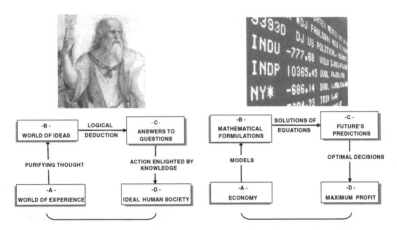

Fig. 9.1 Outline of the relation between progress and knowledge from the point of view of Plato—explained in his *Republic*—and from a modern point of view. The function of knowledge, which according to Plato should lead man to the "*Summum Bonum*", has at present turned into a specialised instrument for solving "*bene*" (*i.e.* well) particular problems

logical-mathematical procedures, allow Man to illuminate his actions, which are intended to create a perfect society (D).

Remote from all philosophical controversies, this methodological system has always been considered valid. Even the system in place today is essentially "isomorphic" to that of Plato, but the terms have changed.

Economy, meant in a general sense, has replaced the world of experience and reason is called upon to produce mathematical models, whose predictions are used in order to make decisions aimed at obtaining maximum profits.[6] The essential difference of the modern position is given by a reversal of the hierarchical position of the terms: if the ideal world (B) held the first hierarchical rank in the past, today this has been supplanted by experimental data, represented by world (A). The consequences are far-reaching. The demand to create trustworthy models of phenomena is based solely on employing these models to manipulate phenomena.

This model has no pretensions of seeking truth, but only that of providing detailed lists of sufficiently reliable predictions and instructions. There is no doubt the method works, but its new asset has implied that man renounce visions that extend beyond immediate experimental data.

---

[6] The binomial relation economy/profit here does not merely refer to concepts of wealth/gain, but, generally, to an initial state (a) with which we have a relation, and to a corresponding state (d), in which this relation is improved in quality. Note that this scheme has also been discussed with regard to the progress of modern mathematics according to the expectations and to the effective creativity of future generations (see W. Haken [75])

If we consider modern research both in its structure and in its results, we can actually realise that knowledge is increasingly considered as something extemporaneous and provisional, and definitively degraded from being intimately implicated in man's final aim to the rank of instrument of production and profit.

9.2 Reality and Utopia

Today, when we examine current views on the future of mankind, apart from utopian visions we also encounter necessary projects which, however, can only be realised after solving a number of serious problems that demand decision-making power and economic means, which, however, are currently not available, neither in public nor in private hands.

In the seventeenth century, meditating on the possible scientific discoveries which were expected to occur, Francis Bacon produced a vision of the future, described in his "*New Atlantis*", where a perfect society was prefigured, in which peace and well-being were rendered possible by the conquests of science: "knowing is power", he claimed. Bacon produced a list of technical and scientific innovations supposedly achieved in Atlantis, where scientific knowledge represented the hinge, around which swung the lives of its fortunate and happy citizens.

Viewed from a distance of three centuries, most of the discoveries and innovations imagined by Bacon have been effectively realised, without, however, producing the harmonious society envisioned by the English philosopher. On the contrary, conflicts among peoples have multiplied and have become more serious.

The root of these conflicts lies in the drama of real society. In fact, perfect utopian societies of any sort are always governed with an iron hand and the happiness of the citizens is due to the fact that the existence of unhappy dissidents has been preliminarily excluded. If it is true that the dream of Utopia has gained impetus from critics of the real, imperfect society, from an ethical aspect, however, this dream can be appropriated by radicalised intellectuals, who substitute trumped up evidence of abstract principles for the incoherent diversity of concrete situations. At this point eliminating dissidents is not merely a lack of concern on the part of supposedly unconvinced and unpersuadable people, but becomes an active repression of all "*Andersdenker*" justified on a pragmatic basis. Therefore, when Utopia is assumed as a political objective, its missionary phase always entails violence and blood—as demonstrated, for example, by Jacobinism in France, Bolshevism in Russia, radical Communism in China, *etc.*, (a complete list would extend over several pages).

A positive aspect of utopia is the desire to fill the gap between the real social order, and the ideal one. Utopian visions can be taken as goals, but the problem, perhaps insoluble, consists in finding the path to achieve them pacifically.

Let us examine Bacon's vision, updated to fit our present world, by compiling a new list of discoveries and implementations we may think necessary, if not for the purpose of re-founding a new perfect society, at least for the survival of the actual one. Without any pretence at thoroughness, the modern list of "expected conquests" of science could be imagined as follows:

Advanced sources and reserves of energy, largely exceeding current levels.

Development and implementation of recycling processes in the extraction/ production of raw materials.

Reduction of the environmental impact of all industrial processes.

Regulation of the purity of air and water on earth.

New design and reconstruction of most of the present urban areas.

Rationalisation of transports with rapid methods of underground transportation.

Widespread circulation of information and standardised quality of contents.

Total automation of repetitive or risky labour.

Effective instruments of defence against natural catastrophes, such as extreme climatic variations, storms, volcanic eruptions, sea- and earthquakes, and meteorite impacts.

Construction of nano-machines for medical operations and micro- mechanical and biological applications.

Control of epidemic diseases and development of preventive medicine.

These are very ambitious goals, but, in the light of existing knowledge, not unattainable. However, if we examine the conditions under which this scientific and technological progress is expected, we realise that if the entire world does not benefit from them, a sort of Pandora's Box will be created. By this I mean that together with the obvious benefits an even greater number of conflicts and disasters will take place. The ethical ambivalence of knowledge will reveal itself with repercussions so serious today we can only vaguely guess at their magnitude, considering, for instance, the quantity and power of the weapons currently in circulation, the fragility of advanced civilisations in the face of natural catastrophes and wars, incurable social conflicts and terrorism, overpopulation and unlimited industrial production. Even freedom of thought and action can assume insidious forms of anarchy and become a source of physical and psychological violence.

On the other hand, all utopian societies conceived by philosophers, from Plato to Marx, have as a common foundation the consensus of the people prevailing on diverse groups, an agreement which renders democratic government possible and prevents authority from turning into tyranny.[7] Whether and how harmony can be reached in human society, preserving the necessary diversities and freedom of choices, represents a central problem and the dividing line between Utopia as a dangerous mirage and Utopia seen as a possible historical horizon.

For centuries an uninterrupted philosophical tradition has asserted the ethical value of knowledge and the essential equivalence of Truth and Good, which can be reached by all men through the use of reason. This axiom has been reaffirmed by Christian theology, which, however, pointed out the—not obvious—correct use (*rectus usus*) of reason, since imperfect knowledge does necessarily lead to questioning its acceptance as a common good and may entail deadly differences and strife. Therefore, in the

---

[7] In the Roman Empire, beginning from Galba until the Antoninian dynasty, the image of *Concordia Provinciarum* is sometimes represented on the back of coins (see D. R. Sear [84]). Whatever the political vision of the leading class was, it is a remarkable fact that at that time one conceived a unity of ideals and intentions for all peoples, from the north of Europe until Africa and Asia.

Christian society of the Middle Ages religion claimed for itself the role of advocate of the principle of supremacy and the subordination of temporal power to the supreme authority of God, the "*Summum Bonum*", a principle condensed in the lapidary maxim "*Per Me Reges Regunt*" engraved under the frontal cross of the crown of the Holy Roman Empire.

This principle survived amid false interpretations and abuses which occurred in the history of Western civilisation. It has been reinterpreted by Illuminism, albeit with anti-Christian aspects, as well as by positivistic and philanthropic movements of the nineteenth century, which made Reason and Progress the criterion of supreme authority. Throughout all these historical vicissitudes this concept has rendered possible the birth of western democracies, which became increasingly distant from the despotic forms of government of the great civilisations of the Orient.[8]

Yet the cultural crisis of the last century has completely changed the terms of this fundamental assumption. The downfall of metaphysics has undermined the very concept of an objective truth, degrading knowledge to a psychological mechanism. The wheels of progress became detached from the two axles of metaphysics and religion, continuing to turn faster and faster, but in mutable, unforeseeable and uncontrollable directions. In the Occident we have born witness to the complete disintegration of the intellectual elite for some decades now. Nihilism and relativism have extinguished a feeling and sense of belonging and participating in a common civil project which used to exist in individuals. Everywhere in Western societies one observes a regression to a culture of tribal character that manifests itself in fractions and sects of any sort and kind, with its own hierarchies and values, incapable of amalgamating or even communicating with one another.

In this scenario we must seriously ask ourselves if it is still possible to envisage human progress leading towards a social model founded on civil harmony. The present situation certainly does not encourage us to devise such a course. On the contrary, scientific and technological progress has created serious dilemmas bound to the pernicious use of new products of knowledge, not only in less evolved countries, but also in advanced societies, where ample levels of the population are avid to receive them, but unprepared to avoid their deviating and devastating uses.

The problem of the necessity of human progress having ethical and social aspects is dramatically and urgently phrased in the face of extending the power of knowledge. It has been observed that, historically, the acquisition of knowledge took place within the intellectual elite, but it was the appropriation of the new discoveries by society that guaranteed the continuity and increased the speed of this very progress. However, very rarely has this happened according to pre-established plans. In the past, the subjects intended for study and research resulted from their social and cultural context, but they were never seen as "projects" in a modern sense. Today research is planned and financed for well-defined intentions. Even if this occurs in a climate of freedom and of continuous discussions fully open to new ideas and views, it is, how-

---

[8] Oriental despotism exhibits fundamental and structural differences not only from democracies, but also from historical forms of Western monarchic absolutism. The argument is developed, from a Marxist point of view, in an admirable monograph by Karl Wittvogel [83].

ever, a matter of fact that the increasing complexity of the problems investigated and the enormous cost of running the experiments render scientific progress impossible outside a mainstream culture. On the other hand, the possibility of promoting and of directing research and acquisition of knowledge in selected areas and directions represents a key factor in solving the present crisis of our planetary civilisation. The problem of subordinating progress of knowledge to human progress does not have fail-safe solutions, but the destiny of humanity is inextricably bound to our ability to find them.

9.3 Towards Sustainable Progress

Industrial progress during the last decades has activated fatal mechanisms that evidently must be stopped as soon as possible, while others are to be promoted with the available means and upgraded in the future with the aid of new knowledge.

A most ruinous strategy of Industry has been to increase all kinds of production beyond any reasonable limit, submerging under-developed countries with a surplus of production. The resulting tendency was to channel the changes in way of living in these countries into the same road that we have travelled on, reckoning, in the best case scenario, on a rapid recovery of their delay in development.

Yet, we are aware that this would exacerbate the problems connected with the sustainable limits of this policy, problems which have already been encountered by advanced countries (energy crisis, deterioration of the environment, widening gap in living standards between upper and lower social classes, *etc.*,). The remedy consists in investing in innovations and new technologies involving *primarily* poorer countries with basic objectives, for instance:

Fair and even distribution of energy at costs comparable with those in the years following the Second World War while its production from advanced sources would be implemented on the basis of international agreement and co-operation.
Constructing towns of the future to replace infernal "megapoleis" having risen from the waste of technological civilisation.
Creating new geographic space by opening a third dimension where automated industrial activities are confined to the underground.
Enabling people's mobility to be supported by innovative means of transportation, and drastically reducing automotive traffic.
Exerting a reasonable demographic control that makes it possible to optimally increment the population, according to the objectives being followed.

It is not a question of compelling the advanced countries to undertake an immense philanthropic action in favour of the third world, but rather to start constructing our common, future planetary *habitat*. This policy is not even imaginable in the context of endemic conflicts between states and ethnic groups.

Considering the trend of the history of the last two centuries, one must recognise that human progress has not been as rapid as scientific and technological advancements have. On the contrary, both rich and poor countries, of ancient and recent civilisations, have abundantly provided proof of our egoism and ferocity, with genocides, wars and slaughters, with warfare between powerful criminal organisations and with social conflicts perpetually being revived through seemingly inextinguishable mutual hatred.

It is no wonder that today many people see in the progress of science and its applications a danger to mankind and advocate a return to a simpler lifestyle, in harmony with nature. From this perspective, the concept of sustainability of any human activity is no longer tied to its technical feasibility and to an appraisal of its risks and benefits, but rather to the possibility of carrying out the activity in question with a minimum of impact on our environment, which is to be maintained in a state of stable equilibrium. The strategy proposed by modern environmentalists is fundamentally oriented toward imposing a collective behaviour with regard to parsimony and saving of resources, control of production and consumption, limitation of mobility and tightened demographic control. Since, however, a return to pre-industrial standards of living is untenable as a realistic global objective, [9] one ends by relying on vague expectations of new, miraculous technologies which would enable us to realise the above-mentioned objectives. Unfortunately, the vision of certain contemporary environmentalists is based on wishful thinking only. Actually, if extended to all peoples of the earth this vision entails the same limits as those of a utopian world, which have been described above: on the one hand, a positive view of an ideal technological civilisation with its own set of values and lines of demarcation, on the other, a real population strongly variegated in culture and quality of life. How would it be possible, for example, to impose a strategy of parsimony, to which poor countries would have to adhere for a long time in order to maintain the *status quo*, if not with coercion? Once again, we are faced with the necessity of distinguishing model from political objectives in a utopian vision.

Fortunately, modern society possesses instruments to regulate these objectives which were completely unknown in the past. Foremost among them is public management, direct or indirect, of research objectives. This power has two concomitant effects: the first one is to concentrate efforts on defined fields; the second is to prevent research from focusing on branches whose fruits may involve too great an imbalance in global society.

In Western culture dangerous ideas have been converging since the seventeenth century into a recursive vision of a social model, which appears to be feasible with the

---

[9] "Zero-growth" socio-political models were very popular in the 1960s especially due to studies promoted by the "Club de Rome". The basic idea was to create conditions for a worldwide uniform distribution of know-how and goods so as to reach a final common standard of living asymptotically established at the level of Western societies of that time. In reality, the state of these societies was already unstable and totally dependent on the cheap exploitation of primary resources located in the Third World. Therefore, "zero-growth" necessarily entailed a controlled regression of the welfare in rich countries: an extremely dangerous policy, since in any system controlling both progression and regression is much more complex than merely promoting growth, and presents serious dangers of destabilisation with respective risks of explosion or implosion.

advancement of science and these same ideas are now gaining ground in a much more insidious manner than that adopted in the recent past by some totalitarian regimes.

For modern man a salutary mental exercise would be reading and meditating on Aldous Huxley's "*Brave New World*" [88]. Written in 1932, the theme of this book is the advancement of science as it affects human individuals. According to Huxley the triumph of physics, chemistry and engineering can be tacitly taken for granted, but crucial advances are those involving the application of the results of future research in biology, physiology and psychology to human beings since these sciences can radically change the quality of life. The world imagined by Huxley is one where society is definitively stabilised once men became happy and satisfied after being genetically adapted to expect only that which is in accordance with available means and resources of science. In the brave new world this state is achieved through a last, ultimate revolution after which freedom, the alleged source of chaos and evil, is suppressed forever.

The ensuing horror story is a sort of prophecy that was promptly received by the intellectual class of that time. Bertrand Russel, who was very familiar with Huxley's arguments had this to say: "*It is all too likely to come true*". [10] Huxley's reaction to this fear led him to irrationalism: he finally embraced Hindu beliefs and habitually made use of psychedelic drugs , anticipating the hippy revolution of the sixties.

It is not easy for modern civilisation to react against scientism without falling into contrary but irrational tendencies. In order to avoid this pitfall, the direction and amplitude of human progress must be continuously adapted and not remained fixed as a result of abstract ideological thinking. Altogether, Francis Bacon's list, updated to meet modern expectations, does sufficiently represent the need to develop new knowledge in tune with a programme of human promotion of all peoples of our planet. Scientists must look at distant horizons, but always proceed with caution on the concrete but slippery ground of historical evidence. Short-sightedness and long-sightedness in science are equally dangerous in a society whose survival by now is irreversibly tied to scientific know-how applications.

But can we look to the future with the same detached optimism of the empiricists in the past? The sentence that we keep hearing over and over again today wherever scientific research is being carried out is: "*we must know more*". Strangely enough, this statement did not appear nearly so often in the past when, in fact, we knew much less. One should say that initially the salient character of our age was to raise doubts and questions of vital importance and that today our civilisation is harassed

---

[10] It is not a mystery that certain circles of biologists openly pursue research to widely implement artificial reproduction of men in view of creating genetic standards, as already is the case for breeding of selected races of animals.

We should at least be able to expect a little caution on the part of these biologists in how they tamper with the statistical stability, including its inherent variance, of our genetic patrimony, which was derived from a variety of events that have safeguarded the persistence of certain features of the human species in the face of external perturbations. However, these arguments do not adversely affect the rationale of these initiatives which is supposedly pure in its ideological character and indeed aims at organising future man into predefined types.

by an urgent need to find adequate answers. We cannot be satisfied with guesses and conjectures to these answers, but neither do we know if science will be ever capable of providing them. Nevertheless, there is no other way than that implying a further advancement of science, but not merely because of its technological applications, or for our health and longevity or, simply, for a comfortable life, but quite simply because we can still hope that from science will ensue wisdom, without which survival would be equivalent to a sentence in hell.

9.4 A Final Parable

Around 1935 Jorge Luis Borges published a collection of writings which he modestly termed "exercises in narrative prose" [85]. Among these, there is a short story, which is only a few lines long, entitled *"On Rigour in Science"*. We quote here the entire text, maintaining the original Spanish with its deep literary resonances and the (important) use of capital letters.

Del Rigor en la Ciencia

En aquel Imperio, el Arte de la Cartografía logró tal Perfeción que el Mapa de una sola Provincia ocupaba toda una Ciudad, y el Mapa del Imperio toda una Provincia. Con el tiempo, estos Mapas Desmedurados no satisficieron y los Colegios de los Cartógrafos levantaron un Mapa del Imperio, que tenía el Tamaño del Imperio y coincidía puntualmente con él. Menos Adictas al Estudio de la Cartografía, las Generaciones Siguentes entendieron que ese dilatado Mapa era Inútil y no sino Empiedad lo entregaron a las Inclemencias del Sol y del los Inviernos. En los Desiertos del Oeste perduran despedazadas Ruinas del Mapa habitadas por Animales y por Mendigos; en todo el País no hay otra reliquia de las Disciplinas Geográficas".[11]

This short story is clearly a metaphor, which is, however, open to a variety inter-pretations of different levels, as we discover the significance of the archetypes that figure in it: Apart from the title, the word Science does not appear in the text, but in its place we find the term Map, produced through an Art that demands Study and Discipline in order to achieve Perfection. The Cartographers find in their satisfaction a motivation for their work that proceeds beyond what can reasonably be measured. Then we encounter the sudden *caesura*: the Following Generations do not aim at Perfection, but appealing to Measure, they pass judgement on the Map: it is Useless.

---

[11] *"...In that Empire, the Art of Cartography attained such Perfection that the Map of a single Province occupied an entire City, and the Map of the Empire an entire Province. With time, these unbounded Maps did not satisfy and the Schools of Cartographers compiled a Map of the Empire that was the Size of the Empire and corresponded precisely to it. Less Addicted to the Study of Cartography, the Following Generations understood that this expanded Map was Useless and indeed not without Impiety, they abandoned it to the Inclemency of the Sun and of the Winters. In the Deserts of the West broken Ruins of the Map survived inhabited by Animals and Mendicants; in all that Country there is no other relic of the Geographic Disciplines."*

The Map is sentenced to be abandoned in the Deserts of the West where its Ruins serve as shelter for Animals and Beggars.

We leave the reader with the liberty and pleasure of fully deciphering the message.

Throughout various points of the preceding chapters a number of problems have been encountered regarding the multifarious relation between science and society. A product of the exclusive intellectual work of a restricted aristocracy, science has nevertheless been protected and its development supported by all historical societies due to a general interest in knowledge and to the benefits of its applications. But interest alone on the part of members of any society can only persist if the scientific contents are comprehensible within the limits of common sense. An abstract science formulated in an incomprehensible language will not be able to sustain common interest for very long. Consequently, its indispensable public support will finally depend only on the material profit science can offer. What this implies can already be experienced in our times: What stands out in the long term is a perversion of science and its exclusive enslavement for indiscriminate technical applications to a cycle of production and consumption, in which the word usefulness is exclusively synonymous with monetary gain, that is to say, with supplying more goods.

I am not talking about a vague hypothetical risk, but a real incumbent danger. The world of science, having rid itself of philosophy and religion, considered itself able to proceed alone in guiding our society. Furthermore, in the modern age, a general consensus has bestowed an aura of impartiality and democracy upon the decision-making power of science. But the price of this consensus is increasing all the time and it is likely that one day science will not be able to pay any longer. We will then face a catastrophic alternative: the ruin of science or its tyranny over our society.

This ominous outcome can only be avoided if science is able to walk once again on the same paths as philosophy and religion, its original road companions in the history of mankind. Actually, man can find only in philosophy and religion an incentive to introspection, from which originates the self-esteem necessary to harmonise the significance of his own existence with that of the Universe, and of his individual life with that of his fellows. These stimulating aspects have shaped during two millennia the culture of the Occident, a culture which in our times is expected to deeply imprint the newborn global civilisation. It has been never an easy companionship; the relations of philosophy and religion with science have often led to conflicts and very rarely to exchanges of opinions that were *immediately* useful to one another. Nonetheless, their opposing views and their taking turns in guiding the course of history have substantiated all three. Ludwig Wittgenstein asserts that they represent three forms of knowledge of a diverse nature which can be above one another, but never together on the same level. The starting proposition of his *Tractatus* [54] asserts that the world is a totality of facts and the facts in logical space constitute the world. All the following, rigorously organised reasoning of Wittgenstein intended to confine human scientific language to this space, the only one in which the purely logical concepts of true and false have a meaning. In this picture it is impossible to speak of free will as a subject with ethical attributes. In particular, ethics, asserts Wittgenstein, is outside logical space and belongs to what he calls "the world of mystics", where nothing can be *said* but only *shown*. There exist, in fact, things that cannot be log-ically expressed with words, but that simply manifest themselves. Scepticism with regard to this world is nonsense, since it is senseless to raise doubts where strictly logical questions and answers cannot even be formulated. It is, however, up to Man to decide and to establish at every instance their hierarchical order.

One can disagree with these conclusions, but one must recognise their validity in defining the limits of logical-mathematical knowledge and in pointing out the fundamental issue of whether man is one of the many facts that fills up logical space or if, being above all forms of knowledge, Man rises in his entirety, with his capacity to construct his own history, both individually and collectively, on the basis of an intricate weaving of subjective and objective conditions, where his freedom of thought and action is realised and from which emerges his ethical responsibility. Here is the inextricable knot of the question that saint Paul picks up in the second chapter of his first letter to the Corinthians (1 Cor. 2,11) at the apex of the tension surrounding a series of questions on the wisdom of world: *"Quis enim hominum scit quae sunt hominis, nisi spiritus hominis, qui in ipso est ?"*. [12]

From the eighteenth century until now the emphasis has been that human reason is the main, if not the sole, source of knowledge and that scientific method is the only criterion for action with a view to the progress of mankind. Actually, the promotion of reason by Illuminists was due to the necessity of breaking off the ties of mental schemes, which were becoming closed and senescent. But the following cultural revolution affected Western civilisation in every way. What started out as a war on freedom of thought ended by becoming a war of conquest and exclusive hegemony of scientism. The people, who in the century of *"Lumières"* liked to be called *"philosophes"*, started a corrosive polemic against all forms of knowledge whose formulation could not be reduced to logic and mathematics. The first enemy to be discredited was religion, from theology to ethics. Then it was the turn for metaphysics and, later on, all forms of philosophy which were not a mere comment or appendix to scientific thought. We can assert that during the last three centuries the final objective of scientism has been to swallow up the whole mental activity of man, exterminating, idea after idea, all forms of thought that it considers foreign to its world.

When, however, even the horizon of science was finally restricted by its own relentless criticism, which is one of its essential features, science found itself facing *the abyss* of the unknown with the desert of scepticism at its back. In this "Desert of the West", science has remained alone, without point of reference, and moves along without knowing to where and why.

The original critics raised by scientism to philosophy and religion can themselves be criticised for their attributing a higher rank to scientific knowledge than to reality. This way of completely "digesting" experience by converting it into mental schemes represents a sort of re-creation of a reality found in the human mind, where science will finally turn out to be the definitive cage of the spirit.

---

[12] *"For who among men knows what is characteristic of man if not the spirit of man, which is in him?."*

Thus Stephen Hawking declared in a recent interview that God is a verbal expression to indicate what we don't know. In other words, God is the complement of our present knowledge system to a final one, where everything will be explained: once man will have achieved it, this complement will be empty. This argument arises from an obsessive need to construct more and more general and comprehensive models of the universe. But let us suppose that one day we shall find what we consider to be a definitive system of cosmological laws. What will be changed in our position in the face of a universe which preceded any attempt to describe it? The answer is: "nothing". We will in fact still be confronted with a sense of our existence and life, where faith and doubt are the two poles of attraction, poles which are, however, both located outside science.

Fortunately, no matter how strong an influence science may have in our society, scientocracy today represents a thread rather than a reality. In our life, most of our choices are conditioned by ethical and religious criteria. Even aesthetic perception and satisfaction, which are both foreign to reason, represent a fundamental factor in our culture. Yet, although one must recognise that logical-mathematical knowledge comes to a halt at the surface of things, in the currently prevailing culture what Blaise Pascal called the "*esprit de finesse*", a trait which penetrates reality in depth, has nearly been extinguished

In his *Treatise of Human Nature,* David Hume attacks this theme and provocatively writes his famous sentence:

"*Reason is, and ought only to be the slave of the passions, and can never pretend to any other office than to serve and obey them* "

There is no irony or blame in this affirmation which is meant to convey that reason alone is completely sterile in a practical world. It is hopeless to think that reason alone can induce us to act and it is doubtful that syllogisms can dominate passion. This does not entail a pragmatic supremacy of irrationality. If reason cannot stop a passion, it can, however, guide it. How many times, under its aegis, have passions, with their variegated facets, exerted a corroborating influence on the human mind? How many times have they had a determining role in scientific discoveries? How many times did a solution of inner contrasts oblige man to gamble all he was and had, prevailing over what he might have considered a definitive truth? How many times was the course of history radically changed by convictions originating from an unfathomable source of the "*raison du coeur*"?

One deals here with questions that, from the point of view of logic and mathematics, do not make any sense unless one conceives of man as the master and not as the slave of reason.

In Plato's *Phaedrus* the myth of the charioteer is represented by a vision ablaze in beauty, in which the soul joins the Gods' cortege in their periodic travel towards the highest celestial spheres. The charioteer directs the speed of the couple yoked to his chariot. The two horses, the passions, tend to follow opposite courses, but it is the soul that gives the chariot the necessary impetus to ascend, solely using the steering force of its wings.

Epilogue

The author and his reader are here at the end of their journey. The road still extends to a horizon beyond our reach, but we sometimes had to stop along the way. Considerations and reflections on human knowledge have kept us company, but we still find ourselves facing the initial questions that continue to amaze us. At this point an Epilogue pertains only to the end of our journey, but not the end of the interminable story of the Tree of Knowledge. Therefore, it is now appropriate to turn our eyes to the salient milestones we have passed.

Having originated from a planetary climatic catastrophe, quaternary man starts thinking and, within a few hundreds of generations, he succeeds in measuring the space of the Universe and the time that preceded his appearance on earth. He perceives the smallness of his living dimension and the tremendous brevity of his lifetime, but, nonetheless, he rises to take control of the earth aiming, in the end, at the entire cosmos. He doesn't merely pursue, as all other living creatures do, his own biological security and physical well-being. Rather, he begins asking himself about the causes of things and trusting his own answers. It is a risk he takes, the importance of which will never be completely perceived or accurately measured. Critical reason and knowledge compel him to clearly recognise his limits, but nonetheless he proceeds beyond them

---

[1] *"Consider how your souls were sown / you were not made to live like brutes or beasts, but to pursue virtue and knowledge"*

[2] *"Therefore, God has hidden in himself certain causes of events, which he has not inserted in the things He created"*

C. Ronchi, *The Tree of Knowledge*, DOI: 10.1007/978-3-319-01484-5,
© Springer International Publishing Switzerland 2014

every single time, building bridges over unmeasured abysses. He feels the presence in the cosmos of a Creating Mind, but he does not doubt for an instant that his mind is a reflection of it. He feels that the ultimate aim of science is the theoretical reconstruction of the order of the universe with respect to his mental, spiritual and bodily structure, but he realises that neither sophisticated physical and intellectual instruments nor a denial of logic can enable him to overcome the limits of this structure.

Therefore, it is perhaps the intimate conviction that knowledge represents a sort of re-creation, that led ancient man to consider it as a transgression, a ὕβρις, whose nemesis consists in having to experience the two-faced aspect of its implications at every instance. In the prologue, we had referred to the disturbing Biblical symbol of the Tree of Knowledge of good and evil with the purpose of examining the fruits of knowledge in three millennia of human history, without asking which constitutes a "good" type of knowledge and which an "evil" one. On the other hand, even if this had been our intention, we could hardly have passed any ethical judgement on knowledge, even when the aims of the people who pursued it were iniquitous and its applications ominous. Whether the matter was the discovery of penicillin or the atomic bomb, the effects were pre-determined by the laws of nature. Once their discovery was made one could not simply turn away: one had irreversibly climbed one step higher and the consequences of a possible fall were aggravated.

If, therefore, the adjectives "good" and "bad" are applied to scientific knowledge, they lose any ethical reference since their meanings refer only to the thoroughness and the reliability of its contents. It is, however, impossible for man to evade his responsibility in upgrading his knowledge and amplifying his own actions by means of science. In this sense, the crescent uneasiness of the last generations in the face of the power obtained by man has led many people to consider science a threat to his very existence: incapable of judging the effects of the inevitable applications of his knowledge, one fears that man is on the way to transforming himself into a voracious and lethal parasite of his planet. The analysis is correct, but the initial premise tends to divert us dangerously from the central frame of reference: the threat of an anti-scientific Manichaeism, the most recent reincarnation of the ancient Gnosis, is as erroneous as is the acceptance of an indiscriminate use of science.

It is, on the other hand, impossible to establish ethical criteria independent of religious motives that comprise a vision of the final outcome of man, which, as we have seen, science is not in a position to predict. Yet, it would be an error to condemn it for this. In an excellent commentary on St. Augustine's *De Genesis ad Litteram* [66], P. Agaesse SJ makes this important observation:

"Distinguishing between notional knowledge and ethical experience, and discovering the principles upon which this difference is founded, Augustine avoids the gnostic-Manichean temptation to attribute a substance to the evil, but he rather situates it in the will of a subject that is essentially in relation with God. The evil is not a negative truth except that as a notion. In reality it is the perversion or the reversal of man's freedom that refuses to exert its relation to God. In other words, the evil is the drama of a created freedom that acts separately from the creating Freedom and must experience its limits and its insufficiency."

The adventure of human history appears extremely narrow and brief when compared with the magnitude of the cosmic theatre in which its performance takes place. Nevertheless it is Man himself who has defined the dimensions of the scene of his drama: proud to be able to explain the birth and motion of the stars, to control the aggregation of the elements of matter and to explain their origin, he feels himself master of this scene and recites "*a soggetto*" his role in the cosmic drama, even if he does not know how it will end. Yet, once in a while, he is seized by a troubling feeling, when, casting a glance at the darkness where the audience can usually be found, he perceives the presence of a silent Spectator, who observes with interest and is in a position to judge. He is the only one who can write the concluding Epilogue ...

... "ac nobis in suspenso manebit".

Glossary

We report here a few fundamental notions of mathematical physics concerning definitions and formulae used in this book, which might be helpful for readers who are not very familiar with them.

A.1 The Physical Quantities

A central topic discussed in Chaps. 2–4 is the relation between physical quantities and numbers. We summarise here the definition of physical quantity and that, consequent, of physical law.

In physics a *quantity* is expressed by a convenient definition of a *standard unit* and a *measurement procedure*, and by a *number* or a *set of numbers* that represent its *measure* or *value*. When a physical quantity varies as a function of other quantities is identified by a *numerical variable*. *Theoretically*, physical variables are represented by real numbers, while *experimental* measures of a physical quantity are always expressed by *rational* numbers, i.e., by multiple or fractions of the standard unit.

A physical quantity, defined by a specific measurement, can be *dimensional* and hence expressed by a *dimensional formula*, for instance, [length/time], [mass/volume], etc. The values of dimensional quantities depend on the *basic units* assumed (such as *time, length, mass*, etc.) *Dimensionless* quantities, are pure numbers (e.g.,the ratio of two lengths) and are independent of the adopted basic units.

The *dimensional formula* of a physical quantity defines its *species*.

The physical laws are expressed by mathematical relations between quantities. Generally, every quantity is defined as an abstract numerical variable, x, pertaining to a *numerical field* and, specifically, to the field of the *real numbers*. This assumption implies that a quantity can be incommensurable with the adopted measurement unit, i.e., that there are cases where it does *not* exist a fraction of the unit such that the measured quantity can be expressed by an integer multiple of this fraction.

A physical quantity defined by a single number is called *scalar*. On scalar variables can be applied the customary arithmetical operations of sum and product and

C. Ronchi, *The Tree of Knowledge*, DOI: 10.1007/978-3-319-01484-5,
© Springer International Publishing Switzerland 2014

their inverse ones, with the following restrictions: (1) only physical quantities of the same species can be added and their sum is a quantity of the same species. (2) The multiplication or division of two dimensional quantities of equal or different species produces a quantity of a species different from the other two (e.g., dividing a length quantity by a time quantity produces a new quantity that we call velocity).

In physics, a dimensional variable can be the argument of sole *rational functions*, i.e., sums, subtractions, multiplications quotients, powers and roots. Other operations (e.g. trigonometric functions, logarithms, exponentials, etc.) can be applied exclusively to dimensionless physical variables. Non-rational functions are in fact defined by series of powers of the variable, violating the prohibition to add quantities of different dimension.

A.2 The Vectorial Variables

A variable defined by an ordered sequence of n numbers, called *components* (e.g., the triplets of space co-ordinates defining a point in the space) is called *vectorial variable*. A vector is usually indicated with a bold letter, x, that indicates the sequence $(x_1, x_2, \ldots x_n)$ of its components. In a general system of co-ordinates operations on vectors are rather complicated, however, in the case of orthogonal Cartesian co-ordinates the formalism of vector operations is remarkably simplified. Therefore we refer here to this system in the following definitions.

- The sum of two vectors x and y with the same number of components, n, is simply defined as a vector z having as components the sum of the components of the the the addends. For instance, if $n = 3$ we have:

$$z = x + y = (x_1 + y_1, x_2 + y_2, x_3 + y_3)$$

- For two vectors x and y two different product operations are defined:
  The first one is called *internal* or *scalar product*, because its result is a scalar quantity. This product is given by:

$$z = x.y = x_1 y_1 + x_2 y_2 + x_3 y_3$$

One can easily see that if vectors x and y are perpendicular their scalar product is zero. The square root of the scalar product of a vector x and itself corresponds to its length, and is called *norm* or *module* of x.
The second product between two vectors is called *vectorial* (its result is a vector):

$$z = x \times y = (x_2 y_3 - x_3 y_2, x_3 y_1 - x_1 y_3, x_1 y_2 - x_2 y_1)$$

It can be easily seen that vector z is perpendicular both to x and y (i.e., to the plane defined by them); z has the direction of the rotation axis that brings x on y. Its module is proportional to the sinus of the rotation angle and hence the product is zero when x and y are *parallel*.

A.3 The Functions and their Differential Operators

A function $y = f(x)$ is a correspondence between two variables, x and y, belonging to a numerical field. In mathematics a function is defined by the sequence of elementary instructions to execute in order to calculate y from any value of x. The functions can involve scalar or vector variables. We have previously remarked that in quantum mechanics a function $f(x)$ can be considered as a generalisation of a vector having the form:

$$f = (f(x_1), f(x_2), f(x_3), \ldots f(x_n))$$

in which the continuum variable x replaces the sequence of the discrete values of x, associated with the indices $1, 2, 3\ldots n$.

- Similarly to vectors, one defines the *internal* or *scalar* product of two functions $f(x)$ and $g(x)$ with the integral:

$$z = \int f(x)g(x)\mathrm{d}x$$

extended to the whole definition field of x. When this integral is zero the two functions are said to be *orthogonal*. Analogous to the vectors, the square root of the internal product of a function for itself is called *norm* or *module*. The functions having a finite module (called square-summable) constitute a metric space, called Hilbert's space. The set of mutually orthogonal functions belonging to this space and having module 1 is said *orthonormal set*. In Hilbert's space these functions are the analogous of the usual versors in an orthogonal Cartesian system of reference.
- In Hilbert's space it is possible to find a countable set of orthonormal functions, ϕ_k, called *base*, which can be used like the versors of a system of co-ordinates to represent a generic function of the space as a *linear combination* of the functions of the *base*:

$$f(x) = \sum_{k=1}^{\infty} a_k \varphi_k(x)$$

where a_k are constant coefficients.
- Analogous to that of vectors, the module of a function is equal to the sum of the squares of the coefficients a_k that represent the analogous terms of the co-ordinates in a Cartesian system.

A.1 Some Basic Differential Operators

An operator A is a mathematical procedure, which transforms a function $f(x)$ into a function $g(x)$, analogously to a function $y = f(x)$, which transforms a number x into a number y, that is to say, a relation of type:

$$y = f(x),$$

is generalised by a relation between two functions $f(x)$ and $g(x)$ indicated as:

$$g(x) = Af(x)$$

An operator A consists of a defined sequence of operations on function f (derivation, integration, *etc.*) that transform $f(x)$ into $g(x)$.

Two operators can be added and also multiplied, but for the product of two operators the commutative property is generally not applicable.

It exists for every operator A an inverse operator A^{-1} for which:

$$A^{-1}g(x) = f(x)$$

- An important operator in physics is the *gradient* of a scalar variable $f(x, y, z)$, that in a three-dimensional space (assuming the more familiar notation $x = x_1$, $y = x_2$, $z = x_3$) is defined as:

$$\text{GRAD}\,(f(x, y, z)) = \left(\frac{\partial f}{\partial x}, \frac{\partial f}{\partial y}, \frac{\partial f}{\partial z}\right)$$

The operator GRAD transforms a *scalar* into a *vectorial* field. We remind that if f represents a physical potential, GRAD (f) is the field of the force acting in the point $P = (x, y, z)$.

- Another important operator is the divergence of a scalar variable, f:

$$\text{DIV}\,(f(x, y, z)) = \left(\frac{\partial f}{\partial x}\right) + \left(\frac{\partial f}{\partial y}\right) + \left(\frac{\partial f}{\partial z}\right)$$

or of a vector variable, f:

$$\text{DIV}\,(\mathbf{f}(x, y, z)) = \left(\frac{\partial f_x}{\partial x}\right) + \left(\frac{\partial f_y}{\partial y}\right) + \left(\frac{\partial f_z}{\partial z}\right)$$

The operator divergence, $\text{DIV}(f)$, transforms a scalar (or vector) field into a scalar field. The divergence represents the average increment of f around point P (x, y, z).

For instance, in the Maxwell electromagnetic equations the divergence of the electric-field vector is proportional to the density of charges in point P.

- The third operator of basic interest is the ROT (or CURL), a vector operator that describes the infinitesimal rotation of a vector belonging to a 3-dimensional vector field.

$$\text{ROT}\,(f(x, y, z)) = \left(\frac{\partial f_z}{\partial y} - \frac{\partial f_y}{\partial z}, \frac{\partial f_x}{\partial z} - \frac{\partial f_z}{\partial x}, \frac{\partial f_y}{\partial x} - \frac{\partial f_x}{\partial y} \right)$$

The operator rotation, ROT (f), transforms a vectorial field into another vectorial field. It can be seen that this operator is the correspondent of the vectorial product between two vectors defined above and represents the axis and the amplitude of "curling" of the vectorial field f around point P.

We have seen that in the equations of Maxwell the rotation of the magnetic field characterises the circular motion of the electrical charges around the lines of the magnetic field and, *inversely*, the circulation of these lines around the trajectory of the moving electrical charges.

A.2 Operators and Quantum Mechanics

In quantum mechanics, the fundamental concept of operator can be introduced in various ways, not always easy to follow. We try here to introduce it in a simpler, even though mathematically not rigorous, manner.

Let us take the case of the one-dimensional space. In classical physics the varying position of a material point of mass m is given by a real number, x, as a function of time, t, a function from which we obtain the point velocity and its energy through the simple operation of derivation.

In quantum mechanics the material point is replaced by a wave function $\Psi(x, t)$ that we suppose here monochromatic and written in a complex exponential form:

$$\Psi(x, t) = e^{i(px - Et)/\hbar}$$

where $p(= mv)$ it is the moment and E the energy of the particle. We remember that the complex exponential function is periodic and the reduced Planck constant, $\hbar = h/2\pi$, defines the well-known relation between the energy and the frequency of vibration, v:

$$E = h v$$

One can see that deriving $\Psi(x, t)$ with respect to x is equivalent to multiplying by p the wave function, whilst deriving it with respect to time t is equivalent to multiplying it by E. Since there is a simple relation between the moment, p, and the kinetic energy, E, given by:

$$E = \frac{p^2}{2m}$$

the derivative of the wave function with respect to t must be equal, except for a multiplicative constant, to function $\Psi(x, t)$ derived two times with respect to x. Thus one obtains:

$$i\hbar\frac{\partial\Psi}{\partial t} = -\frac{\hbar^2}{2m}\frac{\partial^2\Psi}{\partial x^2}$$

that represents the simplest form of Schrödinger's wave equation.

The mathematical formalism shows that the energy of the particle is obtained by applying to the wave function the operator:

$$E \rightarrow i\hbar\frac{\partial}{\partial t},$$

whilst the moment, p, is obtained by applying the operator:

$$p \rightarrow -i\hbar\frac{\partial}{\partial x}$$

Others, more complex operators can be obtained in order to describe physical quantities in less simple contexts than that illustrated here, but any operator, A allows us to write a differential equation of type:

$$A\Psi = a\Psi$$

This equation[3] tells that the unknown wave function $\Psi(x, t)$ must be transformed by operator A into itself, multiplied by a constant a. Function $\Psi(x, t)$ that satisfies to this equation represents what is called an *eigenstate* of A, that is to say a state in which the value of the quantity represented by A is fully determined and is exactly equal to a, which is called *eigenvalue* of A. Since the measure of a physical quantity must always correspond to a real number, the resulting wave may be a complex one, but the *eigenvalues* must be real numbers. Furthermore, the *eigenvalues* are *not* arbitrary since normally the equation $A\Psi = a\Psi$ admits a solution only for defined values of a, and, in some cases, these correspond to a set of discrete values (as, for example, the energy *eigenvalues* of an electron bound to an atom, the vibration frequencies of a harmonic oscillator, etc.).

A.3 Geometric Representation of the Principle of Special Relativity

The old principle of relativity, formulated by Galilei, asserts that, if observer K' is moving with respect to observer K with a constant speed v, the co-ordinates, x' and

---

[3] For simplest cases operator A consists only of partial derivatives, but for more complex systems the resulting wave equation may be very elaborated and can only be solved numerically by means of dedicated computer programmes.

t', of an object (we suppose for the sake of simplicity a point in a one-dimensional space) observed by K' are obtained from the co-ordinates, x and t, of the same object seen by K by applying the self-explaining transformation:

$$x' = x - vt; \quad t' = t$$

These relations obviously require that K and K' possess two perfectly synchronised clocks. Then, if considered from the geometric point of view, the axis of time t' of K' corresponds for K to the straight line $t = x/v$, i.e., to an axis tilted with respect to the time axis t of K by an angle $\theta = \arctan(v)$ (see Fig. A.1).

One first remarks that, if observer K sees two events, P1 and P2 in the *same x*-position, but at *different times*, observer K'—that between these two events has moved a certain distance—sees them in two *different* positions. However, two events, T1 and T2, simultaneous for K, are *also simultaneous* for K'. (Note that the *unit length* of the t'-axis is larger than that of the t-axis: in fact, one can see that OT2 $= \sqrt{(\text{OT1}^2 + v^2)} \geq$ OT1. However, the *time values* of OT1 and OT2 are equal.

In conclusion, the Galileian principle of relativity entails that, in a Cartesian co-ordinates' system, the axis of time of K' is tilted with respect to that of K, but times and lengths are the same for both K and K'.

Einstein's principle of relativity, in its *restricted* form, regards the kinematic properties of motion as referred to different systems of co-ordinates and is based on the following hypothesis:

One cannot prove the existence of a universal time, and hence one must accept that the two abovementioned observers, K and K', are *not* in a condition to synchronise their clocks. The transformation of the Cartesian co-ordinates from K to K' is, therefore, generally written for a 3-dimensional space as:

$$x' = a_{11}x + a_{21}y + a_{31}z + a_{41}t$$
$$y' = a_{12}x + a_{22}y + a_{32}z + a_{42}t$$
$$z' = a_{13}x + a_{23}y + a_{33}z + a_{43}t$$
$$t' = a_{14}x + a_{24}y + a_{34}z + a_{44}t$$

Yet, knowledge of the three components (v_x, v_y, v_z) of the relative speed of K' with respect to K, together with the conditions of space symmetry, is *not* sufficient to calculate the 16 coefficients a_{ik}. The missing mathematical constraint is supplied by Einstein's *postulate of relativity*, which asserts that the speed of light, c, is the same for K and K', no matter how large their relative speed, v, is This condition can be mathematically expressed as:

$$x'^2 + y'^2 + z'^2 - c^2 t'^2 = x^2 + y^2 + z^2 - c^2 t^2$$

This means that the *norm* of vector (x, y, z, ict), where i is the imaginary unit, is *invariant* with respect to the co-ordinates' transformation from observer K to observer K'. This vector belong to a four-dimensional space, called *chronotope*,

Fig. A.1 Geometric representation of the Cartesian systems of reference according
to the postulate of relativity of
Galileo

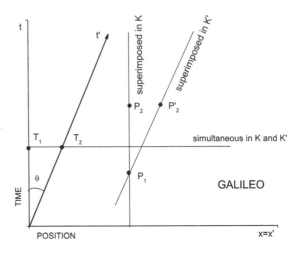

which, in addition to the three spatial co-ordinates, is characterised by a fourth
time-dimension corresponding to the axis -ict. The expressions of coefficients a_{ik} in
terms of v and c can now be obtained with simple algebraic calculations. It is here not
necessary to report them in order to understand their geometric meaning. Actually, the
transformation of co-ordinates operated by a_{ik} is well known in analytical geometry:
it represents a rotation of *all* co-ordinates' axes of an angle $\theta = \text{arctg}\,(v/c)$.
Therefore, the most general proposition of special relativity is as follows:

*"It must be possible to translate all equations of Physics into geometrical relations
within the chronotope, whereby these relations are invariant for any rotation of the
co-ordinates' system."*

Assuming, for the sake of simplicity, a one-dimensional (x, t) system, Einstein's
hypothesis entails that by changing the reference system from K to K$'$ a rotation of
axis x, occurs, in addition to that of axis t, already implied by Galilei's postulate, as
shown, respectively, in Figs. A.1 and A.2.

A perfect symmetry is thus obtained in the co-ordinates' transformation matrix,
where the imaginary axis, ict, whatever might be its interpretation, is dealt with
the same type of mathematical dependence as that of the spacial co-ordinates. This
means that, while in classical physics time represents a privileged variable, which
can be singled out from the process of choice of any reference system, in the relativity
theory time and space co-ordinates are essentially interdependent. Time is, therefore,
to be seen as an intrinsic property of space.

According to Einstein's principle of relativity, the property of Galilei's relativity
to maintain simultaneity conditions between different observers, is *not* valid any
longer. This means that, under certain conditions, an observer K$'$ *sees* two events
in a time sequence opposite to that observed by K. This effect can be measured
experimentally if the relative speed of the observers is sufficiently high to rotate the
axes of a significantly large angle θ (see Fig. A.3). There is, on the other hand, a limit
for this rotation (it cannot exceed $\pi/4$. corresponding to $v = c$), hence there are

Fig. A.2 Geometric representation of the Cartesian systems
of reference according to the
postulate of relativity of Einstein

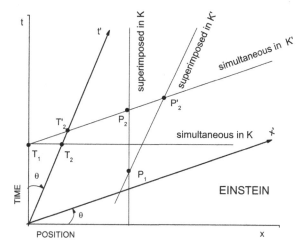

Fig. A.3 The diagram illustrates the subdivision of space-
time according to the model
of the restricted relativity

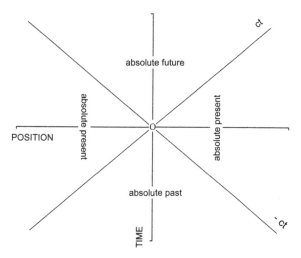

consecutive events whose time distance measured by observer K may appear shorter
to K′, but their sequence can never be inverted. Since for any observer any cause
must necessarily precede all its effects, these events are the only ones which can
be put in a causal relation. The points representing these events are confined in two
distinct geometrical regions of the space-time, respectively called *absolute past* and
absolute future (Fig. A.3). All the other events, falling in the complementary field
of space-time, belong to an *absolute present*, in which their time sequence can be
inverted depending on the speed of the observer and, therefore, they cannot be put
in any causality relation.

References

1. N. Chomsky, *Language and Mind* (Harcourt Brace Jovanovich Inc., New York, 1968)
2. K.G.C. Von Staudt, *Geometrie der Lage* (F. Korn, Nürnberg, 1847)
3. B.E. Schaefer, *The Latitude and Epoch for the Origin of the Astronomical Lore in Mulapin*. Bull. Am. Astron. Soc. **38**, 157 (2007)
4. G.S. Kirk, J.E. Raven, M. Schofield, *The Presocratic Philosophers* (Cambridge University Press, New York, 1984)
5. D.E. Gershenson, D.A. Greenberg, *Anaxagoras and the Birth of Physics* (Blaisdell Publishing Co., New York, 1964)
6. J. Moreau, *Le Temps et l'Instant Selon Aristote, in Naturphilosophie bei Aristoteles und Theophrast*, 4th. Symposium Aristotelicum, Göteborg 1966, ed. by I. Düring, L. Stiehm Verlag (Heidelberg, 1969)
7. E.V. Arnold, *Roman Stoicism: Lectures on the History of Stoic Philosophy with Special Reference to its Development within the Roman Empire* (1911) (Routledge & Kegan Paul Ltd., London, 1958 (reprinted))
8. G. Sarton, *Introduction to the History of Science*, Carnegie Institution of Washington (Williams and Wilkins Publ., Baltimore 1927–1948)
9. K. Popper, *The Open Society and its Ennemies* (Routledge & Kegan Ltd., London, 1945)
10. S. Drake, *Galileo* (Oxford University Press, Oxford, 1980)
11. Simplicius, *On Aristotle's "On the Heavens"* (Cornell University Press, New York, 2005)
12. G. Galilei, *Le Opere di Galileo Galilei*, edn. in 20 volumes ed by A. Favaro, Barbera, Firenze (1890–1909)
13. D. Ulansey, *The Origins of the Mithraic Mysteries* (Oxford University Press, New York-Oxford, 1989)
14. H. Meschkowski, *Was Wir Wirklich Wissen* (München, Piper Verlag, 1984), p. 87
15. A.N. Whitehead, *Dialogues of A.N. Whitehead as Recorded by L. Price*, (A Mentor Book-Penguin Group (USA) Incorporated 1956) p. 277
16. A. Einstein, B. Podolsky, N. Rosen, *Can Quantum-Mechanical Description of Physical Reality be Considered Complete?* Phys. Rev. vol. **47**, 777 (1935)
17. P. Valéry, *Cahiers II* (Gallimard, Paris, 1974)
18. W. Heisenberg, *The Nature of Elementary Particles*, Phys. Today **29**(3), 18 (1976)
19. S. Drell, *Elementary Particle Physics*, Daedalus **I**, 17 (1977)
20. R. Courant, D. Hilbert, *Methods of Mathematical Physics* (Interscience Publ. Inc., New York, 1953)
21. The New York Time, 18 Aug 1901

C. Ronchi, *The Tree of Knowledge*, DOI: 10.1007/978-3-319-01484-5,
© Springer International Publishing Switzerland 2014

22. R. Feynman, *The Pleasure of Finding Things Out* (Cambridge Mass, Perseus Book, 1999), p. 91
23. R. Feynman, *The Feynman Lectures on Physics*, vol. 1 (Addison Wesley, USA, 1964)
24. H. Reichenbach, *Experience and Prediction: an Analysis of the Foundations and the Structure of Knowledge* (University of Chicago Press 1938)
25. C.G. Hempel, *Aspects of Scientific Explanation and Other Essays in the Philosophy of Science* (Collier-Macmillan, New York, 1970)
26. S. Toulmin, *From Form to Function* (in *Daedalus*, J. Amer. Acad. Arts Sci. Vol, I Summer, 1977)
27. K.G. Chesterton, from Daily News, 25 Feb 1905
28. B.B. Mandelbrot, *Fractals: Form, Chance, and Dimension* (Freeman and Company, San Francisco, 1977)
29. I. Prigogine, *From Being to Becoming* (Freeman and Co., San Francisco, 1985)
30. A.H. Compton, *The Freedom of Man* (Yale University Press, New Haven, 1935)
31. K. Popper, *Objective Knowledge: an Evolutionary Approach* (Oxford Clarendon Press, UK, 1972)
32. T.S. Kuhn, *The Structure of Scientific Revolutions* (University of Chicago Press, 1962)
33. J. Ehlers, *The Nature and Structure of Space-Time*, in *The Physicist's Conception of Nature*, ed. by J. Mehra, Reidel (Dordrecht, 1973), pp. 71–91
34. D. Wilkinson, *The Organization of the Universe*, in *The Nature of Matter*, ed by J. H. Mulvey Wolfson College Lectures 1980 (Clarendon Press, Oxford, 1981)
35. S.W. Hawking, *A Brief History of Time* (Bantam Press, London, 1988)
36. M. Paul, *Principles of Quantum Mechanics* (Oxford Press, Oxford, 1951)
37. R. Thom, *Structural Stability and Morphogenesis* (W.A. Benjamin Inc. Publ., London, 1975)
38. V. Braitenberg, *Vehicles* (MIT Press, Cambridge, 1984)
39. P. Medawar, *The Limits of Science* (Oxford University Press 1986)
40. T.R. Malthus, *An Essay on the Principle of Population (1798)* (University of Michigan Press, Michigan, 1959). (Reprinted) ISBN:0472060317)
41. E. Snapper, *Are Mathematical Theorems Analytical or Synthetic?* The Mathematical Intelligencer **3**, 2 (1981)
42. M. Fitzgerald, *Autism and Creativity: Is there a Link Between Autism in Men and Exceptional Ability?* (Brunner-Routledge Publ., London, 2004)
43. M. Fitzgerald, B. O'Brien, *How Asperger Talents Changed the World* (Autism Asperger Publishing Company, UK, 2007)
44. F. Manuel, *A Portrait of Isaac Newton* (Cambridge 1968)
45. B. Hessen, *The Social and Economic Roots of Newton's Principia* in: Nicolai I. Bukharin, Science at the Crossroads London (1931) (Reprint New York 1971) pp. 151–212
46. F. Rosenblatt, *Principles of Neurodynamics* (Spartan Books Publ, New York, 1962)
47. M. Minsky, *Perceptrons: an Introduction in Computational Geometry* (MIT Press, Cambridge, 1972)
48. J.G. Carbonell, T.M. Mitchell, *Machine Learning : an Artificial Intelligence Approach* ed. by R.S. Michalski ISBN-0-934613-09-5
49. R.C. Schanke, R.P. Abelson, *Scripts, Plans, Goals and Understanding* (Eribaum Publ, Hillsdaöle NJ, 1977)
50. J.R. Searle, *Minds, Brains and Programs*, in *The behavioral and brain sciences* (Cambridge University Press, Cambridge, 1980)
51. J. Hadamard, *Essai sur la Psychologie de l'Invention dans le Domaine Mathématique*, Reprint (1993) ISBN 2876470179
52. R. Penrose, *The Emperor's New Mind* (Oxford University Press, Oxford, 1989)
53. N. Chomsky, *Reflection on Language* (Pantheon, New York, 1975)
54. L. Wittgenstein, *Tractatus Logico-Philosoficus* (Routledge and Kegan Paul, London and Henley, 1974)
55. R. Montague, *Universal Grammar in Selected Papers* (Yale University Press, New Haven, 1974)

56. F. Kambartel, in *Semantics from Different Points of View*, ed. by R. Bäerle, U. Egli, A.V. Stechow (Springer, Berlin, 1980), p. 195
57. J. Campbell, *The Masks of God: Creative Mythology* (Viking Press, New York, 1968)
58. J.L. Goff, *Storia e Memoria* (Einaudi, Torino, 1977)
59. J. Good, *Speculations Concerning the First Ultraintelligent Machine*, in *Advances in Computers*, ed. by F.L. Alt, M. Rubinoff (Academic Press, Cambridge, 1965), pp. 31–68
60. N. Bostrom, *Anthropic Bias: Observation selection effects in science and philosophy* (Routledge Publ, New York, 2002)
61. J.D. Barrow, F.J. Tipler, *The Anthropic Cosmological Principle (Oxford University Press* (Oxford University Press, Oxford, 1986)
62. P. Teilhard de Chardin, *Le Phénomène Humain* (Éditions du Seuil, Paris, 1955)
63. P. Teilhard de Chardin, *Je m'Explique* (Jean-Pierre Demoulin, Éditeur, 1966)
64. K. Popper, J. Eccles, *The Self and its Brain* (Springer, Berlin, 1985)
65. O. Spengler, *Der Untergang des Abendlandes* (Beck'sche Verlag, München, 1923)
66. A. Augustinus, *Oevres*, ed. by P. Agaesse and A. Solignac, vol. 49 (Desclee de Brouwer, Paris, 1972) p. 507
67. A.A. VV, *The System Programming Series* (Addyson-Wesley Publ. Co. Inc, Boston, 1984)
68. J.F. Sowa, *Conceptual Structures: Information Processing in Mind and Machine* (Addyson-Wesley Publ. Co. Inc, Boston, 1984)
69. J. Daniélou, *Platonisme et Théologie Mystique : Doctrine Spirituelle de Saint Grégoire de Nysse* (Aubier, Paris, 1944)
70. K. Rossiianov, *Beyond Species: Il'ya Ivanov and His Experiments on Cross-Breeding Humans with Anthropoid Apes*, Science in Context (2002) pp. 277–316
71. E. Whittaker, *A History of the Theories of Aether and Electricity; vol. 2: The Modern Theories(1910–1926)* (Longmans Green Publ., New York-London, 1953)
72. P.K. Feyerabend, *Erkenntnis für Freie Menschen* (Suhrkamp, Frankfurt, 1977)
73. I. Newton, *Philosophiae Naturalis Principia Mathematica*, vol. I, def. VIII, p. 11 and schol. propos. XXXIX p. 464, Geneva Edn. (1739–1740)
74. R.P. Feynman, *The Character of Physical Law* (Penguin Books Ltd, London, 1992), p. 171
75. W. Haken, *Controversial Questions about Mathematics*. The Mathematical Intelligencer **3**(3), 117 (1981)
76. F. Dyson, *Disturbing the Universe* (Harper and Row, New York, 1979)
77. J. Brunowski, *The Ascent of Man* (BBC, London, 1981)
78. P. Davies, *The Mind of God, Science and the Search for Ultimate Meaning* (Simon and Schuster Ltd, London, 1992)
79. H. Frank, *George, Science and the Crisis in Society* (Wiley Interscience Publ., London, 1970)
80. L. Mumford, *The Culture of Cities* (Secker and Warburg, 1938)
81. D. Deutsch, *The Fabric of Reality* (Penguin Books, London, 1997)
82. R.K. Merton, *The Sociology of Science: Theoretical and Empirical Investigations*, ed. by N. Storer (University of Chicago Press, Chicago, 1973)
83. K. Wittvogel, *Oriental Dispotism (A Comparative Study on Total Power* (Yale University Press, New Haven, 1957)
84. R. David, *Roman Coins* (B. A. Seaby Ltd, London, 1964)
85. L. Jorge, *Historia Universal de la Infamia* (Emecè Editores S.A, Buenos Aires, 1954)
86. L. Thorndike, *A History of Magic and Experimental Science* vol. 8 (Columbia University Press, New York, 1958–1960)
87. AA. VV. *p-Adic Mathematical Physics and Analysis*, Proceedins of the Steklov Institute of Mathematics, Vol. 245, Issue 2 of 4 (Malk Nauka Interperiodica Publ. 2004)
88. A. Huxley, *Brave New World* (Chatto and Windus Ltd, London, 1932)
89. D. Barski, G. Christol, *Les Nombres p-Adiques*, la Recherche, 278 (1995)
90. H. Weyl, *Space-Time-Matter* (Engl. Transl.) (Dover Publ. Inc., New York, 1950)
91. G. Friedlein, (ed.), *Procli Diadochi in Primum Euclidis Elementorum Librum Commentarii* (Bibliotheca Scriptorum Graecorum et Romanorum Teubneriana), Teubner, Leipzig (1873) [Reprint Hildesheim: Olms, (1967)]

92. T.L. Heath, *The Thirteen Books of Euclids Elements*, vol. I, Cambridge University Press - reprinted (1925), (Dover Publ. Inc, New York 1956) p. 202

93. J.C. Friedrich Gauss, in a letter to W. Bolyai, Werke, Vol. VIII, (Teubner Publ., Leipzig 1900) Dec. 16 1799, p. 159

94. G. Saccheri, *Euclides ab Omni Naevo Vindicatus* (1733), citation from *L'Euclide Emendato*, abridged Ital. translation by G. Boccardini (Hoepli Publ., Milan 1904)

95. L. Murawiec, *The Mind of Jihad* (Cambridge University Press, New York, 2008)

96. S. Reinstadler, *Elementa Philosophiae Scholasticae* (Herder Publ., Vienna 1904) p. 361

97. G. Cantor, in *Gesammelte Abhandlungen mathematischen und philosophischen Inhalts*, ed. by E. Zermelo (Springer, Berlin, 1932)

98. T. Pesch, *Institutiones Philosophiae Naturalis Secundum Principia* S. Thomae Aquinatis, 2nd. edn. in 2 Volumes, Freiburg i. Br. (1880/1897)

99. P. Florenski, *Werke in zehn Lieferungen*, X. Vol.: *"Meinen Kindern"*, transl. from Russian by O. Radetzkaya and U. Werner, Edition Kontext, (Berlin 1997) p. 240

100. P. Florenski, *Werke in zehn Lieferungen*, V. Vol.: *"Raum und Zeit"*, *"Analyse der Räumlichkeit und der Zeit in Werken der bildenden Kunst"*, transl. from Russian by O. Radetzkaya and U. Werner, Edition Kontext, (Berlin 1997) p. 99–150

101. P.W. Singer, *Wired for War: The Robotic Revolution and Conflict in the 21th Century* (Penguin Publ. New York, 2009)

102. H.A. Lorentz, Proc.of the Royal Netherlands Acad. of Arts Sci. 1, 427–442 (1899)

103. G. Ifrah, *Histoire Universelle des Chiffres* (Editions Seghers, Paris, 1981)

104. A. Einstein, *On the Electrodynamics of Moving Bodies*. Annalen der Physik **17**(10), 891–921 (1905)

105. L.V. de Broglie, *Recherches sur la Théorie des Quanta*, Thesis, Université de Paris, (1924) and Ann. de Physique 3(10), 22 (1925)

Index

C. Ronchi, *The Tree of Knowledge*, DOI: 10.1007/978-3-319-01484-5,
© Springer International Publishing Switzerland 2014

Printed in the United States
By Bookmasters